PHYSIOLOGY OF FUNGI

ADVANCES IN PLANT PHYSIOLOGY SERIES—5

Physiology of Fungi

D. K. Arora
Seema Gupta

Anmol Publications Pvt Ltd
NEW DELHI-110002 (INDIA)

Anmol Publications Pvt. Ltd.
4374/4B, Ansari Road, Daryanganj
New Delhi-110002

Physiology of Fungi

First Edition 1996

ISBN 81-7488-239-1

PRINTED IN INDIA

Published by J. L. Kumar for Anmol Publications, Pvt. Ltd. New Delhi and printed at Chetna Offset Press, Delhi.

Preface

'Physiology of Fungi' offers a comprehensive account of the fundamental principles of plant physiology. It incorporates and organizes the information concerning Physiology of Fungi from relevant authentic sources. It is an humble attempt to present a correct and precise account of the subject matter of this important branch of Botany which forms an integral part of the studies undertaken by the undergraduate and post-graduate students of this subject. The principles have been outlined in considerable detail alongwith this fundamental facts and theories that explain the life processes of plants. The main object of the present book is to provide a comprehensive and well illustrated account of the prescribed subject matter in a very simple and lucid language and concise manner yet—still maintaining its desired academic standard.

Most of illustrations have been taken from professional papers, standard books and journals. Subject matter has also been obtained from a large number of standard books, reviews and research papers by internationally renowned workers and established authors. That is why there can be no claim of originality except in the manner of treatment.

Though the authors have taken special care to present a current account, yet they are fully aware of their limitations, and the readers may come across the mistakes of various types. For all types of mistakes they extend their due apology.

The authors express their thanks to their friends and colleagues whose continuous inspirations have invited them to bring out this text.

Contents include : Fungal Decomposers, Fungal Symbionts in Ectomycorrhizas and Lichens, Symbiotics Relationship with insects, Parasitic Symbionts of Plants, Root Infecting Fungi and so on.

Authors

Preface

Physiology of Fungi offers a comprehensive account of the fundamental principles of plant physiology. It incorporates and organizes the information concerning Physiology of Fungi from relevant authentic sources. It is an humble attempt to present a correct and precise account of the subject matter of this important branch of Botany which forms an integral part of the studies undertaken by the undergraduate and postgraduate students of this subject. The principles have been outlined in considerable detail along with the fundamental facts and theories that explain the life processes of Plants. The main object of the present book is to provide a comprehensive and well illustrated account of the prescribed subject matter in a very simple and lucid language and concise manner yet still maintaining the desired academic standard.

Most of the illustrations have been taken from the standard journals, standard books and manuals. Subject matter has also been obtained from a large number of standard books, reviews and research papers by internationally renowned workers and established authors. That is why there can be no claim of originality except in the manner of treatment.

Though the authors have taken special care to present a correct account, yet they are fully aware of their limitations and the readers may come across the mistakes of various types. For all types of mistakes we extend them our apology.

The authors express their thanks to their friends and colleagues whose continuous inspirations have invited them to bring out this text.

Contents include:— Fungal Decomposers; Fungal Symbionts in Ecosystems and Lichens; Symbiotic Relationship with Insects; Parasitic Symbionts of Plants; Plant Infection; Light and so on.

Authors

Contents

1

Fungal Decomposers

Leaves of all manner of types form suitable substrates for many fungi. Long before any leaf fails, the complex process of its decay is initiated. The greater part of this process takes place above ground in the intercovering the sun surface. In this chapter some facets of this are discussed but particularly the epiphytic leaf surface or phylloplane micro-flora, the leaf surface as a habitat niche for fungi, colonization by the common primary saprotrophs as leaves senesce and die, the attributes which make these primary saprotrophs such widespread and successful colonizers, and some of the subsequent events occurring in the litter.

The leaf as a spore trap

As any leaf unfolds it is a relatively clean sheet which immediately provides landing sites for air-borne particles such as bacteria, yeast cells and fungal spores but also pollen. Spore trapping by leaves is a natural phenomenon of nature. Spores may reach leaves in three main ways : wind-borne and deposited by impaction or by sedimentation under gravity ; in falling rain drops ; or in rain splash droplets. Air-brone spores are usually dry and often rough or spiny and readily detachable from their stalks, excellent examples being the urediospores of rust fungi. They are readily washed out of air by falling rain drops. Rain splashed spores tend to be wet or slimy and borne in a sticky liquid. Adaptations facilitating deposition are far less obvious than in spores of aquatic fungi. Amongst dry spores, a larger size, as seen in the powdery and downy mildews, favours impaction and sedimentation. Rain splashed spores tend to be smaller and spherical. But these are only generalizations and there are many anomalies. The most ubiquitous and by far the most numerous of the phylloplane fungi are members of the Sporobolomycetaceae, the shadow yeasts. They produce air-borne spores which in relative terms are quite minute.

Leaf surfaces are differential spore traps. Their efficiency as traps depends upon whether they are horizontal or vertical, wet or dry, hairy or glabrous, glossy or mat, waxy or non-waxy and so on. Not all spores that land become securely attached. Some are washed off by rain, blown off by wind or redistributed by dew. Some have a two phase dispersal system. For example, the large sporangia of pathongenic species of *Phytophthora* and some other Oomycete Peronosporales are wind-borne and normally impacted onto leaf surfaces. Under moist conditions the impacted sporangia may germinate directly by a germ tube or indirectly to produce motile zoospores which may swim about in moisture or be redispersed further to other leaves in rain splash droplets.

Virtually any spore which may become air-borne can be found on leaves. If leaf surfaces are washed and the washings plated out onto nutrient agar, numerous yeast and filamentous conidial Ascomycotina, some Zygomycete Mucorales and the occasional Mastigomycotina and Basidiomycotina develop on the plates. But microscopic examination of stained leaf surface impressions or peels reveals the presence of not only these but also spores of many other Ascomycotina and Basidiomycotina, including those of agarics, polypores and Gasteromycetes. They just do not grow or grow too slowly on the culture medium used. These impressions of peels can be made by spraying leaves with cellulose acetate in amyl acetate or painting with nail varnish or molten 1% agar, leaving to dry and then stripping off.

Many of these fungi and an even larger number of bacteria actively grow on the surface of the living leaf and have been called 'resident inhabitants' in contrast to 'casual inhabitants' which are unable to grow in such an environmental because of the lack of essential nutrients, unfavourable physical factors, competition with or antagonism by others or some combination of these factors. This rather simple distinction can be extended by dividing the epiphytic fungi into three categories : non-pathogenic epiphytes ; pathogens ; and exochthonous or casual inhabitants. Exochthonous is a fitting, if somewhat clumsy, term as it is used for fungi found on or in a substrate which is not their habitual one.

Phylloplane inhabitants

Amongst the non-pathogenic epiphytes, two main groups, the phylloplane inhabitants and the common primary saprotrophs, can be recognized. The phylloplane inhabitants are able to complete their life cycle or a significant part of it on the living leaf without damaging it. *Sporobolomyces roseus* (Fig. 1.1*a*) not only is a very good example of such a fungus but is virtually omnipresent, being found on leaves of grasses, dicotyledonous herbs, trees and shrubs, wherever they grow. Its cells multiply very quickly by budding when conditions are favourable, forming distinct yeast-like colonies on the leaves. Budded cells can be redistributed on an individual leaf or from leaf to leaf by rain splash and are similarly locally dispersed to leaves of other plants. It also reproduces by ballistospore formation. The ballistospores are very effectively wind dispersed. They are produced under high humidities at night, as are the budded cells, and constitute the major component of the air-spora at that time. Other members of the Sporobolomycetaceae are also common. Species of *Bullera* behave similarly, whereas members of the genera *Tilletiopsis* and *Itersonilia* produce a sparse mycelium from which ballistospores arise (Fig. 1.1*b*). Other yeasts, especially members of the Cryptococcaceae, the non-sporing yeast, exist in the budding phase. They all complete their life cycle in the phylloplane. Such fungi are often called 'shadow yeasts'. Their presence can be demonstrated by suspending leaves from the underside of a Petri dish lid for about 12 h over 2% malt

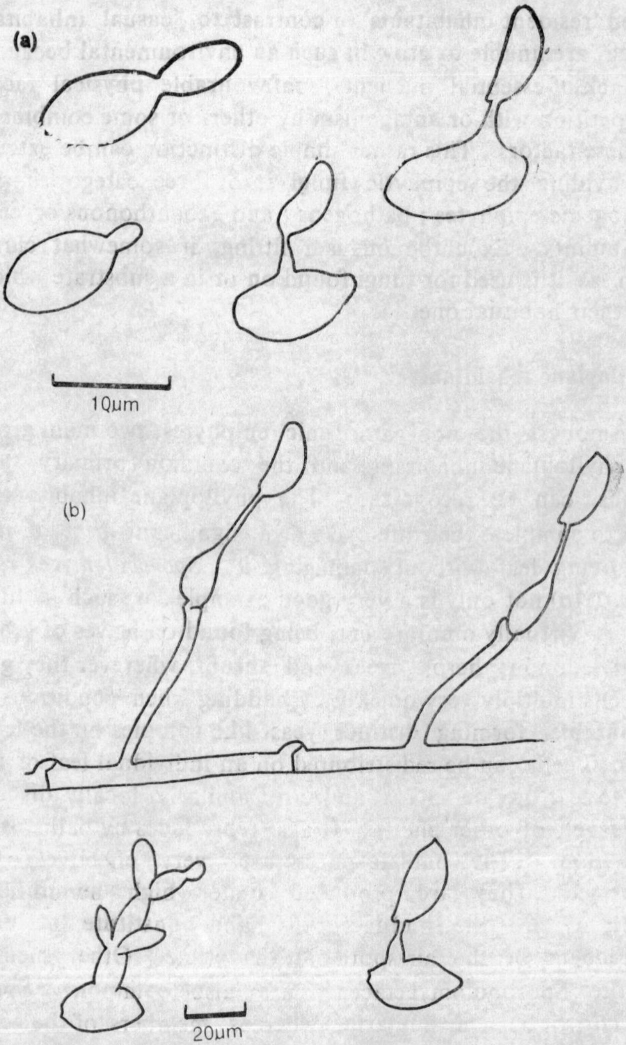

Fig. 1.1. (*a*) Budding and ballistospore formation in *Sporobolomyces roseus*. (*b*) Ballistospore formation in *Itersonilia perplexans*, ballistospores germinating by budding and ballistospore formation.

extract agar. In such a humid and still atmosphere, ballistospores are produced and discharged. They fall vertically onto

the agar below and start budding. Tiny colonies, pink in *S. roseus*, become visible after 2-3 days. These form a mirror image of the distribution of the cells of the leaf.

Two conidial Ascomycotina also grow in the phylloplane. The conidia of *Aureobasidium pullulans* and several species of *Cladosporium* may germinate after impaction and develop into hyphae forming quite extensive colonies under favourable conditions. *Aureobasidium* more often grow by yeast-like budding with minimal hyphal growth (Fig. 1.2*a*). This is a modification of its normal cultural form. The budded cells may again be redistributed in moisture films or by rain splash droplets. The conidia of *Cladosporium* may germinate and produce secondary conidia from short germ-tubes rather than grow as hyphae (Fig. 1.2*c*). These conidia are dry and become air-borne. So both reproduce rapidly and complete a signi-

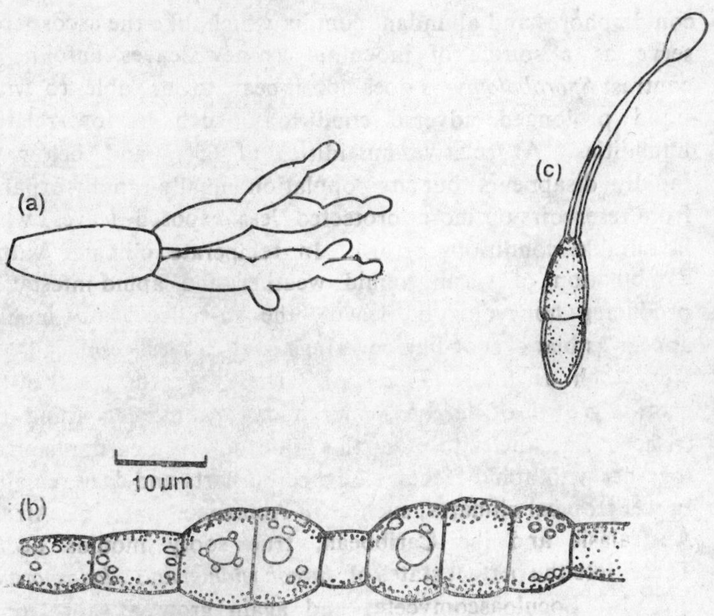

Fig 1.2. (*a*) Yeast-like budding by *Aureobasidium*; and (*b*) chlamydospores of *Aureobasidium* on a leaf surface. (*c*) Conidium of *Cladosporium* germinating to produce a secondary conidium.

ficant part of their life cycle in the phylloplane. Both eventually produce ascocarps to complete their life cycle. These are found only in the spring on overwintered fallen leaves in temperature climates. Ascospores are discharged from these as new leaves unfold.

Aureobasidium and *Cladosporium* are also both very well-adapted to survive in this rigorous habitat. Their hyphal walls rapidly become thickened and melanized. This enables them not only to survive exposure to damaging ultra-violet light from the sun but may also help to prevent excessive desiccation and make them more resistant to bacterial lysis. *Aureobasidium* produces dark, thick-walled multicellular chlamydospores in chains or in clumps (Fig. 1.2*b*). *Cladosporium* produces more distinct microsclerotia, compact spheres of 10-100 cells with an outer layer of thick-walled cells with heavily melanized walls. Under favourable conditions, these produce clusters of conidiophores and abundant conidia which, like the ascospores, serve as a source of inoculum as new leaves unfold. In contrast *Sporobolomyces* does not appear to be able to withstand prolonged adverse conditions, such as low relative humidities. At relative humidities of 65% and below, it rapidly disappears but its population equally rapidly expands from reservoirs on more protected less exposed leaves, when favourable conditions return. In temperate climates with a combination of warm humid weather and aphid infestation producing honeydew on leaves, the so-called sooty moulds appear as black, soot-like coverings over leaves, especially of trees such as limes (*Tilia* spp). These are the result of the profuse growth of *Aureobasidium* and *Cladosporium* using the trisaccharide melezitose in the honeydew as a carbon-source, together with aphid faeces, sloughed off parts and dead remains. In wet tropical climates, such as in Amazonia, parts of Africa, Australasia and the Caribbean, true sooty moulds occur. These, like the perfect states of *Aureobasidium* and *Cladosporium*, are also Loculoascomycetes and again grow as saprotrophs associated with honeydew from aphids. A wide range of species from several fungal families, especially the Capnodiaceae and Chaetothyriaceae are involved. They form distinct, dense, dark hyphal networks on leaves often in the form of a thick felt and each fungus produces abundant conidia of often two or even three types as well as ascocarps.

Nutrient sources

All the phylloplane inhabitants, the yeasts, the filamentous fungi and the bacteria, are chemo-organotrophs requiring organic nutrients for growth. Some of their nutritional requirements may be met by organic substances absorbed or deposited onto leaves such as detritus trapped in their superficial wefts of hyphae, as also happens with fungi living on paint films or glass. Most of their nutrients, however, must be derived directly or perhaps indirectly from the host. A great variety of substances exude or leak out of leaves. These include free sugars, amino acids and inorganic ions which are all essential for fungal and bacterial growth. For example, water droplets placed on leaves exhibit an increase in conductivity indicating exudation from the leaf; increased growth of some fungi in these drops shows that certain of these exudates are of nutritional value. Two sources of added nutrients are from pollen and other spores. Nutrients also leak out of these. Pollen added to leaf surfaces stimulates the development of *Sporobolomyces* and *Cladosporium* and probably accounts for the sudden increase in their population shortly after flowering on leaves of plants such as rye. Conidia of *Botrytis cinerea* placed in droplets on leaves leak out amino acids and sugars in sufficient quantity for phylloplane bacteria to develop in such numbers as to inhibit germination of the conidia themselves. This all occurs on the intact surfaces of healthy leaves On aphid infested leaves nutrients may be derived indirectly from the host. Host sucrose is converted to melezitose in honeydew and this is used by the phylloplane inhabitants.

The number of phylloplane inhabitants inceases with the age of the leaf. This association between population density and age of the leaf is usually explained by increase in leaf exudates with ageing. It is also assumed that the restricted availability of the nutrients is one of the main causes of the relatively poor development of the phylloplane inhabitants on immature leaves. There is also evidence that some of these fungi can slowly degrade the surface waxes and cuticle and so gradually increase the permeability of the epidermis. Their numbers are also far greater on leaves infected by pathogenic fungi, such as rusts and mildews. Four to five times as many colonies of *Sporobolomyces* can be isolated from mint leaves infected with the rust fungus, *Puccinia menthae*, as from healthy mint leaves. Here the injurious effect of the pathogen, especially perhaps the changes in cell permeability, cause an outflow of additional nutrients.

The phyllophane inhabitants are also not uniformly distributed over the leaf surface. Most more prevalent on the upper surface and are usually more predominant along the veins, frequently with their cells oriented to lie parallel with the vein axis. They also tend to align themselves along the anticlinal walls, as with veins there is a slight depression there. They could be washed into these positions but there may also be more exudates released along the veins ; also vein sheath cells may bring nutrients nearer the surface and thus facilitate exudation.

Common primary saprotrophs

The common primary saprotrophs are unable to grow to their full extent in the phylloplane until the onset of senescence. Their pattern of development is restricted until senescence and several rarely on never grow on the green leaf. Their spores accumulate on the leaf prior to senescence and remain dormant until the death of the tissues. If they do germinate they do so only to a limited extent. On senescence they very quickly take advantage of the changing conditions. Sporing colonies of these fungi are ubiquitous on newly dead leaves of the majority of plants. The phylloplane inhabitants and the common primary saprotrophs by no means form distinct groups. *Aureobasidium* and *Cladosporium* have to be included in both groups because, although they grow and reproduce by conidia in the phylloplane, they develop to a much greater extent in the dead leaf. Other fungi, all conidial Ascomycotina, in this group include *Alternaria alternata*,

Botrytis cinerea, *Epicoccum purpurascens* and *Stemphylium botryosum*. In the Tropics the list can be extended to include species of *Curvularia* and *Nigrospora*. Spores of a great variety of other saprotrophs may also be present on the leaves. They germinate only on the death of the leaves or sometimes thereafter.

Pathogens

Two distinct catagories can be recognized amongst the pathogens found on leaves. There are those from the Plectomycete Erysiphales, the powdery mildews, which are wholly restricted to the phylloplane except for haustoria in the epidermal cells of the host leaf. All their very extensive mycellium, conidia and ascocarps are borne on the leaf surface. The second category, covering virtually all other pathogens, infect leaves and grow almost entirely within them with only their reproductive structures having access to or being produced on the outside. These latter exhibit all gradations, from those which produce an appressorium from their spore and penetrate immediately, to those which have a prolonged and relatively extensive phase of epiphytic non-parasitic hyphal growth on the leaf surface before they penetrate (Fig. 1.3). The spores of many of these pathogens remain dormant for considerable periods only germinating as host resistance begins to fall prior to senescence or following a suitable change in the weather. Again no clear distinction can be drawn between pathogens of this latter type and some of the common primary saprotrophs. *Botrytis cinerea* is one of the latter but is also a necrotrophic parasite of some hosts under particular environmental conditions, such as prolonged very high humidities which favour it but not its host. Its conidia may then germinate on the leaf surface and after a phase of epiphytic growth penetrate and bring about a soft watery rot by means of its pectolytic enzymes.

Exochthonous fungi

Spores of pathogens which are unable to infact the leaves on which they have landed may also be present. They may remain decimant or they may germinate before they recognize

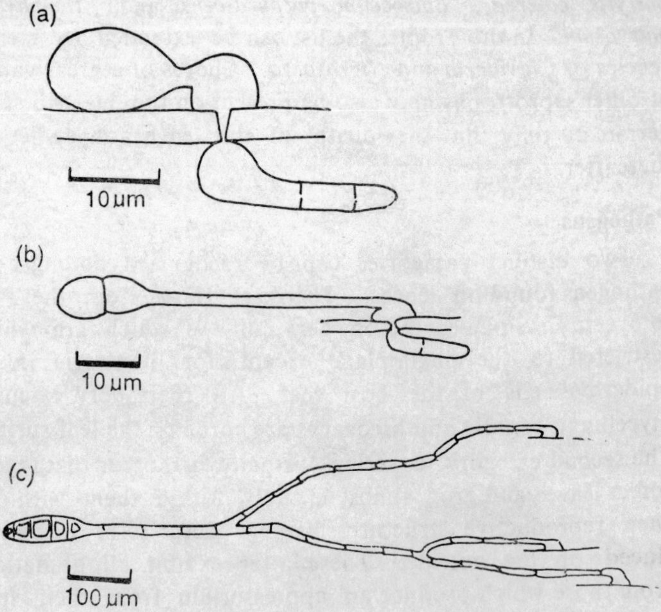

Fig. 1.3. Three different types of growth shown by pathogenic leaf-inhabiting fungi on leaf surfaces. (*a*) *Botrytis fabae*. (*b*) *Mycosphaerella ligulicola*. (*c*) *Cochliobolus sativus*.

that they are on the wrong host. They may contribute, like pollen, to the nutrients available on the leaf surface. They could be included with the exochthonous or casual fungi as they are found on leaves but do not grow there. The latter are unable to gain any nutritional advantage from the habitat which is clearly a dead end for many but by no means all. Any soil fungi with air-borne spores may be trapped on leaves and later washed off by rain onto the soil beneath and so they are successfully dispersed. Spores of many coprophilous fungi on herbivore dung are discharged onto grass leaves surrounding the dung and remain there until the grass is eaten by herbivores. Passage through the gut of a herbivore may be necessary to trigger-off their germination. Direct dung to dung dispersal is abortive. Thus impaction onto leaf surfaces is important if they are to complete their dispersal and life cycle.

The leaf surface as a habitat for fungi

The leaf surface is a most inhospitable niche in both physical and chemical terms for fungi. Although transpiration may mitigate against extreme low levels of relative humidity, the fungi are repeatedly dried by the sun and wind and re-wetted by rain and dew. They are not insulated against temperature fluctuations and as such are subjected to marked and very rapid variations in temperature. Even in temperate climates in relatively still air, leaf surfaces may be 10-12°C above ambient in the sun at one moment and in the next 2° below ambient as a cloud passes over the sun. They are exposed to the harmful ultraviolet component in daylight. Nutrient sources must always be fluctuating and low competition for them must be severe.

Microbial interactions in the Phylloplane

Interest in the leaf surface as a habitat for fungi has centred mainly on the fact that it is here that any pathogen must spend a critical period of time until it can establish and infect. During this time it is not only subjected to such environmental stresses but it may also be subjected to antagonism from the phylloplane inhibitants as well as from the host itself. A great variety of microbial interactions occur in the phylloplane and in the applied field thoughts are turning to consider the possibility of achieving biological control of some leaf pathogens by building up sufficiently large populations of phylloplane inhibitants. This may be an accordingly difficult objective to achieve but a consideration of some of the research which led to the development of such ideas gives further insight into the biology of phylloplane fungi.

Pollen as a nutrient source and competition for nutrient

The fact that phylloplane inhabitants such as *Cladosporium* and *Sporopolomyces* could benefit from nutrients leaked from pollen grains was convincingly demonstrated by Fokkema (1971). Rye leaves from two separate plots, in one of which the plants had their inflorescences removed or covered, so that no pollen fell onto the leaves below, were taken from plants and washed, twice a week from early June to September, in 1968 and 1969. The washings were plated out onto nutrient agar and the colonies of *Cladosporium* spp. which developed were counted, the assumption being made that each colony arose from a single spore (Fig. 1.4). In 1969, the number of

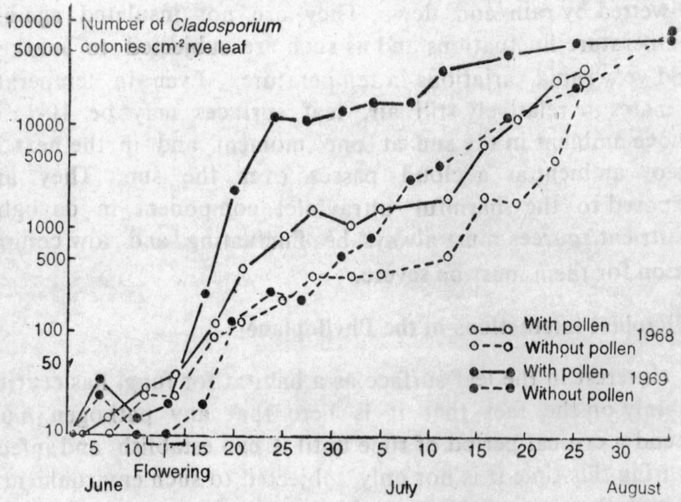

Fig. 1.4. The successive changes in the number of *Cladosporium* spp colonies per cm² ery leaf during the season. The data are the means of the numbers of colonies from washings of eight penultimate leaves.

colonies from leaves with pollen rose from 15 to 13,000 cm² two weeks after flowering. On leaves without pollen the numbers were 10 and 550 respectively. On leaf senescence the colonies recorded from all leaves reached the same levels. The leaves at this stage leaked more nutrients and the stimulating effect of the pollen wore off. The larger number of colonies of *Cladosporium* recorded in 1969 was shown to be due to more pollen on the leaves, a mean of 3450 cm^{-2} as against 300 cm^{-2}. There was frequent rain after flowering in 1968 so that the pollen was washed off. Stimulation in the presence of pollen was not restricted to *Cladosporium*. *Aureobasidium pullulans* and *Sporobolomyces roseus* were also stimulated. For the latter, two weeks after flowering, 33,600 colonies cm^{-2} were recorded from rye leaves with pollen and only 3800 colonies cm^{-2} from rye leaves without pollen.

The Loculoascomycete *Cochliobolus sativus* is a leaf pathogen of rye which makes a variable amount of epithytic growth before penetration into the leaf (Fig 1.3*c*). Fokkema inoculated rye leaves with conidia of *Cochliobolus*, together with pollen and without pollen. The effect of the pollen on successive stages of the infection process and the resultant necrosis is given in Table 1.1. Leaves inoculated with *Cochliobolus* and pollen had a significantly larger percentage of necrotic areas. He attributed this to the pollen leaking out nutrients which greatly stimulated the superficial

Table 1.1. Effect of pollen on successive stages of the infection process of rye leaves by *Cochliobolus sativus*.

		Time after inoculation			
			2-3 days		*7 days*
Experiment	*Pollen addition*	*Mean number of germs tubes/ 100 spores*	*Mean mycelium length in μm mm⁻²*	*Mean number young lesions 10 cm⁻²*	*Mean % nectrotic area*
1	—	23	0	17	2
	+	137	3600	113	58
2	—	67	250	22	2
	+	102	3250	56	35

growth of the mycelium. The increased the inoculum and so more young lesions unit area were obtained and hence eventual necrosis. In these experiments both the phylloplane inhabitants and the pathogen were relying, in part at least, on the same nutrient source and in nature they might well compete for such a source. Some degree of biological control could be achieved if the phylloplane inhabitants could markedly neutralize the stimulating effect of the pollen when inoculated with both it and the pathogen. This is exactly the effect that Fokkema later observed. The relative inhibitory effect on surface mycelial development of *Cochliobolus* on leaves was depressed by 72% and the extent of necrosis by 75% on rye leaves inoculated with *Cochliobolus*, pollen and *Aureobasidium*, as compared with leaves inoculated with *Cochliobolus* and pollen and *Cochliobolus* and *Aureobasidium* only. Effective competition for nutrients by *Aureobasidium* appears to be an adequate explanation for these reductions.

Antagonistic reactions

A number of such interactions has been reported but competition for nutrients is not the sole explanation for some of these. In some cases inhibitory substances produced by the phylloplane inhabitants may also be involved and normal leaf exudates rather than pollen may be the nutrient source. Pace and Campbell (1974) found that *Aureobasidium pullulans* and *Epicoccum purpurascens* were common in the phylloplane of *Brassica* spp. and that they were antagonistic to the wound parastic *Alternaria brassicicola* in culture. Their growing colonies inhibited mycelial growth of *A. brassicicola* and their germinating conidia inhibited the germination of its conidia. They inoculated leaves of cabbage and Brussels sprout after wounding with *Alternaria brassicicola, Aureobasium pollulans* and *Epicoccum purpurascens* separately and with *A. brassicicola* plus *A. pullulans* and *A. brassicicola* plus *E. purpurascencs*. There was an 80-100% infection with *A. brassicicola* alone but none with the two saprotrophs alone. The percentage of successful infections by *A. brassicicola* was reduced where it was inoculated with either of the two saprotrophs (Fig. 1.5). The reduction was greater when the saprotrophs were inoculated 14 h prior to *A. brassicicola*. Both *A. pullulans* and *E. purpurascens* were capable of active growth on the leaf surface so might compete with the pathogen for nutrients in the form of leaf exudates. This could explain why they were more effective when inoculated 14 h prior to the pathogen. But they may also produce inhibitory substances. Some evidence for this is that a 50% reduction in successful infection was obtained when conidia of *A. brassicicola* were suspended in a culture filtrate of *A. pullulans* and used instead of water for inoculation. The culture medium itself enhance infection.

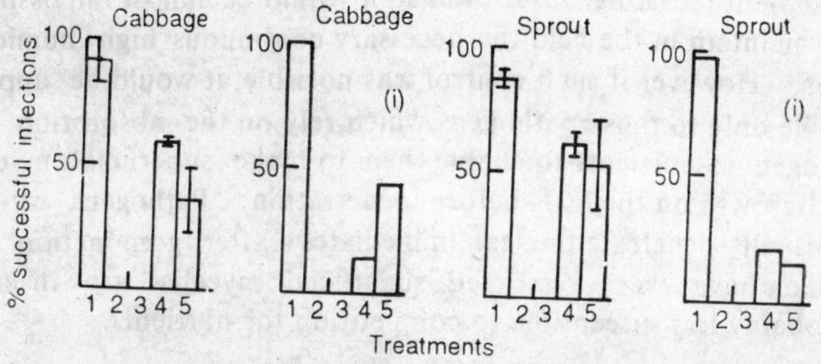

Fig. 1.5. Percentage (mean of three values) of successful infections obtained on cabbage and Brussels sprout leaves. Treatments : 1=*Alternaria brassicicola* alone ; 2=*Epicoccum purpurascens* alone ; 3=*Aureobasidium pullulans* alone ; 4=*A. brasicicola* + *E. purpurascens* ; 5=*A. brassicicola*+ *Au. pullulans*. The vertical bars are the standard deviations from the mean. (*i*)=after pre-incubating the antagonist for 14 h. Reproduced by permission of the British Mycological Society.

Towards biological control

These sorts of antagonism must obviously be having some effect in the field but the question to be answered in terms of achieving biological control is how can their effects be maximized ? One approach is to manipulate the system to stimulate the phylloplane inhabitants. Biological control using the indigenous population, rather than introducing others, would be possible if a sufficiently large population of antagonistic phylloplane inhabitants could be built up. Bashi and Fokkema (1977) have shown that continuous high humidities and nutrients in excess of those exuded by leaves are necessary to maintain a phylloplane population of *Sporobolomyces* dense enough to have sufficient antagonistic potential to control *Cochliobolus*. *Sporobolomyces* is particularly sensitive to low relative humidities. Populations on leaves decreased markedly when maintained at 65% RH. Any added nutrients would have a stimulatory effect on the pathogen as well. To be effective in stimulating only the saprotroph nutrients would have to be added just prior to any significant build up of spores of the pathogen on the leaf surface. This would require accurate disease forecasting. But even so it would be almost impossible to maintain in the field the necessary continuous high humidities. However if such control was possible, it would be applicable only to those pathogens which rely on the absorption of exogenous nutrients to enable them to make superficial mycelial growth on the leaf before penetration. Pathogens which normally penetrate the leaf immediately after germination of which have every restricted superficial mycelial growth are probably less susceptible to competition for nutrients.

The alternative approach is to allow the natural control measures to proceed and to avoid the indiscriminate use of fungicides which may affect the phylloplane inhabitants more than the pathogen. Pace and Campbell (1974) noted that the systemic fungicide Benomyl gave a good control of many diseases but not the leaf spot of brassicas caused by *Alternaria brassicicola.* The pathogen is resistant to it. The two antagonistic saprotrophs which they used. *Aureobasidium pullulans* and *Epicoccum purpurascens* are inhibited by Benomyl. Therefore use of this fungicide could make the disease worse. Fokkema found that *Cochliobolus* is also relatively resistant to Benomyl, and that inoculation of rye leaves with *Cochliobolus* just after flowering (*i.e.* with pollen), resulted in 60% less necrosis on water.sprayed leaves than on Benomyl-sprayed leaves. At the time water sprayed leaves had a natural phylloplane population of 10,000 spores cm^{-2} and Benomyl-sprayed leaves only 1200 spores cm^{-2}. This implies that the Benomyl had reduced the antagonistic capacity of the phylloplane inhabitants but it also provides direct field evidence for naturally occurring biological control.

Antagonism via lysis, antibiotic production or pH changes

Other forms of antagonism are exhibited in the phylloplane. Bacteria may lyse fungal spores. Chitinolytic enzymes are usually involved. Lenne and Parberry (1976) noted clusters of bacteria surrounding lysed conidia and germs tubes of the pathogen *Colletotrichum gloeosporioides* on leaf surfaces. Appressoria are necessary for penetration to occur. The bacteria failed to lyse these. They have melanized walls and there are numerous reports of melanized structures resisting the lytic action of bacteria. The production of appressoria was enhanced in the presence of bacteria but was reduced by added nutrients, such as 1% glucose peptone solution. The stimulated production of appressoria in the presence of bacteria is a normal response of the fungus to a hostile environment. Desiccation and starvation also cause appressorial formation. This response serves as an important short term survival role during the infection phase. It should be noted that in this particular case added nutrients increased germ tube growth but fewer appressoria were formed. Since the latter are necessary for penetration, added nutrients may in this case enhance disease control.

Several phylloplane inhabitants, such as *Aureobasidium* and *Sporobolomyces* have been shown to produce antibiotics in culture although there is no direct evidence that they play a role *in vivo*. Although antibiotic production by bacteria on leaf surfaces does not appear to be very widespread, some bacteria have been shown to produce antifungal peptides which, under experimental conditions at least, reduce incidence of disease caused by a number of species of *Colletotrichum*.

Some fungal leaf pathogens are very sensitive to pH changes. In *Septoria nodorum*, for instance, spore germination is inhibited below pH 6. Conidia of this fungus, placed around the edge of a growing colony of *Botrytis cinerea*, failed to germinate. The pH of the medium fell to below 6 in advance of the hyphal tips of *Botrytis*. Such a mechanism could operate in the phylloplane.

Fungistatic substances produced by leaves

In addition to nutrients, leaves of many plants may exude fungistatic substances which cause inhibition of spore germination or restriction of germ tube growth. Phenols are the most widely known fungistatic substances produced by leaves. They are responsible for the inhibition of spore germination of the apple scab fungus, *Venturia inaequalis*, on some apple cultivars. Gallic acid has been identified as an antifungal component in droplets of dew obtained from sycamore leaves. Apart from substances formed within the leaf cells and exuded onto the surface, some constituents of the cuticular waxes may also be fungistatic. An acidic ether-soluble fraction from the wax of apple leaves inhibits the growth of the apple mildew fungus, *Podosphaera leucotricha.* The properties of waxes on leaves will also affect the exudation of both nutrients and antifungal substances. Waxes with high proportions of more hydrophobic constituents will tend to limit the movement of exudates to the surface.

Thus at the leaf surface a series of complex interactions occur between pathogen/host/phylloplane inhabitants/environment. Numerous aspects and the outcome of many of these interactions are still to be discovered, but it is evident that the phylloplane inhabitants act in some sort of buffering capacity against some pathogenic fungi at least.

Distribution of the common primary saprotrophs

Eventually the leaf senesces either naturally or prematurely after supporting, in some cases, one or more pathogens. Of the multitude of fungal spores of a vast array of a species which are impacted onto leaf surfaces only relatively few succeed in colonizing the leaves as they senesce and grow as active saprotrophs within the leaf tissues after death. These common primary saprotrophs are virtually ubiquitous colonizers. On most leaves such as those of deciduous trees, shrubs, herbs, grasses including cereals and even bracken, most are usually present and exceptions are difficult to find. Pine needles are a particularly selective substrate and of these fungi only *Aureobasidium pullulans* is ever at all common. In the tropics, *Alternaria alternata* is less common and is replaced by *Nigrospora* spp., especially *N. sphaerica,* and *Curvularia* spp., especially *C. lunata,* as is evident from examining senescent leaves of guinea grass (*Panicum maximum*) and banana (*Musa sapientum*). The differences are also reflected in the comparison of the dry air-spora of tropical and temperate climates. The association of these particular fungi has been noted on other substrates, such as cereal stubble and cotton fabrics exposed to the weather. The blackening of the ears of cereals in a damp season is caused mainly by *Alternaria*, *Cladosporium* and *Epicoccum*. Christensen and Kaufmann (1965), in their studies on the deterioration of grain, designated these and others, such as *Chaetomium*, *Fusarium* and *Rhizopus* spp., as 'field fungi'. This is an appropriate term as they are almost always and constantly associated with exposed freshly decaying green parts of plants. On leaves they are usually associated with one or more other saprotrophs which are more restricted in the range of leaves which they colonize. These restricted primary saprotrophs may be confined to a particular host genus or a related group of plants. *Readeriella mirabilis* and *Piggotia stellata* appear to be restricted to *Eucalyptus*. Several species of *Leptosphaeria*, such as *L. microscopica*, are restricted to the Gramineae and *Fusicoccum bacillare* and *Sclerophoma pithiophila* are both very common on pine needles but the latter, at least, is also found on other coniferous leaves. In many of these substrate specificity might be synonymous with and explained by, host specificity. Many of these, although very active saprotrophs, may have an additional advantage in that they can gain access as parasites, *S. pithiophila*, for instance, has been associated with the defoliation of the current year's needles of *Pinus sylvestris*.

As a group these common primary saprotrophs may be well-established in leaves long before leaf-fall. For example *Cladosporium herbarum* often colonizes and produces conidia on damaged necrotic parts of beech leaves in June, within two months of their unfolding. The duration of their persistence on leaves once they are in the litter is dependent upon many variables, one of which is the texture and another the composition of the leaves. In general, free leaves, such as those of ash and sycamore, which decompose and disappear rapidly from the litter, support a more substantial growth of these common primary saprotrophs for a shorter time than do leaves of beech and oak which persist much longer in the litter. On beech leaves, for example, *C. herbarum* persists in high frequency through the winter after leaf-fall until the following June and disappears after September.

Similar sequences can be found on other substrates. Primary saprotrophs are the first fungi to appear on flowering stems of cocksfoot, *Dactylis glomerata*. They are present on the basal leaves in early summer and progress up the stems as successive leaves senesce. They are well-established by July and August on the upper leaf sheaths and intermodes of stems which flowered in late May and June and they persist there until the following summer. On nettles, *Urtica dioica*, primary saprotrophs colonize the upper leaves in August or September of the year of flowering at the onset of basipetal senescence. They again persist throughout the winter until the following spring and summer.

Attributes of the common primary saprotrophs

The intriguing aspect of this particular fact of fungal ecology is to ponder why so few of all the fungi use equipped to assume this role of primary saprotrophic colonizers of such exuberantly plentiful substrates.

Nutrients

In the well-known scheme for fungal successions proposed by Garrett (1963), the primary saprotrophs to invade are 'sugar fungi'. They are not-cellulolytic and rely upon readily such as available sugars, hexoses and pentoses, and other carbon sources simpler than cellulose such as pectins and starch. These fungi also normally possess a high mycelial growth rate and a capacity for rapid spore germination. The classic example of such fungi is the Zygomycete Mucorales, common on herbivore dung. Primary saprotrophic sugar fungi are usually very ephemeral because of the transient nature of their substrate. The persistence of the common primary saprotrophs for months on leaves would suggest that they are not confined to such ephemeral substrates. The ability to utilize cellulose is often regarded as essential for saprotrophic fungi and the majority, except most Zygomycotina and Mastigomycotina, can do this. Of this particular group of leaf saprotrophs only *Aureobasidium pullulans* is non-cellulolytic. It probably relies on pectic substances for its carbon sources and this ability is often used to explain its role as a primary colonizer. None of the others is markedly cellulolytic when compared with some of the Basidiomycotina which later colonize leaves in the litter layer. For instance, Hering (1967) inoculated oak leaves sterilized by γ irradiation with *A. pullulans*, *Cladosporium herbarum* and *Mycena galopus*, a Hymenomycete agaric from the leaf litter, and measured loss in mass after six months at 9-15°C. The two former brought about a loss of 2 and 4% respectively and the latter 15—20%. Not all the loss in mass was of cellulose but in the latter case the loss corresponded with the utilization of about one sixth of the total cellulose present. On filter paper cellulose, various isolates of *Alternaria alternata* brought about losses of 4—8% in 14 days and *Epicoccum purpurascens* about 4%. For comparison, under the same conditions the vigorously cellulolytic *Chaetomium globosum* brought about a 10% loss. It must be remembered that mass loss methods measure only the amount of substrate, in this case cellulose, respired and lost as carbon dioxide and water and not the amount incorporated into fungal material. The cellulytic ability of these fungi thus varies and they all probably use simpler carbohydrates, such as sugars and starch, as long as they last and then go on to utilize cellulose even if to a limited extent and slowly. Thus they persist. This may be placing undue emphasis on their carbohydrate nutrition to the neglect of their nitrogen requirements. The nitrogen supply might be extremely critical in determining their distribution. In culture they can all use nitrate, ammonia or amino acids as their sole nitrogen source but nothing is precisely know as to what sources are available to them within the leaf. Indications of the over-riding limitations of their nitrogen supply are seen when leaves are amended with an available source. Foliar applications of 5% urea solution, after harvest but before leaf fall prevent ascocarp development in the apple scab fungus, *Venturia inaequalis*, on the overwintering leaves and is used as a control measure to limit the ascospore inoculum available to infect the newly emerging leaves in the following spring. Birchill and Cook (1971) in studying the mode of action of the urea demonstrated that both chemical and microbial changes occurred in the leaves after treatment. Marked alterations occurred in the composition and density of fungal and bacterial populations present on treated leaves. The urea in particular enormously increased the relative abundance, as assessed by the number of conidia produced, of both *Cladosporium* spp. (Fig. 1.6) and *Alternaria*

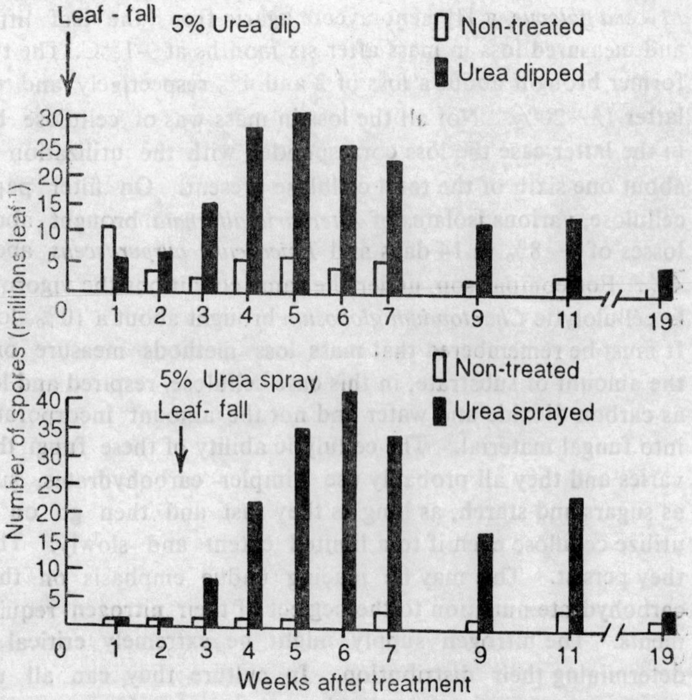

Fig. 1.6. Effect of urea dip and urea spray applied 17 October 1967 on numbers of *Cladosporium* spores washed from overwintering leaves. (After Birchill and Cook, 1971).

spp. So marked was the development of the conidia and conidiophores of *Cladosporium* that they could be seen with the naked eye as olive green lawns. Many times more conidia were produced overall on the treated leaves suggesting that the added nitrogen enabled them to utilize more carbon sources and thus outcompete *Venturia* for substrate. Application of urea to fallen pine needles also dramatically changes the fungal succession. *Cladosporium herbarum*, rather than being an occasional inhabitant, is again stimulated to develop to such an extent that its conidiophores may cover the needles as a dense felt and other common primary saprotrophs, such as *Epicoccum purpuruscens*, which does not normally occur on pine needles, are stimulated to develop by the urea. The mode of action of the urea is not known in this case out is complex ; its property of acting as an alkali may be one important aspect of its effect. For example, several agarics which have not been recorded from pine litter appear when plots are treated with urea or alkalis. An example is *Myxomphalia maura* which is characteristically found on the alkaline ash of bonfire sites on acid soils in coniferous woods. It is not found on woods on alkaline soils so is not a calcicole. *M. maura* is markedly encouraged by the addition of lime to pine litter. Pine needles treated with sodium carbonate are also colonized by *C. herbarum* and *E. purpurascens*. Urea and alkalis both produce similar effects on the litter. They both cause it to darken, become water-soaked and raise the pH from about 3.5-4.0 to 5.5-6.0. They both bring about the release of ammonia from the litter and its use as a nitrogen source may be another important factor inducing these changes.

These saprotrophs are not primary colonizers solely because they grow faster than any other would-be colonizers. Their only attribute with regard to growth rate is their great variability. *Aureobasidium pullulans* produces its slimy conidia very rapidly, but its yeast-like colonies are relatively slow growing. *C. herbarum* also sporulates rapidly but its growth rate is even slower. *Botrytis cinerea* grows very rapidly and *Alternaria alternata* and *E. purpurascens* not so rapidly but faster than *C. herbarum*.

Tolerance to desiccation

Senescing leaves on the tree and recently fallen leaves are very prone to drying out and also subject to strong sunlight. Webster and Dix (1960) compared the growth rates, latent period for germination, germ tube growth rate at 100% RH and the lowest RH which spore germination occurred in three primary colonizers with two later secondary colonizers. *Torula herbarum* and *Tetraploa arictata*. They found that there was little difference between the capacity of the mycelium of the various colonizers two grow at low humidities and the primary colonizers did not make better growth at low humidities. But it can be seen from Table 1.2 that under favourable humidities (100% RH), *A. alternata* and *E. purpurascens* not only grew

Table 1.2. Mycelial growth rate, latent period for germination, growth rate of germ tubes at 100% RH and lowest RH at which spores germinated.

Primary colonizers	*Mycelial growth rate (mm day⁻¹)*	*Latent period (h)*	*Germ tube growth rate (gm ⁻¹)*	*Lowest RH at which germination occurs*
Cladosporium herbarum	2.96	6—12	4.2	89%
Alternaria alternata	6.41	3—6	29.1	89%
Epicoccum purpurascens	6 45	0—3	31.6	92%
Secondary colonizers				
Torula herbarum	1.41	12—18	0	Water
Tetraploa aristata	3.34	12—18	2.4	98%

faster than the secondary colonizers but also had a shorter latent period before germination and their germ tubes grew faster. These features, coupled with the fact that their conidia can germinate at lower relative humidities, would give them an advantage over the others in that their conidia would germinate under less ideal conditions of humidity and they would quickly exploit, by virtue of their more rapid growth rate, any changes to more humid conditions.

Because of the rapidly fluctuating conditions on the leaf surface, germinating conidia may rapidly dry out before penetrating into the leaf. Diem (1971) has investigated the survival at low humidities of germinating conidia of *C. herbarum, A. alternata* and some casual inhabitants of the phylloplane. He found that germinating pigmented conidia, such as those of *Cladosporium* and *Alternaria*, were more resistant than germinating colourless conidia of *Aspergillus* and *Penicillium*. The germ tubes of *Cladosporium* were remarkedly resistant. Some 90% grew on at 100% RH after 8 h in a desiccator over anhydrous calcium chloride and 99% did so after being kept at

40% RH for 8 h. This would indicate that if they germinated in the more humid conditions of the night and had not penetrated into the leaf by the morning they could survive the drier conditions of the day. Germ tubes of the conidia of *Alternaria* were equally resistant but some failed to grow on after periods at relative humidities below 65% but the conidia either germinated again from another cell or from a lateral branch from below the damaged part of the germ tube. In contrast, the germ tubes of the conidia of *Aspergillus* and *Penicillium* were no longer viable after periods at 85% RH. Species with coloured conidia are thus more likely to be successful in the phylloplane and as subsequent primary colonizers of the leaves. But it should be noted that the conidia of *Aureobasidium* and *Botrytis* are not pigmented. The biotrophic Erysiphales which produce their mycelium on leaf surfaces also have colourless hyphae and conidia. Thus pigmentation is a useful but not an essential attribute to possess.

Primary saprotrophs also show an equally remarkable tolerance to desiccation in their hyphal tips as distinct from germ tubes. Hyphal tips are very delicate structures but in these fungi they survive periods of extreme desiccation, some as long as three weeks, above a saturated solution of potassium nitrate (a_w 0.45). Thus they can rapidly exploit the return to favourable conditions of humidity with no apparent loss of previously synthesized biomass. Other fungi do not show this ability. This attribute may be critical in enabling such fungi to tolerate cycles of wetting and drying.

Survival structures

Once established on the leaf surface, most of the common primary saprotrophs produce some form of pigmented survival structure : *Cladosporium herbarum* minute micro-sclerotia ; *Botrytis cinerea* and *Epicoccum purpurascens* sclerotia ; and *Aureobasidium pullulans* aggregates of chlamydospores (see Fig. 1.2*b*). All have a pigmented mycellium. Such structures and pigmentation protect against desiccation, ultra-violet light and microbial lysis.

It is thus clear that the common primary saprotrophs possess a multiplicity of attributes by which they have become successfully adapted to this relatively inhospitably niche with each fungus possessing its own particular complex of attributes, not all necessarily the same.

As suggested, in temperate climates some of the common primary saprotrophs produce ascocarp initials in the late autumn in the year of leaf-fall as do a number of leaf pathogens, such as *Apiognomonia errabunda* on beech and *Venturia inaequalis* on apple. Ascospores are discharged from these over the period early April to early June. This is the time when the next crop of leaves is unfolding. The initially spore-free leaves become impaction sites for air-born ascospores and under favourable conditions infection occurs. Such a life history is of particular significance in leaf pathogens with restricted periods of spore formation and release and where the host virtually frees itself of infection by shedding all its leaves prior to its dormant season. The requirement for an overwintering phase, a period of low temperature (5-8°C), before ascocarp initials mature is very common in these fungi. The maturation and release of the ascospores thus coincide with the breaking of bud dormancy of the host. Thus telemorphic states of *Aureobasidium pullulans* (*Guignardia fagi*) and *Cladosporium herbarum* (*Mycorsphaerella tassiana*), which are common on fallen leaves, may be regarded as additional survival structures adding an ascospore inoculum to the conidial inoculum available in the spring.

Subsequent colonizers and leaf decay

These initial colonizers gradually disappear, being replaced by other leaf-inhabiting saprotrophs which begin to reproduce in the late summer of the year after leaf-fall, reach a maximum in the autumn and persist over the winter, until the spring. These include a very wide variety of conidial fungi, such as *Polyscytalum fecundissimum* and *Chalara cylindrospora*, Ascomycotina such as *Microthyrium fagi*, and *Helotium caudatum*, and Basidiomycotina with minute basidiocarps, such as *Lachnella villosa* and *Pistillaria pusilla* on beech leaves. With fragmentation in the final stages of decomposition, the fungal flora becomes dominated by typical soil-inhabiting fungi, mainly Zygomycete Mucorales, especially species of *Mucor* and

Mortierella, and conidial fungi, such as species of *Penicillium* and *Trichoderma*, together with litter-decomposing Hymenomycete Agaricales, such as species of *Collybia* and *Mycena*.

The soil-inhabiting fungi grow up from the soil *via* the continuum of organic debris. The role which they play in the decomposition process has not been fully elucidated. At this stage the Mucorales are certainly not using any simple carbohydrates initially present in the leaves as these would have been utilized already. They could be living in association with the cellulolytic Agaricales as commensals by taking a share of the hydrolytic products of cellulose and thus acting as secondary saprotrophic sugar fungi rather than primary ones. Alternatively, they could be primary colonizers of the wealth of faecal pellets produced by the micro-fauna, especially mites, as they are on pellets of *Glomeris*. The H layer of the soil is particularly rich in chitin in the form of hyphal wall fragments and exoskeletons of insects or other chitinized remains of the micro-fauna. Species from several common genera of soil-inhabiting fungi, including *Mortierella*, *Penicillium* and *Trichoderma*, have the ability to break down this very resistant substrate and their activity may well represent one of the final stages of the mineralization of primary and secondary organic materials in the soil.

Decomposition of pine needles

The time period between leaf-fall and the final decomposition of a leaf varies enormously. In cool North temperate pine forests it may be 10 years or more, in ash and sycamore under 1 year and in tropical forests mere weeks. Pine needles are extremely durable and decay very slowly. Their decomposition most often results in the formation of a mor type of soil. The needles are shed mainly in August and September and there is an accumulation of considerable bulk of leaf litter, each successive leaf-fall burying the previous one so that a stratified litter layer is produced. The animals in the litter are sufficiently small not to disturb this stratification and such a litter layer well illustrates the diversity of organisms, fungi and animals, involved in the decomposition process. A very considerable amount of potential energy is available for micro-organisms in this litter. In *Pinus sylvestris*, production of needles accounts for about one third of the total productivity and accounts for about 60—80% of the total litter. Although the decay process is a continuous, if fluctuating, one, it is convenient to recognize a number of stages. The A horizon may thus be divided into L, F_1, F_2 and H layers. The L layer consists of freshly fallen, undecomposed needles, light brown to buff in colour and others somewhat darker in colour, which have fallen earlier. Needles remain in this layer for about six months. They all have a high tensile strength, a relatively low but fluctuating moisture content and form a loose, uncompacted layer on the litter surface. In this layer the needles are very susceptible to drying out and conditions are unfavourable for continuous fungal growth. In the upper parts of the F_1 layer, the needles are grey, becoming dark brown with depth but recognizable as needles. Their tissues become softened and they have a low tensile strength and a high moisture content. They remain in this layer for about two years. Below, in the F_2 layer, the character of the needles again changes. They are greyish, fragmented and compressed, but again still recognizable as needles. The mesophyll collapses and most bear dark amorphous faecal masses of the micro-fauna. Eventually the remains of the needles enter the H layer which consists of an amorphous mass of faeces and the remains of both the micro-fauna and fungi, the needles having undergone complete physical reduction. Below this layer is an intimate mixture of humus and mineral soil.

Two factors that may greatly influence the sequence of decomposer fungi on the needles are the time at which they fall and their previous history. Pine needles are far from being a homogeneous entity and at needle fall vary in age, physical structure, nutrient content and the presence or absence of fungal colonizers in or on the needles. The needles have a very thick, waxy cuticle and support a much sparser population of phylloplane inhabitants. *Sporobolomyces roseus*, although present on most attached needles, occurs in very low frequencies only. This contrasts markedly with its abundance on leaves of deciduous trees and herbaceous plants. It decreases rapidly on needle fall, whereas some other yeasts, such as *Bullera* spp., increase in frequency and persist. A number of other mycelial fungi such as the conidial *Sclerophoma pithiophila* may grow and sporulate on the leaf surface. Vigorous ones, such as *Lophodermella sulcigena*, an apothecial Ascomycete, and *Coleosporium senecionis*, a rust, cause premature needle cast, either directly or by predisposing first year needles to infect'on by secondary pathogens. For example, *L. sulcigena* infects young first year needles and predisposes them to infection by *Hendersonia acicola* or *Lophodermium pinastri* and finally *Naemocyclus niveus*, which cause the needles to fall in their first summer. Such weak pathogens may colonize the needles directly but spread very little until senescence. They may also gain access via tissues damaged by insect pests. Living needles may also be colonized by *Fusicoccum bacillare* or *Sclerophoma pithiophila*. Needles infected by either of these two conidial fungi soon die and turn brown but remain attached to the tree. Such needles again fall in the summer. *S. pithiophila* is also a frequent colonizer of needles containing high nutrient levels, such as first year needles, shed while still green and of needles of felled pines. Clearly pine needles can fall at varying times of the year and may already be colonized by a variety of fungi which have already initiated the process of decomposition *Lophodermella sulcigena* actively decomposes the mesophyll tissue and *Hendersonia acicola* may remove much of the cellulose, reducing the needle to a skeleton of epidermal waxes and lignified tissues. *Lophodermium pinastri* produces pigmented diaphragms across the needles delimiting the extent of its colonization Fig. 1.7*a*). Such parts later escape extensive internal attack by saprotrophic needle-inhabiting fungi. This is often attributed to their inability to penetrate the melanized diaphragms but in culture at least *Lophodermium* produces powerful antifungal antibiotics and they may also play a part in restricting saprotrophic colonization. Such parts of the needles decay more slowly than uninfected parts and as-a consequence accumulate in lower layers of the litter. The saprotrophic colonization of naturally fallen needles and needles shed after parasitic attack may thus be distinct.

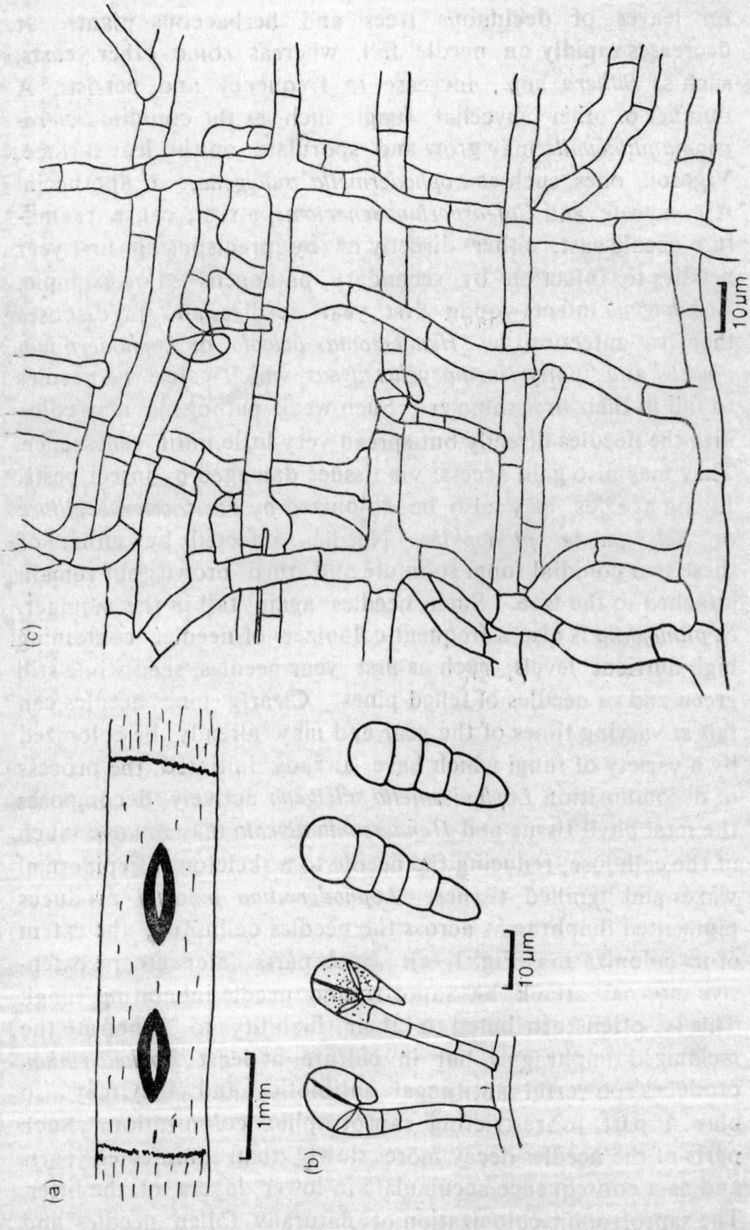
(a)
1mm
(b)
10 μm
(c)
10 μm

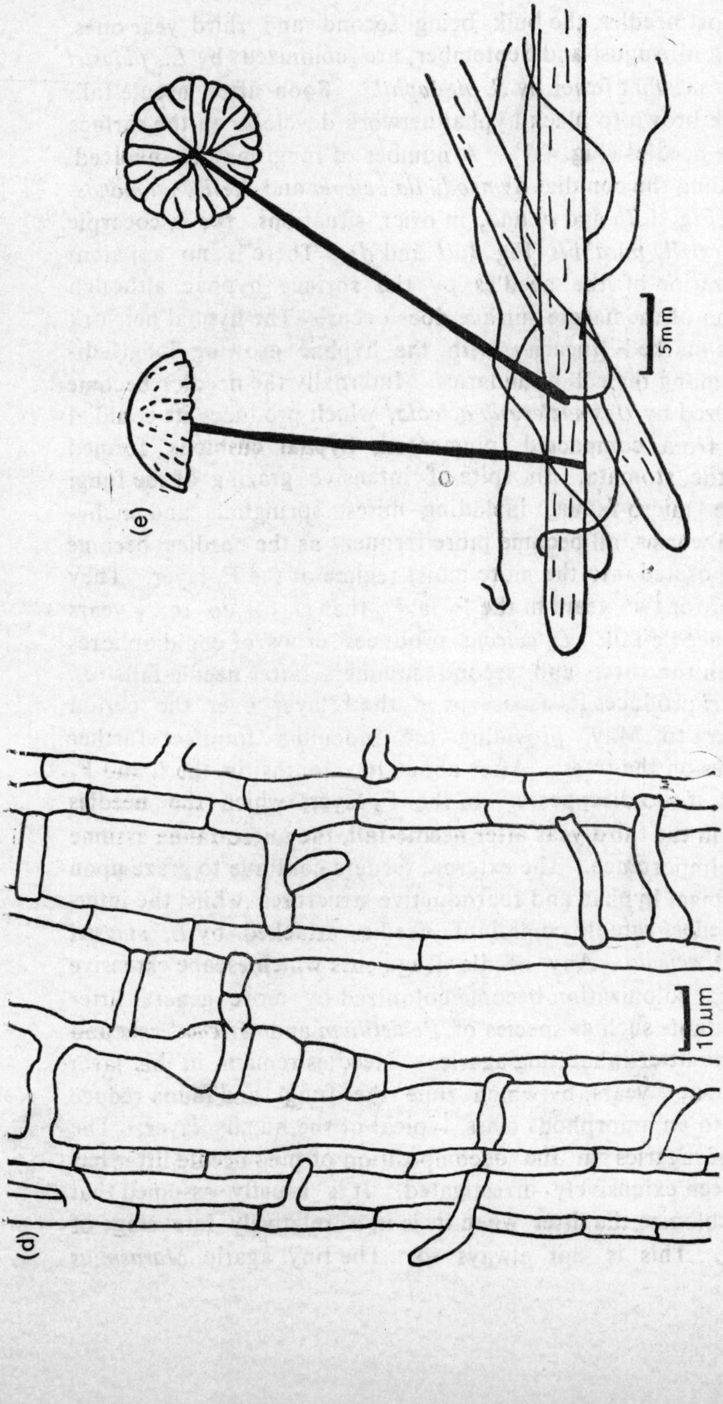

Fig. 1.7. (*Left*) (*a*) Two asoocarps of *Lophodermium pinasta* and diaphragms across a pine needle (*b*) Conidia of *Helicoma monospora* and ascospores of *Kriegeriella mirabilis*. (*c*) Conidia and hyphal network of *Helicoma monospora* on the surface of a pine needle (*Above*) (*d*) Ascospore and hyphal network of *Kriegeriella mirabilis* on the surface of a pine needle (*e*) Thread like thizomorphs and basidiocarps of *Marasmius androsaceus* an pine needle litter.

Most needles, the bulk being second and third year ones, falling in August and September, are colonized by *L. pinastri* and somewhat fewer by *S. pithiophila*. Soon after needle-fall, a dark brown to black hyphal network develops on the surface of the needles (Fig. 1.7). A number of fungi may be involved, including the conidial *Sympodiella acicola* and *Helicoma monospora* (Fig. 1.7*b* and *c*) and, in drier situations, the ascocarpic *Kriegeriella mirabilis* (Fig. 1.7*b* and *d*). There is no apparent penetration of the needles by the surface hyphae although erosion of the needle surface does occur. The hyphal network shows marked linearity with the hyphae growing longitudinally along the cell boundaries. Internally the needles become colonized by *Desmazierella acicola*, which produces its conidial state from compacted, pigmented, hyphal cushions formed over the stomata. In spite of intensive grazing of the fungi by the micro-fauna, including mites, springtails and enchytraeid worms, all become more frequent as the needles become incorporated into the more moist regime of the F_1 layer. They persist for two years in the F_1 layer, that is for up to 2½ years after needle-fall. *D. acicola* produces crops of conidiophores in both the first and second summers after needle-fall. *L. pinastri* produces its ascocarps in the L layer over the period January to May, providing the inoculum to infect further needles on the trees. After about 10 months in the L and F_1 layers, it too disappears. In the F_2 layer, which the needles enter in the third year after needle-fall, the micro-fauna assume more importance. The external feeders continue to graze upon the fungal hyphae and reproductive structures whilst the internal feeders rapidly comminute needles attacked by *L. pinastri* and *D. acicola*. Any needle fragments which escape extensive internal colonization become colonized by more general litter inhabitants such as species of *Penicillium* and *Trichoderma* and by pine litter-inhabiting agarics. Needles remain in this layer for about 7 years, by which time the fungi and fauna reduce them to an amorphous mass, typical of the humus layer. The role of agarics in the decomposition of pine needle litter has not been extensively investigated. It is usually assumed that they colonize the litter when it is in a relatively late stage of decay. This is not always so. The tiny agaric, *Marasmius* *androsaceus*, is very common in pine needle litter. It is often called the 'Horse hair fungus' because its stalk is shiny black, like horse hair and is of about the same diameter. *M. androsaceus* colonizes the needles very shortly after needle-fall. Its delicate black, cotton-like rhizomorphs grow up from previously colonized needles below, binding them together in a loose tangle. Dense masses of basidiocarps may appear on the needles in the litter, any time from May to November (Fig. 1.7*e*). It is both strongly cellulolytic and ligninolytic and causes very extensive internal decomposition. The role of such Basidiomycotina should not be underestimated. Their mycelium is often prolific in both the L and F layers. Long lists of agarics have been recorded from pine woods. Richardson (1970) has estimated the total productivity of these in a woodland of *Pinus sylvestris* in Scotland, to be between 0.25-0.5 million basidiocarps 10^4 m^{-2} y^{-1}. The majority are produced from August to September. However, because of our inability to distinguish species of Basidiomycotina from their mycelium and because we know insufficient about the biology of some of these, the problem is to assess the relative contributions of the litter decomposers and the mycorrhizal fungi. Since many agarics in the litter decompose both cellulose and lignin, it is probably delignification that reduces the needles to a greyish colour in the F_2 layer.

The role of the litter micro-fauna

As in other litter systems, the micro-fauna are important agents in the decomposition process. Mites and springtails cause considerable comminution of the needles and in so doing convert them to faecal pellets. It has been estimated that a pine needle with a surface area of 180 mm^2 would have a surface area of 1.80 m^2 after comminution to faecal pellets by micro-arthropods. Such comminution would present a much larger surface area to microbial enzymes and thus be expected to increase the decomposition rate. This may not always be so. Orobatid mite pellets persist longer than the source from which they are derived. This may be due to the nature of the substances cementing the particles of the pellets together and digestion by the animal of the more easily decomposable components of the litter. It is clear, however, that the water holding capacity of the pellets is higher and the rate of evaporation from the pellets decreased. This creates a higher and more stable moisture regime which again should favour microbial activity.

The enchytraeid worms, which are very abundant is podsols in northern coniferous forests, in addition to their grazing activities, play a vital role in the absence of earthworms in this mor type litter, in mixing the amorphous remains with the mineral soil. The major group of animals involved in the decomposition of pine needle litter are mites, many of which are strictly mycophagous, with some showing marked preferences for particular fungi. They are most abundant in the moister F layers where fungi are also more active. Protozoa and nematodes also occur in pine litter and feed on the contents of living fungal hyphae. There is no clear evidence to support or refute the hypothesis that mycophagy by the microfauna stimulate the growth of fungal mycelia.

Perhaps the most important effect of the micro-fauna is that they act as a reservoir of plant nutrients as they do in decaying wood, gradually making available the minerals which have become immobilized in the fungal hyphae. The large number of different species involved and their varying life spans mean that the nutrients contained in their tissues are only gradually mineralized.

FUNGI AS DECOMPOSERS OF WOOD

Perennial woody plants are the predominant vegetation on earth. Forests form the climax vegetation of all parts of the world except where temperature and moisture extremes limit plant growth. Forests also contain the greatest biomass varying from 500 Mg 10^4 m^{-2} in tropical rain forests to 100-300 Mg 10^4 m^{-2} in northern temperate coniferous forests. Perennial woody parts above ground make up about three-quarters of this biomass. Woody tissues thus provide the bulk source of organic carbon for decomposer heterotrophs. Fungi are the major group of organisms responsible for wood decay and a number of groups of fungi are solely wood-inhabitants. They exist entirely on the components of wood. A detailed consideration of these fungi will further illustrate the versatility of fungi as saprotrophs.

The structure and components of wood

Technically, wood is the xylem cylinder inside the break of trees. In many trees it consists of an outer, light coloured sapwood and an inner, darker heartwood. The bulk of the wood consists of dead and empty lignified vessels and/or tracheids and fibres but it also contains xylem parenchyma. Much of the parenchyma in the sapwood remains alive and unlignified and acts as a food store, mainly for starch but soluble sugars, proteins, peptides, and amino acids, lipids, nucleic acids and vitamins, such as thiamine, are also present. Once the tree is dead, these afford substrates for a wide variety of fungi but are all relatively minor and ephemeral components.

Wood consists of three major components—40-60% cellulose, 10-30% hemicelluloses and 15-30% lignin. Although the biological decomposition of lignin is of critical importance in the continuous cycling of carbon, its degradation is incompletely understood. This can be attributed to many facts ; not least of these are our lack of understanding of its precise chemical structure, the diversity of its structure in different woods, our inability to produce a pure form of lignin for cultural studies and a suitable assay for lignin degradation, the availability of a very potent lignin degrader, and the general cellular and chemical complexity of wood.

Lignin, in addition to making up about one-quarter of the dry mass of wood, is undoubtedly the structurally most complex of all the polymers and the most resistant of all to microbial decomposition. It is a three dimensionally branched aromatic polymer, formed by the oxidative polymerization of three different building blocks, not just one as in cellulose. The building blocks are the phenyl propanes: coumaryl, coniferyl and sinapyl alcohol (Fig. 1.8*a*). The lignin of different plants may contain different proportions of the three building blocks. Conifer lignin consists of mostly coniferyl alcohol, with small amounts of coumaryl alcohol and minor amounts of sinapyl alcohol. In angiosperm lignin there are approximately equal amounts of coniferyl and sinapyl alcohol and minor amounts of coumaryl alcohol. These phenyl propane units are built up into a branched polymer by covalent bonding involving three major linkage types. By far the commonest, making up 40-60% of the total and most important structures form 10-20% of the linkage types and biphenyl structures another 10-25% (Fig. 1.8*b*). Thus in lignin there are three functional monomers, varying in proportions in the various lignins, and three major linkage types, but also other minor ones. There is no regular repeating unit as there is in starch or cellulose, nor are there bonds which are easily hydrolysed. Because of this structural complexity, decomposition must necessarily differ from that of most natural polymers where there is usually straight cleavage, often by hydrolysis, to produce the monomers. There is also the possibility of microbial enzymes bringing about a variety of limited changes to the intact molecule and only partially degrading it to substances which pass with little further change into humic materials. It appears that only the so-called white-rot fungi can completely decompose lignin to carbon dioxide and water. Lignin imparts rigidity and resistance to mechanical stress in woody plants and also resistance to microbial attack. Nevertheless it degraded in natural environments but degradation is a very slow process.

Types of wood decay—white, brown and soft rots

As with other decomposing substrates, the form in which wood is presented to micro-organism and the environment in which it occurs have a major influence on the path that degradation takes. The degradation of wood in the form of trunk and large branches of trees above the soil may be very different from that of the woody tissues of leaves and small roots in the soil. This may be quite different again from logs submerged in the sea.

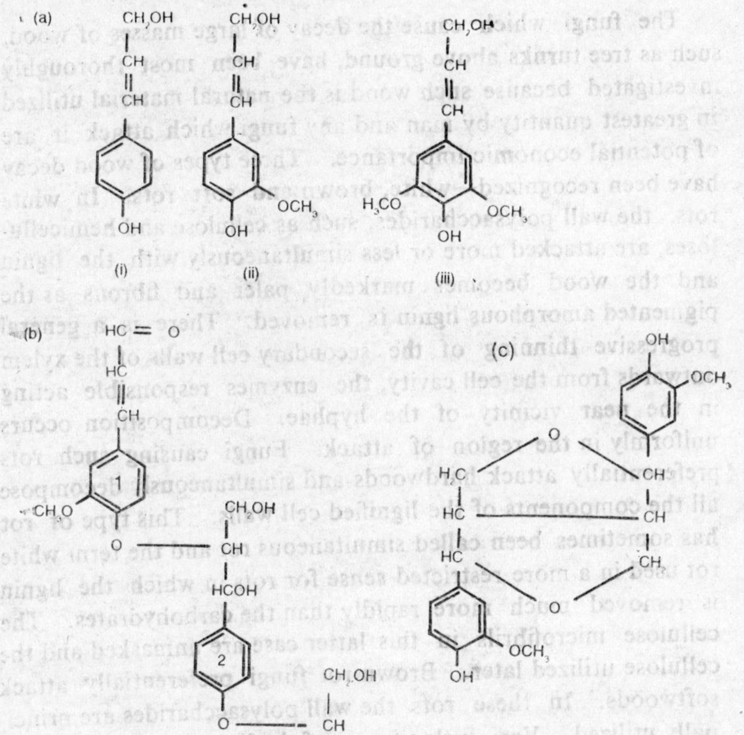

Fig. 1.8. (*a*) Building units of lignin ; (*i*) coumaryl ; (*ii*) coniferyl ; and (*iii*) sinapyl alcohol ; (*b*) Major linkage types of lignin (*c*) Pinoresional, a dilignol.

The fungi which cause the decay of large masses of wood, such as tree turnks above ground, have been most thoroughly investigated because such wood is the natural material utilized in greatest quantity by man and any fungi which attack it are of potential economic importance. Three types of wood decay have been recognized—white, brown and soft rots. In white rots, the wall polysaccharides, such as cellulose and hemicelluloses, are attacked more or less simultaneously with the lignin and the wood becomes markedly paler and fibrous as the pigmented amorphous lignin is removed. There is a general progressive thinning of the secondary cell walls of the xylem outwards from the cell cavity, the enzymes responsible acting in the near vicinity of the hyphae. Decomposition occurs uniformly in the region of attack. Fungi causing such rots preferentially attack hardwoods and simultaneously decompose all the components of the lignified cell walls. This type of rot has sometimes been called simultaneous rot and the term white rot used in a more restricted sense for rots in which the lignin is removed much more rapidly than the carbohydrates. The cellulose microfibrils in this latter case are unmasked and the cellulose utilized later. Brown-rot fungi preferentially attack softwoods. In these rots the wall polysaccharides are principally utilized. Very little, if any, of the lignin is used, although it may be altered structurally as, for instance, by the removal of methoxyl groups. With decay the wood becomes darker brown. There is no thinning of the walls. The enzymes responsible diffuse away from the hypha and act on the entire cell wall, often at some distance from the hyphae. The structural polymers are removed, leaving a framework of lignin to maintain the general cell shape so that there is little apparent damage until the cell walls collapse. Decomposition occurs in irregular patches in the attacked wood. This leads to the cubically cracked appearance of brown-rotted wood. It also crumbles readily to a powder when rubbed between the fingers. In both white and brown rots, the hyphae grow and branch in the cell cavities and penetrate the walls mechanically via pits or the surfaces in general, by coupling penetration with enzymic erosion, producing bore holes somewhat wider than the hyphae. In both, they penetrate deeply into the wood. Soft rots, on other hand, are more conspicuous near the surface, advancing inwards after destroying the outer layers of the wood. They occur only in wood of unusually high moisture content, such as water-logged river and marine timbers. The soft-rot fungi again principally utilize the cellulose and the hemicelluloses of the walls, but their hyphae penetrate and grow within the secondary cell walls ; here they enzymatically create chains of typically rhomboidal or elongated cylindrical cavities, with conically tapering ends (Fig. 1.9). Decomposition is restricted to the immediate neighbourhood of the hyphae.

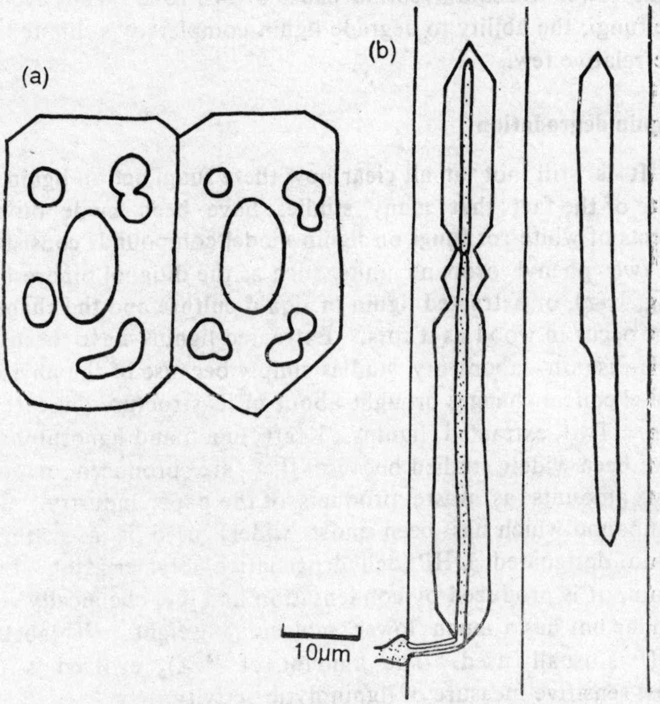

Fig. 1.9. Diagram of (*a*) transverse and (*b*) longitudinal sections of tracheids of *Pinus* with soft rot cavities in the secondary walls.

Whereas soft rots are caused by Ascomycotina, such as species of *Chaetomium* and *Ceratocystis*, and anamorphic states such as *Alternaria* and *Phialophora*, white and brown rots are caused mainly by Basidiomycotina. Two good examples of these are *Coriolus versicolor* and *Piptoporus betulines* respectively. The former is one of the commonest polypores and is found on a great variety of hardwoods whereas the latter is a facultative wound parasite restricted to birch (*Betula* spp.). *Coriolus* can degrade over 90% of the lignin in wood. Several hundred species in the Hymenomycete Agaricales, but more so in the Aphyllophorales, an order almost entirely confined to wood, cause white rots. But apart from these, only a very few Ascomycotina, including *Xylaria polymorpha* and *Ustulina deusta*, can cause a white rot. Somewhat fewer Basidiomycotina cause brown rots. Thus even in the fungi, the ability to degrade lignin completely is limited to the relative few.

Lignin degradation

It is still not at all clear how these fungi act on lignin, in spite of the fact that many studies have been made on the effects of white-rot fungi on lignin model compounds consisting of two phenyl propane units, such as the dilignol pinoresinol (Fig. 1.8*c*), or extracted lignin in liquid culture and the changes that occur in wood as it rots. Extracted lignins have been of little use in laboratory studies simply because of the physical and chemical changes brought about in its structure on extraction. Two extracted lignins, Kraft lignin and lignosulphate, have been widely studied because they are produced in such vast amounts as waste products of the paper industry. The compound which has been most widely used is a synthetic lignin designated DHP (dehydrogenative polymerizate). Like lignin, it is produced by condensation and it is chemically very similar but has a much lower molecular weight. ^{14}C labelled DHP is usually used. The amount of $^{14}CO_2$ evolved is the most sensitive measure of ligninolytic activity.

Role of extracellular phenolases

When cultured on agar containing phenolics, such as gallic or tannic acid, most—over 90%—of the white-rot fungi produce extracellular phenolases, such as laccase, peroxidase and tyrosinase, which oxidize these acids ; a brown coloured diffusion ring appears around the colony margin. These catalyse the removal of electrons from phenols. They have long been considered as being involved in lignin degradation because lignin is a phenolic and so a substrate for these and lignin degradation is certainly an oxidation. Lignin is resistant to decomposition in anaerobic conditions. Also white-rot fungi produce these enzymes but the closely related brown-rot fungi, which do not decompose lignin, do not. Thus there is the apparent correlation between the ability to degrade lignin and the production of extracellular phenolases. The ability to degrade implies the formation of smaller compounds, yet it is usually considered that these oxidases act by coupling and polymerization to form compounds of higher molecular weight. There is, however, some scant evidence from experiments using white-rot fungi and model compounds that these enzymes can bring about limited depolymerization. It has also been shown that the continued action of these enzymes on wood itself lead to some degradation of the lignin. It is doubtful, however, whether they play significant part in lignin degradation. In any case they can only be part of the enzyme complex which attacks lignin. It has been suggested that they have an indirect role in polymerizing and so detoxifying any toxic phenolics released during degradation—that is to suggest that they act after the monomers have been cleaved off. Simple phenolics are often toxic to fungal growth and they may have an important function in coupling these. This would be comparable to depside and depsidone formation from monocarboxylic acids in lichens.

Cleavage of major linkage groups

One obvious step towards decomposition would be to cleave any of the major linkage groups between the phenyl propane units to release the C_6-C_3 monomers. *Coriolus versicolor* and a number of other white-rot fungi can cleave lignin models bonded by the arylglycerol-β-aryl ether bond. Although oxidative cleavage of the β-ether linkage occurs, there is again no convincing evidence that any great part of the lignin molecule is cleaved by white-rot fungi to produce the single monomers.

There is also evidence that the monomers may be attacked while still bonded in the polymer, not by breaking the bonds between them but by directly attacking the aromatic rings, by either ring cleavage and/or demethylation of the methoxyl groups to hydroxyl ones. Demethylation may also be coupled with side chain oxidation. These are oxidized by the loss of two carbon atoms and the formation of new carboxyl groups. Both these, the formation of-OH and-COOH groups, would lead to increased solubility. Support for demethylation and side chain oxidation comes from two sources. Lignin degraded by white-rot fungi contains less carbon, slightly less hydrogen, fewer methoxyl groups but more oxygen and carboxyl and hydroxyl groups. Culture filtrates from white-rot fungi grown on extracted lignin contain small amounts of vanillin, vanillic acid and syringaldehyde. There is also slightly less vanillin in rotted wood compared with sound wood. This is taken to indicate that some phenyl propane units, either in the lignin or after cleavage, have their side chains oxidized with the loss of two carbon atoms. Alternatively, it has been argued that vanillin and vanillic acid are attached as side groups along the main polymer and are released on hydrolysis.

A hypothetical scheme for lignin degradation

Although white-rot fungi unquestionable can use lignin as a sole carbon source and completely decompose it, we are by no means certain as to how lignin is degraded. A number of very hypothentical schemes have been put forward. These usually assume intial cleavage of the arylglycerol-β-aryl ether bounds between the monomers (Fig. 1.10). This is followed by oxidative cleavage of the side chain with the loss of two carbon atoms and the formation of a carboxyl group to give vanillic acid is demethylated to protocatechuic acid. Ring cleavage then occurs to keto-adipic acid and this is used in the tricarboxylic acid cycle. It is most likely that these reactions occur simultaneously over the surface of the polymer with the oxidative cleavage of the side chains being centrally significant for fragmentation of the polymer.

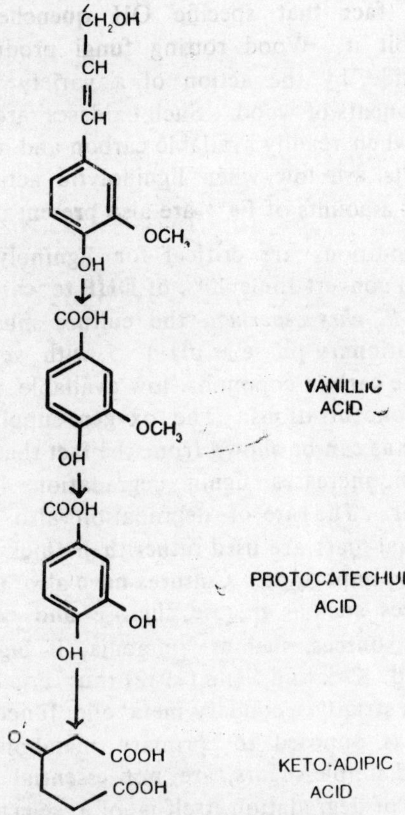

Fig. 1.10. Scheme for ligning degradation.

Role of agents other than enzymes

Evidence is accumulating that agents other than enzymes, such as the hydroxyl radical (OH), may be involved in lignin degradation. In cultures of *Phanerochaete chrysosporium*, one of the most widely used white-rot fungi, hydroxyl dependent formation of ethylene coincides with ligninolytic activity. The radical is probably formed from hydrogen peroxide in the so-called Haber-Weiss reaction which is catalysed by iron and requires the superoxide radical (O_2^-)

$$O_2^- + Fe^{+++} \rightarrow Fe^{++} + O_2$$
$$F^{++} + H_2O_2 \rightarrow Fe^{+++} + OH^- + OH$$

The dependence of ligninolytic activity on the radical is verified by the fact that specific OH quenchers, such as mannitol, inhibit it. Wood rotting fungi produce sufficient hydrogen peroxide, by the action of a variety of oxidases, from the components of wood. Such oxidases are synthesized most rapidly when readily available carbon and nitrogen containing nutrients are low, when ligninolytic activity is at its peak. Sufficient amounts of Fe^{++} are also present in wood.

Cultural conditions are critical for ligninolytic activity. For example, to convert some 40% of DHP to carbon dioxide and water with *P. chrysosporium*, the culture must be maintained in the stationary phase at pH 4—5, with very low levels of metabolizable carbon copounds, low available nitrogen and high oxygen concentrations. The oxygen supply is a very critical variable as can be shown from the fact that incubation in pure oxygen increases lignin degradation 10-fold over incubation in air. The rate of degradation also increases if very thin mycelial mats are used rather than thick ones, diffusive supply being important. Cultures have also to be starved of carbon sources, such as glucose, sucrose and cellobiose, as well as nitrogen sources, such as ammonia, if high rates are to be maintained. Kirk and Fenn (1982) thus argue that lignin degradation is a strictly secondary metabolic function in that the products, as opposed to primary metabolites such as amino acids and simple sugars, are not essential for growth. But the process of degradation itself is of a selective value to the fungi. It gives such fungi a competitive ecological advantage in providing access to the cellulose and hemicelluloses masked by the lignin. White-rot fungi clearly produce a very elaborate and complex ligninolytic system, in part enzymatic and in part associated with the hydroxyl radical, to be able to degrade lignin completely to carbon dioxide and water. Brown-rot fungi lack the complete system. The most that many of these can do is to bring about a limited attack on the lignin molecule and cause such effects as demethylation.

Lignin as a physical barrier to cellulase

Both the brown-rot and soft-rot fungi decompose the carbohydrates, especially the cellulose of the wood. In the soft rots, the characteristic cavities are caused by decay being restricted to the immediate neighbourhood of the hyphae. The diffusion of their cellulase is definitely restricted. This situation contrasts markedly with the brown-rot fungi where the cellulase diffuses freely into the walls, hydrolysing the cellulose throughout and leaving a skeleton of predominantly lignin. It has often been suggested that the cellulolytic enzymes of the two groups differ in size and shape and that the lower diffusibility of the cellulase produced by the soft-rot fungi indicates larger molecular dimensions but this is not so. The cellulases produced by the two groups have similar dimensions and properties. Many actively cellulolytic fungi may be restricted in their ability to utilize cellulose in wood by virtue of the intimate nature of the association between the cellulose and the lignin. A particularly good example is *Chaetomium globosum* which rapidly degrades cotton and filter paper cellulose completely, but only attacks wood of high moisture content and merely produces soft rot cavities in the cell walls. Lighin appears to act as a physical barrier that prevents the cellulase from reaching sufficient glycosidic bonds in the cellulose to permit any large scale hydrolysis. Thus the accessibility of the cellulose to the degrading enzymes is a most important factor. The evidence for this comes from a number of sources. Increased accessibility can be achieved by breaking down the wood to much finer particles before adding cellulase. This exposes larger surface area of the cellulose free of its association with lignin. For example, in experiments using sawdust and ball-milled sawdust, increased hydrolysis occurred in the latter when cellulase was added. It thus appears that brown-rot fungi possess some system—a pre-cellulolytic phase—which enables the cellulase to get at cellulose in wood. In the cell walls of wood, the cellulose microfibrils are encrusted with and surrounded by lignin and hemicelluloses. One suggestion that has been made is that brown-rot fungi produce enzymes which the soft-rot fungi lack. Some of these degrade the hemicelluloses and others disrupt the links between the cellulose and the lignin. There is also some evidence that one does not necessarily have to postulate enzymic dissociation of the lignin from the cellulose. Brown-rot fungi growing in wood, develop and maintain their own pH of between 2—4, whereas soft-rot fungi develop best in near neutral conditions. It may well be just that acidic conditions are necessary to disrupt the association between the lignin and the cellulose. If wood is treated initially with acid and, after removing the acid, soft-rot fungi allowed to attack it, the fungi bring about a greater loss than in untreated wood.

NATURAL RESISTANCE OF WOOD TO FUNGAL DECAY

Lignification

Lignification of the cell walls is obviously a very important factor that contributes to the natural resistance of wood to fungal decay. This is more important in soft rots. Softwoods are more resistant to these than hardwoods. This is usually attributed to the higher degree of lignification and the higher density of cell walls in conifers. It seems unlikely that mere abundance of lignin can account solely for the difference in resistance. The different proportions of the various phenyl propane units in the lignin, the degree of cross linkage with the cellulose and the nature of the hemicelluloses must also be important. Nevertheless lignification must act as some sort of physical barrier. Many very actively cellulolytic fungi and bacteria cannot attack wood because the lignin prevents their cellulase from reaching sufficient glycosidic bonds to permit hydrolysis on such a scale that they can grow on the proceeds.

Refractivity of cellulose

Many other factors contribute to decay resistance. The cellulose in wood tends to have a higher degree of refractivity or crystallinity than in the cell walls of herbaceous plants. The microfibrils are more highly ordered and there are correspondingly less amorphous or more randomly organized areas. The higher the refractivity, the smaller the surface immediately accessible to the components of cellulase.

Nitrogen content

In addition to being distinguished by its high lignin content, wood can also be distinguished from other plant materials by its very low nitrogen content. This also increases its resistance to decay. Woody tissues contain 0.03—1.0% nitrogen as compared to 1.0–5.0% in herbaceous tissues. The carbon : nitrogen ratio in most woody tissues is thus high, in the order of 350—500 : 1 and may exceed 10(0 : 1. For most fungi a substrate with such a high carbon : nitrogen ratio would be nitrogen dificient and growth limiting. Wood-decay fungi are unusual in that they can grow in such substrates. They metabolize large amounts of carbohydrates (and lignin in white rots) in the presence of very small amounts of nitrogen. The mycelium of most fungi, grown on nutrient media, contains about 5.0% nitrogen and has a carbon : nitrogen ratio of about 10 :1. The nitrogen content of the medium may fall, under starvation conditions, to around 1.0% before growth stops. The white-rot fungus *Coriolus versicolor* is unusual in that on high carbon : nitrogen containing media, the total nitrogen in the mycelium may fall as low as 0.2% before growth rapidly declines. The ability to grow under such conditions suggests a greater efficiency in its nitrogen metabolism. This may be achieved in a number of ways. Mycelial nitrogen may be re-used either by internal translocation from old to young hyphae or by autolysis and re-use. Extracellular lytic enzymes may be secreted which break down the old hyphal walls making the constituents, especially the nitrogen in the chitin, available for re-assimilation. Preferential allocation of available nitrogen to nucleic acids and enzymes may occur. For example, when growth of *C. versicolor* on media containing low and high carbon-nitrogen ratios was compared, the total nitrogen, expressed as percentage dry mass of the mycelium, fell from 4.4 to 0.2 but the percentage nitrogen in nucleic acids rose from 4 to 25 and the amount of cellulase produced per unit of mycelium was comparable in each. In fungi in general, cellulolysis diminishes with increase in the carbon-nitrogen ratio. White-rot fungi are unique in being able to produce cellulase at a carbon : nitrogen ratio of 2000 : 1 whereas in most other fungi this ability is negligible at a ratio of about 200 : 1.

Moisture content

Wood-decay fungi have higher moisture requirements for growth than fungi which attack most other plant materials. Their growth rate is very sensitive to changes in the water activity (a_w) of the medium. Whereas cotton is susceptible to fungal attack when it has a moisture content of more than 10% on a dry mass basis and cereal grains more than 13%, wood decay can be initiated only at moisture levels of about 26—32%. In standing trees and freshly felled timber, most of the cell cavities in the wood are water-filled. Such wood may have a moisture content of well over 100%. Such completely saturated wood is quite immune to fungal attack, presumably because the oxygen tension is too low to support active hyphal growth and the carbon dioxide content raised considerably above atmospheric levels. Many wood-decay fungi are very tolerant of high carbon dioxide concentrations. Whereas litter-inhabiting Basidiomycotina may be inhibited from growth by a partial pressure of 10 kPa, wood-decay species still grow at 30 kPa and some, including *Piptoporus betulinus*, still grow at 70 kPa. It appears that air equivalent to something more than 20% by volume of the wood is necessary for actual decay to take place. The existence of intact wooden galleys, submerged since Roman times, is adequate proof that completely waterlogged wood does not decay. The point at which all the free water has disappeared, but the cell walls are still fully saturated, is known as the fibre saturation point. For most woods this is around 26—32% (c. 0.3 gg^{-1}, equivalent to an a_w of 0.97). Wooddecay fungi begin to grow at around this level and make optimum growth at about 40%. Tresner and Hayes (1971) tested just over 100 species of Basidiomycotina and found that 94 were unable to grow at an a_w of 0.97 and below and only one species grew down to 0.94. Other evidence suggests that the lower limit for growth of wood-decay fungi is about 0.97, with the linear growth rate reduced to about half normal even at 0.989.

Worked wood that has been thoroughly air-seasoned contains 15—18% moisture, which is far too low to support any fungal growth. Requirements for such high moisture contents, and thus water activities, obviously contribute to the resistance of wood decay. Dry rot, caused by *Serpula lacrimans,* is an exception. Wood with a moisture content as low as 20—24% becomes liable to attack by *S. lacrimans.* Furthermore, as a brown-rot fungus, it produces metabolic water during cellulose degradation which considerably raises the moisture content of the wood on which it is growing. Once established on a small pocket of damp wood it can continue to colonize dry wood in this way. The exact relationship between water activity and wood decay is difficult to obtain because all, like *S. lacrimans,* degrade the cellulose in the cell walls. The complete degradation of 1.0 g cellulose liberates 0.56 g metabolic water. This is sufficient to alter the a_w of the wood significantly.

Toxic substances

All these factors contribute to decay resistance but the principal sources of such resistance in wood are toxic substances deposited during the formation of the heartwood. These are synthesized in the senescing parenchyma cells and diffuse out into the walls of the adjacent xylem elements. The distribution of decay resistance within a tree has been correlated with both the distribution and the nature of these toxic substances. They have been studied most in Gymnosperms, Most are phenolics. They fall into four main chemical groups, terpenoids, tropolones, flavonoids and stilbenes (Fig. 1.11). Of these the thujaplicins (tropolones) are the most inhibitory.

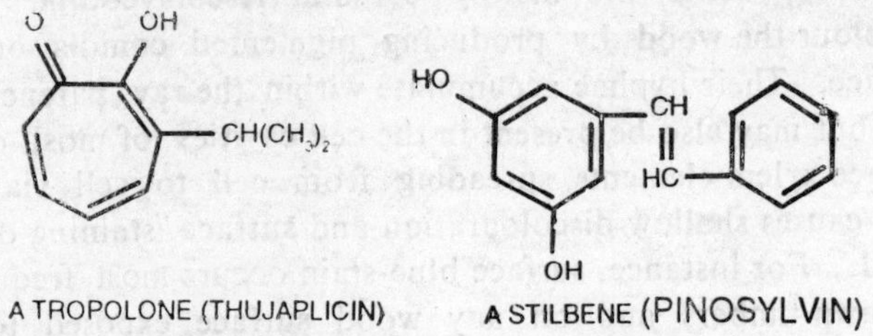

A TROPOLONE (THUJAPLICIN) A STILBENE (PINOSYLVIN)

Fig. 1.11. Structure of two toxic chemicals extracted from gymnosperm wood.

They all provide protection from decay for many years but with time they may become lost by leaching or become inactivated. In spite of their toxicity, it is well-known that several fungi are able to destroy the heartwood, even in living trees, and also timber impregnated with similar phenols, such as pentachlorophenol and 2, 4-dinitrophenol, which are used to protect less durable timbers and the sapwood of conifers. Such heartwood rotters are not insensitive to these toxins. They use their phenolases to oxidize them and polymerize the products to non-toxic melanins. Tannins are very common in the heartwood of Angiosperms and they play a similar role in decay resistance there. They inhibit fungal phenolases but decreased toxicity of the heartwood occurs with time by auto-oxidative polymerization of the tannins.

Sapwood is ordinarily very susceptible to decay but the resistance of different heartwoods is very variable. Trees with very resistant heartwoods include many oaks, cedars and the redwoods and those with non-resistant or only slightly resistant heartwoods include alders, beech, elms and poplars. A durable heartwood may be of survival value to the tree itself. Cedars live 2000 years or more whereas any of those in the slightly resistant category rarely live as long as 500 years.

OTHER WOOD-INHABITING FUNGI

Other fungi which inhabit wood occur chiefly in the sapwood where they obtain their food supply from the contents of the dead xylem parenchyma cells. These are the so-called moulds and stain fungi.

Mould fungi are mainly conidial Ascomycotina. They discolour the wood by producing pigmented conidia on the surface. Their hyphae accumulate within the ray parenchyma cells but may also be present in the cell cavities of most of the surface xylem elements, spreading from cell to cell via pits. This causes shallow discolouration and surface staining of the wood. For instance, surface blue-stain occurs most frequently on sawn timbers and on any wood surface exposed to the rain. It is caused by the surface growth of common airborne fungi, but especially *Cladosporium* spp., with dark brown hyphae and coloured conidia. Such staining is easily removed during planing treatments.

Blue-stain fungi

Typical blue-staining is caused by pigmented hyphae that grow in the wood, whereas other stains, such as brown ones, are caused by chromogenic substances actually excreted by the hyphae into the wood. At least one fungal stain has been used commercially. The mycelium of the Ascomycete *Chlorosplenium aeruginascens* permeates the dead wood of oak and beech on the woodland floor and colours it a brilliant green. Such 'green oak' has been used for inlays and decoratively as Tunbridge Ware. The wood is unaltered in texture and resists decay.

Blue-stain fungi are common in coniferous sapwood, especially pines, but are also found in hardwoods. They are non-cellulolytic 'sugar' fungi, in that they utilize only the more readily assimilable carbon compounds, such as sugars and starches, which occur in the ray parenchyma cells of freshly killed wood. They do no structural damage to the wood as they move across it mainly through the pits. Blue-stain is thus not the first stage of a form of rot, but its occurrence does indicate the wood has been kept moist and exposed to conditions favourable to the development of decay fungi. Although the structural properties of the wood are unaltered, blue staining of coniferous sapwood is responsible for large financial losses to the timber producer. The mere discolouration of the wood makes architects disinclined to use it and it is less acceptable to the manufacturers of packing cases and paper. In vigorously growing pine trees, the moisture content of the sapwood is too high to permit growth of blue-stain fungi. The low oxygen tension again appears to be the major limiting factor in the growth of these fungi in wood with a very high moisture content. In nature they may colonize standing pine trees which have been killed either by root-rot caused by *Heterobasidion annosum*, or some other disease, or by suppression. They are much more common on felled pine logs and will soon appear on these if they are left on the forest floor for any length of time. However they are rapidly replaced by wood-decay fungi. Both death and felling cause the wood to dry but progressively and such wood will support the growth of blue-staiu fungi unless the moisture content falls below about 27%.

Blue-stain fungi are Ascomycotina, mainly of the genus *Ceratocystis*, most of which have been perithecial and conidial states. The majority present their spores for dispersal in the form of stalked spore drops, the spores, in this case being insect dispersed. The perithecia of *Ceratocystis* have a swollen base and a very long, slender neck, some often 1 mm or more in length (Fig. 1.12*a*). The ascospores are not violently discharged but the asci break down within the ascocarp and the ascospores are forced up the neck ; they are extruded in a mucilaginous drop at the apex where they are held in place by a fringe of hair-like hyphae lining a pore. A variety of conidial states are produced. The *Graphium* state has a thick sheath of dark hyphae forming the stalk (Fig. 1.12*b*). The component hyphae branch at their tips and produce masses of sticky conidia, whereas, in the stalked spore drop of the *Leptographium* state, the stalk is a deeply pigmented and very wide single hypha which branches profusely at the apex (Fig. 1.12*c*).

Blue-staining is usually associated with attack by bark beetles. Species of *Hylastes*, *Myelophilus* and others introduce spores in making their brood chambers at the interface of sapwood and bark. The spores germinate and grow radially and longitudinally in the sapwood forming wedges of bluestained timber and then produce their conidia projecting into the brood chambers. These adhere to the young beetles as they emerge and are dispersed to other logs as they in turn make brood chambers.

Blue-staining becomes a problem where felled pine logs are left in piles on the forest floor for 2-3 months before being removed to timber depots. There are a number of ways of treating the problem, such as the use of insecticides and fungicides, but the most successful method of control, widely used in Europe, is to remove the bark immediately on felling. This not only prevents beetle attack but assists rapid drying

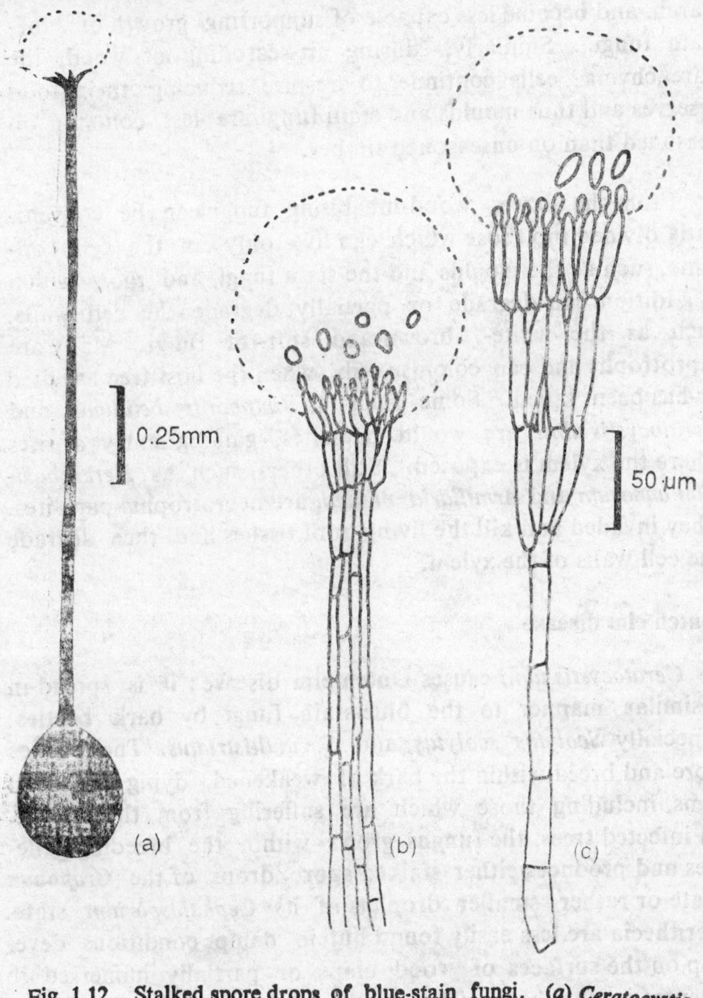

Fig. 1.12. Stalked spore drops of blue-stain fungi. (*a*) *Ceratocystis*. (*b*) *Graphium*. (*c*) *Leptographium*.

out to moisture contents below those which will support fungal growth.

Depletion of the food reserves of the xylem parenchyma cells during ageing is one nutritional factor that tends to limit the susceptibility of sapwood to staining fungi. During ageing and the transition from sapwood to heartwood the parenchyma cells gradually die and become depleted of reserves, especially starch, and become less capable of supporting growth of blue-stain fungi. Similarly, during air-seasoning of wood, the parenchyma cells continue to respire reducing their food reserves and thus moulds and stain fungi are less common on seasoned than on unseasoned timber.

From the above, wood-inhabiting fungi can be conveniently divided ino those which can live only on the cell contents, such as the moulds and the stain fungi, and those which in addition can degrade or partially degrade the cell walls, such as the white-, brown-and soft-rot fungi. Many are saprotrophs and can colonise only when the host tree has died or has been killed. Some such as *Piptoporus betulinus* and *Ceratocystis ulmi* are would parasites, gaining entry at sites where the xylem is exposed. Still others, such as *Heterobasidion annosum* and *Armillaria, mellea*, are necrotrophic parasites. They invaded and kill the living root tissues and then degrade the cell walls of the xylem.

Dutch elm disease

Ceratocystis ulmi causes Dutch elm disease ; it is spread in a similar manner to the blue-stain fungi by bark beetles, especially *Scolytus scolytus* and *S. multistriatus*. The beetles bore and breed within the bark of weakened, dying and dead elms, including those which are suffering from the disease. In infected trees, the fungus grows within the breeding galleries and produces either stalked spore drops of the *Graphium* state or rather smaller droplets of its *Cephalosporium* state. Perithecia are less easily found but in damp conditions develop on the surfaces of wood chips or partially immersed in fissures in the bark. The young adults emerge in May to October and the sticky conidia may adhere to their bodies. They fly immediately too feed on young, healthy, elm twigs and in doing so may introduce the conidia into the xylem in wounds made as they feed. Beetles thus spread the fungus from branch to branch and tree to tree. The fungus enters the xylem and grows in a yeast-like form. It can be carried up in the xylem in the traspiration stream as such or as conidia. Infected trees soon show signs of wilting and yellowing or drying out of the foliage. Fungitoxins may be involved but part of the symptoms may be explained by the occlusion of the xylem of the current year's growth by gums and tyloses.

Environmental factors and the decomposition of wood

The decomposition of wood under natural conditions is an exceedingly protracted process. Whereas leaves of the majority of northern temperate deciduous trees may decompose in one, two or three years, a tree trunk under the same conditions may take a decade or even two to do so. The low level of available nitronen may be the overriding factor contributing to its slow rate of decay. The addition of organic nitrogen to wood blocks inoculated with various Basidiomycotina has been shown to increase their decay rate by over 60%. Other minerals, especially phosphorus and potassium, may also be limiting. The relatively high demand for such mineral nutrients combined with their relatively low availabilily places a limitation on the amount of fungal mycelium such a substrate can produce. Fluctuations, both diurnal and annual, in temperature content must also be important. However, in aseptic laboratory experiments decomposition of wood by a single species of decay fungus may be relatively rapid. The white-rot fungi *Lenzites betulina*, *Coriolus hirsutus* and *C. versicolor*, inoculated onto small blocks (20 mm^3) of birch wood kept in sterile moist oil at 22°C, caused more than a 75% loss in dry mass in three months. *Piptoporus betulinus* and a number of other brown-rot fungi caused mass losses of 50—70% over the same period. These are substantial losses, the more so when it is borne in mind that birch wood contains some 20% lignin which is not avaiable to brown-rot fungi. These facts may be contrasted with the observation that *P. bitulinus*, which had killed birch trees in East Angila, was still producing basidiocarps on these at least five years after they had fallen. The time period from infection to falling was not known but it may have been at least another five years. Even if the variable temperature and moisture regimes are taken into account, the decomposition of one of the least durable woods is very much slower in nature than in laboratory tests.

Most wood-decay fungi are mesophiles in terms of their temperature requirements, although some come into the category of cold-tolerant ones. The optimum temperature for the growth of most lies between 25 and 30°C. *P. bitulinus* has an optimum at 25°C and its growth falls off very rapidly above and ceases at 30°C. Its minimum temperature for growth, it must often experience in the field, lies between 7 and 9°C. For many others, the minimum lies below freezing point but decay at such temperatures would be very slow. The geographical distribution of a number of species is related to their temperature requirements. *Serpula lacrimans* has a low maximum of 25-26°C. It is absent from the tropics and other parts of the world with high summer temperatures.

In bulk wood, temperature is probably a more important variable than moisture content. With reference again to *P. betulinus*, it is able to decompose birch wood with a moisture content within the range 35-100% on a dry mass basis, although near maximum decomposition rates only occurred between 60 and 120% in laboratory experiments. Birch logs, stored outside in Central Sweden, had a moisture content of 85-91% on felling and after three years the moisture content was still 51-67%. Fluctuations did occur, with some drying in the summer and some water uptake in the winter, but over the whole period the moisture content was somewhere near the optimum for decay. However, if the bark peels off the position is quite different. In summer, rapid drying out may occur to moisture contents below those which will support growth and equally rapid soaking will occur in rain. A thick, highly suberized outer bark is not only a structural deterrent to fungi, but because of its high content of tannins, phenols and the like, also a chemical one. However, given this, if it remains intact after death of the tree, it helps to maintain a more equitable moisture regime within the wood and this will favour any decay fungi.

Habitat relations and specificity of wood-inhabiting fungi

The habitants of wood-inhabiting fungi vary from minute twigs, small and large branches to the most massive of tree trunks and stumps and from minute rootlets to major roots and include such man-made habitats as fencing post, house timbers, sawdust and chip piles. Any one of these substrates is particularly complex in more ways than one. The trunk of any one tree will have varying proportions of bark, sapwood and heartwood along its length. The proportions will differ from those in the trunk of another species. The wood from different tree species differs structurally as can be seen by contrasting ring-porous with diffuse-porous types. Further marked differences occurs between softwoods and hardwoods. Over and above these differences as already indicated the composition of the lignin varies in different wood. This complexity and heterogeneity make it difficult to generalize about the decomposition process. A number of successional studies have been made on woody substrates but these have to be interpreted with caution. A succession can be defined as the appearance of different fungi in sequence on the same part of the substrate. The fact that one fungus appears on one part of a log at one time and another fungus on another part, even an adjacent part, at another time does not necessarily prove a succession. Their habitat niches may be quite different, one growing on the sapwood and one on the heartwood or the latter may be colonizing a part of the heartwood not colonized by the former. For example, basidiocarps of *Daedaleopsis confragosa* or of *Hypholoma fasciculare* may appear on a birch trunk which has, for a number of years, supported basidiocarps of *Piptoporus betulinus*, but the mycelia of these would almost certainly be growing on parts of the wood not colonized by *P. betulinus*. The latter is specific to birches and is a wound parasite gaining entry where a branch has been fractured. Infected trees are usually killed by the fungus and the trunks of these often break off remarkably cleanly and transversely at a height of about 3 m in high winds. The structural polysaccharides in the walls are rapidly and completely removed leaving a cellular framework of amorphous lignin which has insufficient tensile strength to withstand to bending strains incurred. The fungus then continues to grow and to produce its characteristic kidney-or hoof-shaped basidiocarps on the fallen and standing parts of the tree. As with many woody substrates, *P. betulinus* is the primary and sole colonizer. It may completely permeate the wood of the whole trunk and persist there, virtually in pure culture, for several years, by which time the wood is in a very late stage of decay and extremely friable. In such a state the wood is unlikely to be capable of supporting fungi such as *D. confragosa* and *H. fasciculare.* They would not succeed *P. betulinus* but would be growing on parts not colonized by it.

Wood-decays fungi exhibit all degrees of specificity. Considering the white-rot and the brown-rot fungi as two groups, there are many more of the former than the latter. Those of the write-rot group primarily attack hardwoods and those of the brown-rot group softwoods. Some of these may be restricted to a single host genus. *P. betulinus* is a good example. *Fistulina hepatica* which causes a serious decay of the heartwood of oaks is another. The causes of such marked specificity are obscure. Other fungi may be restricted to the wood of a relatively small number of trees *Polyporus squamosus* is, like *P. bitulinus,* a wound parasite, in this case of elm in particular but it is often found on other trees, such as ash and sycamore. It causes a white-rot of the heartwood and may persist for a number of years on fallen trees which it has killed or which have been wind-blown, as a consequence of the rot. On elm trunks. *P. squamosus* is replaced, but only in a temporal sense, by a number of other wood-decay fungi. Two in particular, *Auricularia mesenterica* and *Pleurotus cornucopiae* are rarely found on other wood. Basidiocarps of the former soon appear on any felled elms and production of these continues for up to eight years. Spatially it utilizes the bark and surface layers of the sapwood so it does not succeeded *P. squamosus.* Basidiocarps of *P. cornucopiae* appear on elm trunks only some 3-10 years after they have fallen. The fungus then fungus then persists until the wood is well-decayed. Its mycelium appears to be confined to the sapwood not utilized by *P. squamosus* and again it does not actually succeed the latter. Wheareas *P. cornucopiae* is most common on fallen elm trunks, *Flammulina velutipes* is most often found on standing dead elms, especially those killed by *Ceratocystis ulmi* and which have lost their bark.

Still other fungi, such as ***Cariolus versicolor*** and ***Stereum hirsutum***, are much less discriminating and grow on a wide range of hardwoods. The former is entirely saprotrophic and is one of the commonest fungi found on fallen twigs, branches, trunks and dead stumps of hardwoods where it produces the most rapid of white rots but, *Stereum*, is confined to the sapwood. It can actually replace, and therefore succeed, other established and less aggressive white-rot fungi. *Xylaria hypoxylon* and *Daldinia concentrica*, two Ascomycotina, produce black lines, zone lines, in the sapwood of ash delimiting areas which they have colonized. The hyphae of *C. versicolor* will penetrate these and grow on to replace them. Other fungi show a preference for coniferous wood. *Heterobasidion annosum*, *Paxillus atrotomentosus* and *Tricholomopsis rutilans* are characteristic of conifer stumps, *Hirschoporus abietinus* and *Stereum sanguinolentum* of coniferous twigs and branches and *Auriscalpium vulgare* of pine cones. These are all Basidiomycotina but similar examples can be found in the Ascomycotina. For example, *Daldinia concentrica* is very common on ash but is occasionally found on other hosts, especially beech and birch *D. vernicosa* occurs on gorse, especially bushes which have been burnt and subsequently weathered. *Ustulina deusta* causes a white-rot of lime and beech, whereas *Xylaria polymorpha* and *X. hypoxylon* are very common on a wide variety of dead hardwoods.

Each tree species thus may have, within limits, its own particular wood-decay fungi. A number of these may enter as necrotrophic parasites at wounds above ground or along roots below ground and then persist as active saprotrophs after death. They would thus have a competitive advantage over purely saprotrophic fungi in being established first. As parasites they may only be overcome the host resistance of one or a few species of trees. This might account for some of the specificity noted. It may well be that the different naturally occurring tannins, terpenoids, etc., present in the different heartwoods further help to determine specificity. Only fungi which can tolerate or degrade these are able to become established.

Ecological studies on decaying wood

Numerous ecological studies have been made of fungi colonizing specific woody substrates including wounded living tree trunks and fallen dead ones, tree trunks after insect attack, fire-killed trees, branches and slash on the ground, tree stumps, fence posts, beech capules, etc. Changes with time in the fungal communities on these have been recorded and described, accurately or inaccurately, as successions. As might be expected, the sequences of fungi observed on these considerable variation depending upon the species of wood, the type of substrate and, in addition, the environment in which decomposition is occurring. But fungi are not the only organisms found in decaying wood. A very wide variety of invertebrates and bacteria also occur and they, too, may play an important role in its decomposition.

Swift (1977) recognized three stages in the decay process—the pioneer colonization stage, the major decomposition stage and the incorporation stage, in which the products of decay are incorported into the soil.

Pioneer colonization stage

Patterns of colonization may vary. In some cases, the Basidiomycotina which are going to dominate the decomposition stage are the primary and sole colonizers. In others, their colonization is preceded or accompanied by a variety of decay or non-decay fungi or bacteria. This may be illustrated with some specific examples. *Heterobasidion annosum* is a white-rot fungus causing butt- and root-rot of conifers. It may colonize via roots or the surfaces of freshly cut stumps. Infection of a healthy living root almost invariably occurs as a result of mycelial transfer from another infected root coming into contact with it. From the root the fungus grows up to the base of the stem and colonizes and kills the combaium, thus effectively girdling and so killing the tree. It then progressively rots the roots and the stem base. Rapid desiccation of the wood after death usually prevents extensive spread up the stem. In this case it is the sole colonizer. Alternatively it may colonize the surfaces of freshly cut stumps via its air-borne basidiospores. These stump surfaces are highly selective substrates and are initially colonized by a relatively small number but, nevertheless, a variety of fungi. These include, in addition to *H. annosum*, non-cellulolytic blue-stain fungi utilizing the contents of the parenchyma cells, cellulolytic fungi, such as *Phialophora* and *Trichoderma* spp., utilizing cell contents and may easily accessible cellulose, and other wood-decay fungi such as the white-rot fungus *Peniophora gigantea*. This is a much more competitive situation ; whether or not it emerges as the major decomposer will depend upon a multiplicity of factors, including its ability to compute with these for the more readily available nutrients which are necessary if it is to become established. Similarly patterns of colonization can be seen in the initiation of decay in trunks following, woundings, such as by the branches breaking off in high wind. This applies to most species of *Sterium*. They invade only freshly exposed tissues and are inhibited by the presence of other pioneer micro-fungi and bacteria. In other cases, such as with *Phellinus igniarius* invading wounds on poplars and other hardwoods, prior colonization by bacteria and micro-fungi such as the stain, mould and soft-rot fungi generally occurs and may even be a prerequisite if it is to attack and cause of progressive rot of the heartwood.

Insects, especially members of the Ipidae and Scolytidae, may attack living trees and introduce bacteria, yeasts, blue-stain or ambrosia fungi below the protective bark. The combined activities of the insects and the fungi may weaken or kill the tree. The wood-decay Basidiomycotina then follow. The attack of *Scolytus scolytus* on elms introducing *Ceratocystis ulmi*, followed by *Flammulina velutipes*, is a case in point.

Decomposition phase

The decomposition phase is dominated by the white- and brown-rot fungi but wood-boring beetles (Coleoptera) and wood-eating termites (Isoptera) may also contribute to decay. In many woody substrates ony one fungus may be involved in the decomposition phase ; examples of *Piptoporus betulinus* on birch, *Heterobasidion annosum* on pines and *Coriolus versicolor* on hardwoods in general, have already been given. In others, a number, but usually a very limited number of fungi are involved. Each of these occupies discrete volumes of wood which are often clearly demarcated from each other by distinct dark lines. These colonies may intricately interlock but their mycelia do not intermix. They remain isolated by zone lines into virtually pure cultures. This balanced state may persist for a number of years, but, depending upon the relative competitive ability of adjacent mycelia, there may be eventually some replacement of one fungus by another, or aggressive saprotrophs, such as *Hypholoma fasciculare*, *Phallus impudicus* and *Phlebia merismodes*, may colonize from the surrounding litter and replace them (Rayner and Todd, 1979). For example, both *C. versicolor* and *Stereum hirsutum* are susceptible to replacement by any of these three in hardwood trunks and branches and *Heterobasidium annosum* by *Peniophora gigantea* in pine stumps. *H. annosum* is particularly sensitive to hyphal interference caused by the latter and this may be one factor involved in its replacement.

The role of animals in degradation

Many wood-boring beetles and their larvae and termites are wood feeders depending upon microbial symbionts in their guts to semi-digest the wood. Some feed on sound wood, others on decaying and well-rotted wood. In the latter case, the fungi growing in the wood may be an important component of their food. Many termites are polyphagous. When *Kalotermes flavicallis* is fed on wood, it decomposes 94—95% of the cellulose, 60—70% of the hemicelluloses and 3—4% of the lignin. Thus large populations of these wood-boring beetles would almost certainly contribute substantially to wood decay. Another important aspect of their degradative activity is the comminution of the wood as they attack it.

As the white- and brown-rot fungi exploit the wood, it softens and becomes friable and as such is more attractive to animals as a food source, as somewhere to live and as a breeding ground. Their access to it, if it is bulky, may be dependent upon the prior activity of wood-boring animals, such as the beetles. Their bore holes afford ports of entry for a very great variety of animals generally common in litter and soil. These include micro-arthropods, such as Acari and Collembola, and macro-arthropods, such as Diptera and Isopoda, as well as Oligochaetes, such as Enchytraeid and Lumbricid worms. Many of these, such as the Mycetophilid dipterous larvae feed mainly on the mycelia of the fungi. They all accelerate the process of comminution and carry spores from the surrounding litter and soil into the wood and thus inoculate it with common soil fungi. Many Zygomycete Mucorales appear for the first time on the decaying wood, along with a variety of conidial fungi including species of *Penicillium*, *Scytalidium* and *Trichoderma*. These now have a quite wide choice of substrates on which to grow. They may utilize the partly degraded wood, the dead hyphae of the wood-decay fungi, dead fauna or their faecal remains. Some may live as commensals sharing the hydrolytic products of the enzyme systems of the major decomposers. This is the incorporation stage. As the wood becomes more extensively decayed, the activity of the wood-decay fungi may decline and they are eventually replaced by such soil-inhabiting fungi. With time, the wood disintegrates and as it does so it is incorporated into the soil.

Wood decay and the cycling of mineral nutrients

Wood decay is important in regulating the cycling of mineral nutrients in the woodland ecosystem and contributes to the process of soil development there. Over the period of fungal decay, virtually all the important minerals, but in particular nitrogen and phosphorus, become immobilized in an organic form in the fungal hyphae and their reproductive structures, such as basidiocarps. Although the mineral content is low per unit volume, the sheer volume of decomposing wood means that it forms a very substantial part of the total minerals in the woodland ecosystem. Comminution of the decaying wood by the animal invaders leads to a release of some minerals. The small particulate form of the frass or faecal materials means that they are more effectively leached. As with the fungi the animals themselves act as a further reservoir of plant nutrients in a considerably more concentrated form than in the wood itself. Their wanderings, after feeding, lead to some redistribution of minerals but by far the more major redistribution and export from the wood occurs when adult stages emerge from the broods reared in and on the decaying wood. But this is essentially only a redistribution. The adults eventually die and their tissues are mineralized elsewhere in the ecosystem, while any fungal remains are mineralized *in situ.*

DECOMPOSITION OF LIGNIN AND HUMUS IN THE SOIL

The white- and brown-rot fungi are, more often than not, associated with relatively large masses of wood, such as the dead tree trunk and decaying stump. This may be because only these contain enough energy resources for these fungi to amass sufficient to produce their relatively massive and conspicuous basidiocarps. Vast quantities of lignin are incorporated into the soil in the vascular network of the leaves, fine rootlets and so on. These are very different substrates for fungi and other micro-organisms and they are in a vastly different environment. The substrate is richer in terms of associated readily available carbon and nitrogen sources, as tissues other than the highly lignified xylem are present in relatively larger proportions than in bulk wood. It is also presented to a much more varied population of micro-organisms. As such it would support a more diverse micro-flora and any lignin decomposers would be competing for, not necessarily lignin, but other more generally assimilable components, which are necessary for establishment, and would also be exposed to antagonism by others. This situation is markedly different from the decaying tree trunk with its one or few decomposer fungi in isolation. Further large quantities of lignin may be introduced into the soil in the form of organic residues from wood decay, especially from brown rotted wood. The process of lignin degradation in the soil may be quite different from that occurring in a tree trunk. There is very little direct evidence that any Basidiomycotina degrade such lignin in the soil. This may be because we are ignorant of the facts. In studies on soil fungi, Basidiomycotina are only rarely recorded. They tend to be slower growing and so are easily overgrown on most widely used culture media. They are often very sensitive to antagonism by others and so suppressed. Most do not produce spores or possess any other readily recognizable feature so could easily be overlooked. Nevertheless, they may be equally important as lignin decomposers in the soil itself as they are in the litter and decaying wood. The number of soil-inhabiting micro-organisms which have been reported as being able to utilize lignin is very small. These include a few aerobic, Gram-positive, non-sporing, red-shaped bacteria, in the genera *Bacillus* and *Flavobacterium*, and a few conidial fungi, in genera such as *Humicola* and *Phialophora*. The evidence for their ability to utilize lignin has again been obtained from the use of extracted lignin and lignin model compounds. These have been used incorportated in Kaolin pellets to enrich soil and fungi subsequently isolated from them and tested for their ability to utilize such compounds as vanillic acid and syringaldehyde. The ability to grow on and utilize these should not be taken as an ability to utilize lignin itself just as the ability to utilize carboxymethylcellulose is not taken as an ability to utilize native cellulose. They are partial degradation product and these fungi should be regarded as occupying a similar niche with regard to lignin as secondary sugar fungi do to cellulose. The latter do not possess the whole enzyme system necessary to hydrolyse cellulose (Fig. 1.13). They lack the C_1 component but possess the C_x component and β-glucosidase so that they can utilize the hydrolytic products. Some also lack the C_x component as well. Similarly with lignin, the fact that a fungus lacks one component of the multi-enzyme system does no necessarily debar it from participating in lignin degradation. Fungi may co-operate, sometimes synergistically, in the degradation of both cellulose and lignin. For example, it has been shown in experimental systems that a mixture of the C_1 component from one fungus and the C_x component of another is as efficient at cellulolysis as when both components are derived from the same fungus. It has also been shown using lignin preparations in culture tests that, in many cases, when two wood-decay fungi are grown

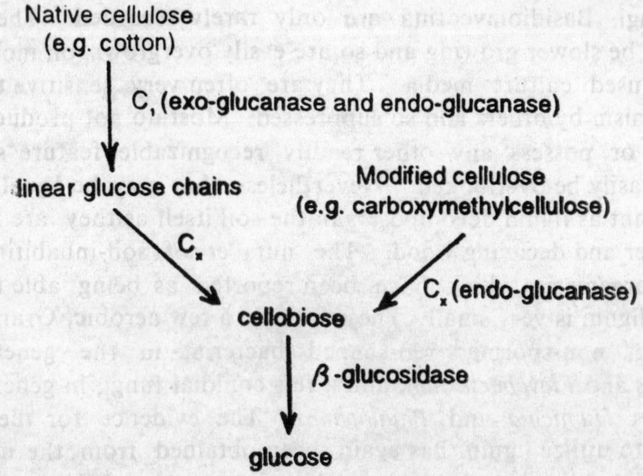

Fig. 1.13. A schema for the degradation of cellulose.

together in mixed culture, degradation is more pronounced than when both fungi are grown apart.

Lignin in the soil decomposes very slowly and its degradation there is more of a joint effort. There may well be a pooling of enzymes from a variety of fungi and perhaps bacteria and actinomycetes—some enzymes capable of cleaving bonds between monomers, others of demethylation and still others of side chain oxidation and so on until the final products enter the respiratory pathways of one organism or another.

The nature of humus

With this breakdown there is a gradual accumulation of dark, amorphous, organic humus. The chemistry of humus has by no means been fully elucidated. It is very heterogeneous and can be separated into a number of molecular categories using extraction techniques. It forms a very dark solution in dilute NaOH and a black precipitate called humin. Acidification of the solution with HCl to between pH 1-2 precipitates out a fraction called humic acid, leaving fulvic acid. The humic acid fraction is the major molecular category and forms from between 50 to 80% of the soil humus. It usually contains about 5% nitrogen, mainly in the form of bound amino acids but also in amino sugars and heterocyclic purine or pyrimidine derivatives. The most favoured idea is that humic acid has a heterogeneous aromatic core with carbohydrates, peptides and proteins, phenolics and metals, attached peripherally. In some soils, especially under woodlands, the humic acid may originate from lignin residues, possibly the end products of the brown-rot fungi, which have been considerably modified by microbial action. Syringic and vanillic residues can often be detected in the degradation products of humic acid and the distribution of these residues is consistent with the composition of the lignin found in the vegetation above. Work with tracers has shown that as much as one third of the humic acid in the soil is derived from lignin and only about one twentieth from cellulose. Reductive cleavage of most humic acid fractions shows that they also contain units based on phloroglucinol. This suggests that seed plant flavonoids also contribute to humus. Flavonoids are phenolics with two aromatic rings and include pigments, such as anthocyanins. The phenolics are degraded to simple phenols which become polymerized into the humic acid fraction. But humus is not solely a product of degradation of the more resistant parts of seed plants. It is also in substantial part a product of microbial synthesis. When ^{14}C labelled glucose is added to the soil, 40—80% of the carbon is lost as carbon dioxide within a few days, but even after two years, about 5—10% is still present in the soil humus. Intracellular transformation of carbohydrates and other simple organic organic substances occurs to produce phenols, quinones and other aromatic substances. These are oxidatively polymerized and combined with peptides and other cell constituents to form humic-like pigments, melanins, inside or outside the cell. These serve several functions. They may be deposited in the walls of hyphae, spores or ascocarps, to protect against excessive ultra-violet light or as a water-proofing to prevent water loss. Eventually, on the death of these structures and with time, they become variously transformed and incorporated into the humus fraction. The existence of significant amounts of amino sugars and non-protein amino acids, such as diaminopimelic acid, also suggests that residues of bacterial cell walls may form part of the humus.

Turnover of humus in soil

Humus is extremely resistant to microbial degradation but nevertheless there is a very slow turnover with the rate depending upon the soil type. A sample from a chernozem soil from the USA was ^{14}C dated as 990 ± 60 y old.

In other soils, humus is less stable *e.g.* humus from a coniferous forest soil in Sweden was dated as 370 ± 100 y. A number of fungi have been found to decompose humic acid in laboratory test. Humic acid was extracted from a Canadian soil. It contained 26% of the total soil carbon and was dated as 785 ± 50 y old. It was supplied as the sole carbon and nitrogen source as a 0.2% solution to a number of microorganisms, isolated by direct plating of the soil onto humic acid containing media. Four bacteria, in the genera *Bacillus* and *Pseudomonas*, and two conidial fungi, *Penicium frequentans* and *Aspergillus versicolor*, could utilize the humic acid as a sole carbon and nitrogen source but no actinomycetes could. *P. frequentans* made the best growth and it appeared to utilize the humic acid by initially reducing carboxylic groups to aldehydes and then alcohols. Salicylaldehyde and salicyl alcohol appeared in culture filtrates. A number of Basidiomycotina, including *Coriodus versicolor*; *Hypholoma fasciculare* and *Trametes suaveolens*, all active white-rot fungi, can also utilize humic acid and this ability is always associated with the reduction of carboxylic acids to alcohols. This suggests that one of the first steps in the degradative process is an aerobic reductive one. Subsequent steps have as yet to be elucidated. Very little can be concluded from such studies about the process of degradation in soils. Resistance to degradation may not be so much that it is not susceptible to microbial enzymes but that its multi-dimensional complex structure physically restricts the access of such enzymes.

2

Fungal Symbionts in Ectomycorrhizas and Lichens

A wide variety of fungi are apparently unable to exist either as free-living saprotrophs or as necrotrophic or biotrophic parasites and have developed various kinds of mutualistic associations with plants and animals. The majority of those associated with plants exist in association with roots as mycorrhizas or with algae and cyanobacteria (blue-green algae) as lichens. A very diverse array of mutualistic associations exists between fungi and the roots of plants. On morphological and anatomical features such mycorrhizas can be divided into two groups, ecto-and endo-mycorrhiza. In ectomycorrhizas, the ultimate absorbing rootlets of the root system are completely surrounded by a distinct mantle or sheath of fungal tissue from which hyphae penetrate between the outermost cell layers of the root. In endomycorrhizas, there is no such well-developed sheath. Most of the fungus is within the root and intracellular, as well as intercellular, penetration is common. Intermediates between these two groups are recognizable and have been termed ectendomycorrhizas.

ECTOMYCORRHIZAS

Ectomycorrhizas, aptly called sheathing mycorrhizas, are the minority group. Some 3% of all seed plants have ectomycorrhizas. They are particularly common in northern temperate forest trees, especially in the families. Fagaceae and Pinaceae but also occur in others such as the Betulaceae and Tiliaceae. However, they also occur in families well-represented in the tropics, especially some members of the Myrtaceae and Dipterocarpaceae.

Structure

The sheathing mycorrhizas of the european beech, *Fagus sylvatica* and various species of *Pinus* have been most thoroughly investigated. In *Fagus* the ultimate lateral rootlets are differentiated into 'long' and 'short' roots. The long roots are of indefinite growth. They are the main extending roots and their branches are the short roots which are of restricted growth and short lived. These are the absorbing roots and are usually modified into mycorrhizal organs. They are thicker, more brittle and coloured differently from uninfected roots. They develop many branches also of restricted growth so that a complex racemose branching system is formed. The mycorrhizas of *Pinus* are exceptional amongst sheathing types in being once or several times dichotomously branched (Fig. 2.1*b*). There is also a much more marked differentiation into long and short roots. The whole of these short rootlets, including the apex, are bound by a sheath of fungal pseudoparenchyma. In *Fagus*, this is some 20—40 μm thick and comprises some 20—30% of the total volume of the root (Fig. 2.1*a, c*), In transverse section the sheath can usually be differentiated into two layers—an outermost layer of thicker walled hyphae with no or very few interhyphal spaces and an innermost layer of loosely packed thinner walled hyphae. Anything absorbed from the soil must pass through this thick and somewhat differentiated sheath. There are no root hairs and the meristematic tissues are reduced in comparison with nonmycorrhizal rootlets. From the outside of the sheath, hyphae may spread to varying distances and in varying densities into the soil ; from the inside of the sheath, hyphae penetrate between the cells of the epidermis and occasionally further between one or two cell layers of the cortex forming a hyphal net work called the Hartig net. The cell layers penetrated by this Hartig net may also be radially elongated. In

Pinus with narrower roots, the net may penetrate as far as the endodermis. The hyphae never penetrate the endodermis or stele. The only other visible change in such roots is that the epidermal layer of cells adjacent to the sheath is frequently impregnated with dark staining tannins.

Fungi involved

The majority of the fungi involved are members of the Basidiomycotina especially Hymenomycete Agaricales. Well over 100 speeies of common toadetools have been shown to form mycorrhizas. The majority of these are in the genera *Amanita*, *Boletus* and *Tricholoma* but there are also many species in the genera *Cortinarius*, *Lactarius* and *Russula*. Some of these are relatively specific in their associations, for example *Boletus* (*Suillus*) *elegans* and *B.* (*Suillus*) *aeruginascens* associate only with larches, *Pseudotsuga* and a few species of *Pinus*. Others may associate with 20 or more hosts. *Amanita muscaria*, for instance, is in Britain chiefly associated with birches, *Betula* spp., but is also known to be mycorrhizal with various pines, larches and Norway spruce, *Picea abies*, amongst others. Similarly many host species are capable of forming mycorrhizal associations with several fungi. Professor E. Melin, to whom we owe much of our knowledge of the identity and cultural characteristics of sheathing mycorrhizal fungi, has shown by experimental inoculation that 40 species are capable of forming mycorrhizas with *Pinus sylvestris*. It is also clear that any one tree may associate with several fungi at any one time. Other Basidiomycotina involved are Gasteromycetes in such genera as *Rhizopogon* and *Scleroderma*. Still others are Ascomycotina, such as *Gyromitra esculenta*, all species of *Tuber*, such as *T. melanosporum*, the perigord truffle, and the sterile form *Cenococcum graniforme*.

Although all sheathing mycorrhizas have a number of common characteristics, in reality they show quite wide diversity in both their morphology and anatomy. *C. graniforme*, which is perhaps the least specific of all mycorrhizal fungi, produces one extreme type of mycorrhiza. The fungus has a black mycelium which produces variable sized black sclerotia. The

mycorrhizas which it forms with beeches, birches, limes, pines and spruces are also black and brittle due to dense deposits of melanins encrusting the outer walls of the sheath. Abundant

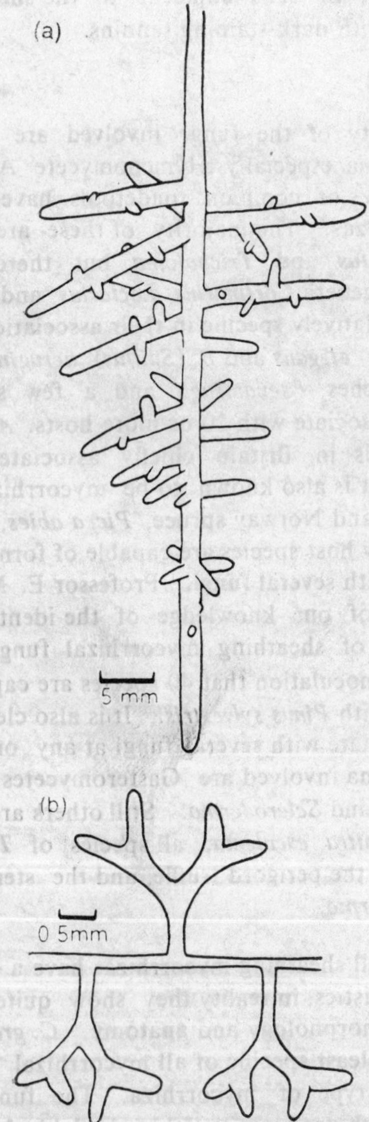

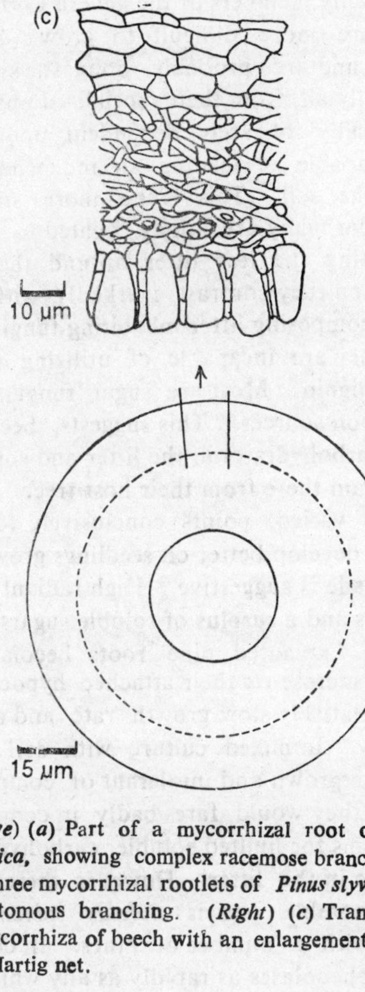

Fig. 2.1. (*Above*) (*a*) Part of a mycorrhizal root of beech *Fagus sylvatica*, showing complex racemose branching pattern. (*b*) Three mycorrhizal rootlets of *Pinus slyvestris* showing dichotomous branching. (*Right*) (*c*) Transverse section of mycorrhiza of beech with an enlargement of the sheath and Hartig net.

very coarse stiff, bristle-like structures grow out from the sheath into the soil. In still others, the sheath is a loose weft of easiiy recognizable, tangled, clamp-bearing hyphae.

Nutrition of Fungi

Many of the agarics and others can be readily grown in pure culture so that they are not obligate symbionts in the sense that they cannot be grown in the absence of the host. Others, especially members of the genera *Cortinarius*, *Lactarius* and *Russula*, are more difficult to grow. Some have never been grown and are probably obligate symbionts. But in nature, virtually all, from their specialised physiology, are probably ecologically obligately dependent upon their host tree. Most are incapable of leading an independent saprotrophic existence in the soil. This is the more striking when it is appreciated that many are closely related to free-living saprotrophs inhabiting the leaf litter around the roots. In their carbon nutrition they contrast markedly with these cellulose and lignin decomposing litter-inhabiting fungi. With very few exceptions, they are incapable of utilizing polymers such as cellulose and lignin. Most are 'sugar fungi' requiring simple sugars as carbon sources. This suggests, because of the scarcity of such carbohydrates in the litter and soil in general, that they must obtain these from their host tree. All the evidence, which is very varied, points conclusively towards this. The fact that they develop better on seedlings grown in full sunlight than in the shade is suggestive. High radiant flux allows rapid photosynthesis and a surplus of soluble sugars for translocation to the roots. Explanted pine roots become mycorrhizal if supplied with sucrose *via* their attached hypocotyl. In culture they have a relatively slow growth rate and a very poor competitive ability. In mixed culture with soil saprotrophs, they are readily overgrown and intolerant of competition. This all indicates that they would fare badly in competition with soil micro-organisms for limited soluble carbohydrates as they become available in the litter. However there are a few exceptions. One notable one is *Boletus subtomentosus* which is mycorrhizal with a number of pines. In culture it produces extracellular phenolases as rapidly as any whiterot fungus. It has also been shown to produce cellulases and to bring about very rapid decomposition of leaf litter. A few others also produce amylases, cellulase and hemicellulases but their ability to do so is often poor compared with litter-inhabiting species such that they would be adversely affected by competition in the free living states ; as symbionts, high sugar levels in the roots would be sufficient to repress synthesis of such enzymes.

Most sheathing mycorrhizal fungi require one or more vitamins, especially thiamine or one of its moieties. Many also require, in addition biotin, nicotinic acid, inositol and others especially of the B complex. Some require specific amino acids. These they also obtain from their host but even so, given these and a suitable carbon source, most grow better if supplied with exudates from living roots. All of these commodities would be very scarce in the litter and soil outside the rhizosphere and there would be very intense competition for them in the rhizosphere. Because the requirements of the rhizosphere micro-organisms and the sheathing mycorrhizal fungi are so very similar and because they both obtain these from the same source, the host roots, one way of looking at the latter is to consider them as a logical development from the rhizosphere where one fungus becomes dominant on the root surface ; even so why a sheath develops has still to be explained.

Physiology, fungus and host benefits

Virtually all the motivation for research into the physiology of sheathing mycorrhizas has stemmed from the inspiration and dedication of Professor J.L. Harley and for a more detailed treatment his monograph with S.E. Smith *Mycorrhizal symbiosis* (1983) should be consulted.

Dependence of the fungi on the host for their carbon sources

Direct evidence for the dependence of the fungi on their host tree for carbon sources comes from tracer experiments. When mycorrhizal seedlings of *Pinus sylvestris*, produced by inoculation with *Boletus variegatus*, were allowed to photosynthesize in ^{14}C labelled carbon dioxide, ^{14}C labelled photosynthate was translocated to the root system and it rapidly appeared in the fungal sheath. In *Fagus* mycorrhiza, host sucrose translocated to the roots is converted into the fungal carbohydrates trehalose, glycogen and mannitol ; none of this can be effectively reabsorbed and utilized by the host. It is a remarkable feature that in all symbioses involving an autotrophic and a heterotrophic partner the heterotroph always converts the products of the autotroph's photosynthesis to compounds which the autotroph cannot utilize. This is as true for the invertebrates which associate with algae as it is for the fungi of lichens and for the fungi of mycorrhizas as it is for biotrophic pathogens such as the rust fungi. This conversion obviously creates a sink for host sugars in the fungus and maintains a concentration gradient in favour of the fungus. It also prevents reciprocal movement. The rate of translocation to the roots of mycorrhizal systems is several magnitudes more rapid than to uninfected ones. The sheath may accumulate as much as 70% of the sugars supplied to the roots. There is thus not only conversion of sugars but also a massive diversion of these into mycorrhizal roots. Because of this it is often argued that mycorrhizal roots are a considerable drain on host photosynthate and that any selective advantage that such mycorrhizas confer on their host must outweigh this large drain. This should be considered in terms of the concept of source to sink ratios. It is probably merely a question of the source, the photosynthesizing leaves, being more efficient in the presence of a large sink. To achieve such efficiency they would probably require only an enhanced mineral supply. This is the contribution of the fungus to the symbiosis.

Enhanced mineral ion uptake

Ectomycorrhizal infection tends to be most intense under conditions of mineral deficiencies or imbalance in the soil ; the greater the availability of mineral nutrients, the less infected the root system tends to be. It is in such conditions as relative mineral deficiency or where there is intense competition from other plants, indeed in most natural habitats, that mycorrhizal infection enhances growth and increases the mineral status of the host. The fact is that most northern temperate trees are normally infected and benefit from infection under natural conditions. The evidence for this is again very varied. Foresters and others who attempt to introduce exotic trees via seed into new localities often find that the seedlings fail to make adequate growth. Conifer seed imported into Western Australia in the late 1920s germinated well and initially grew into healthy seedlings which soon began to show symptoms of mineral deficiency. The introduction into the nursery beds of small quantities of soil in which the same conifers had grown successfully elsewhere was more effective in causing recovery than was adding the deficient minerals. The recovery was associated with mycorrhizal development and failure was due to the absence of the mycorrhizal fungus in the soil. There is a large volume of evidence of this nature. Other evidence comes from a comparison of the growth of inoculated and uninoculated seedlings. Hatch's early work (1937) is often quoted. He compared the growth of inoculated and uninoculated seedings of *Pinus strobus* in prairie soil from Cheyenne, Wyoming, USA. He chose this soil because it was treeless and assumed, from the sort of evidence discussed above, that it would not contain the appropriate mycorrhizal fungus. After one year's growth, inoculated, and thus mycorrhizal, plants were healtheir, taller and heavier. This is indicated in the differences in their dry mass (Table 2.1). Inoculated plants

Table 2.1. Results of experiments by Hatch (1937) on the effect of mycorrhizal infection on growth and nutrient uptake by *Pinus strobus*.

Treatment	*Dry mass (mg)*	*Nutrients absorbed per cent of dry mass*		
		Nitrogen	*Phosphorus*	*Potassium*
Mycorrhizal	404.6	1.24	0 196	0.744
Non-mycorrhizal	202,7	0.85	0.074	0.425

also showed greater absorption of nitrogen, phosphorus and potassium expressed both as a percentage of their dry mass and per seedling. Most striking of all was the increased phosphorus uptake. There was an almost three-fold increase over uninoculated plants whereas nitrogen uptake was less than doubled. In data from other workers, phosphate absorption was always conspicously and consistently increased. In such studies, it has been invariably found that the increased uptake of the three mineral nutrients, nitrogen, phosphorus and potassium, was proportionately higher than the increase in dry mass of the mycorrhizal seedlings. It can be argued from this that the mycorrhizal fungi increase nutrient uptake in general. This can be explained on the increase in absorbing area on mycorrhizal infection, achieved by increase in diameter of the absorbing roots, increase in their number by stimulation of branching and the extension of the hyphae well into the soil. This is offset to some extent by the absence of root hairs on mycorrhizal roots and it assumes that host and fungal surfaces have equivalent absorptive capacity. Mycorrhizal roots also live longer. The absorptive efficiency of mycorrhizal and infected roots is not the same on a unit area basis. By using excised mycorrhizal and uninfected roots in ^{32}P-labelled orthophosphate solution, it has been shown that uptake by mycorrhizal roots is between 2-5 times greater than that of uninfected roots per unit area per unit time.

It is often argued that the enhanced growth of mycorrhizal seedlings is primarily due to increased phosphorus uptake, increase in its uptake leading to increased growth rate and as a consequence increased uptake of other elements. This is a fair argument but it should not be allowed to obscure the fact that phosphates are either deficient or at least in an unavailable form in most soils and in any case are very immobile. It would thus be expected that any increase in absorbing area would markedly increase their uptake.

The sheath as a reservoir of mineral ions

In short term experiments carried out with attached and excised roots at various times during the year, the primary site of absorption in the mycorrizal roots is the fungal sheath. About 90% of the phosphate taken up remains in the sheath (Table 2.2). Even in experiments lasting a few days and in which phosphate uptake was linear with time, a constant proportion of about 10% passed into the host. Only if the mycorrhizas were swamped with unrealistically high levels of external phosphate did more than about one tenth pass to the host. The fungal sheath thus acts as a reservoir of accumulated mineral ions but it then releases these to the host in periods of deficiency. This has been demonstrated in a number of experimental situations. One of these has been to grow mycorrhizal and uninfected pine seedlings in ^{32}P-labelled orthophosphate solutions and then transfer them to phosphate-

Table 2.2. Accumulation of phosphate in the sheath of attached and excised mycorrhizas of nature beech trees at three different seasons.

Condition of roots	*Attached*				*Excised*			
Date	*31 Mar.*	*1 May*	*23 June*		*31 Mar.*	*1 May*	*23 June*	
Condition of leaves	*In bud*	*Expand-ing*	*Fully expanded*		*In bud*	*Expand-ing*	*Fully expanded*	
Conc. H_2PO_4 supplied (mM)	0.074	0.32	0.16	1.6	0.074	0.32	0.16	1.6
Mean percentage phosphate in sheath	88	88	89.8	86.8	91	89	84.9	91.3

free solutions and to follow movement of the label to the shoot. This is steady in both initially while in the phosphate, but both uptake into the plant and movement to the shoot are greater in mycorrhizal plants. The movement continued at a steady rate for 21 days in mycorrhizal plants when placed in the phosphate free solutions but fell rapidly in infected plants. Phosphate appears to be accumulated as polyphosphate granules. These are chains of covalently linked orthophosphate molecules complexed with calcium. These increase in both quantity and size during phosphate accumulation. This is a convenient way to store phosphate as the polymerized form is osmotically inactive. Polyphosphate granules also occur in other types of mycorrhizas and in some lichens. They may be a common characteristic of phosphate storage in such symbiotic systems.

Organic phosphorus and phosphatases

It has often been suggested that sheathing mycorrhizas may actually mobilize insoluble mineral nutrients which are unavailable to uninfected roots thus explaining their efficiency as absorbers. Much of the phosphate in the soil is in an organic form. The average value for mineral soils lies between

50-65%. Of this organic phosphorus, various inositol containing salts of iron, aluminium and particularly calcium, such as calcium inositol hexaphosphate, commonly called calcium phytate, are by far the most important components.

Sheathing mycorrhizas develop best in raw humus where as much as 90% of the phosphorus may be in this organic form. Both mycorrhizal and uninfected roots of beech possess surface bound acid phosphatases which will hydrolyse calcium phytate into free inositol and orthophosphate. The activity of these phosphatases is 2-8 times higher on mycorrhizal roots than on uninfected ones. These enzymes can be found throughout the sheath. Under experimental conditions no phosphatase activity occurs in the external medium. Phosphatases thus appear to be firmly bound to the plasmalemma or cell wall. The ability to hydrolyse such insoluble organic phosphates has been clearly demonstrated experimentally and it is often assumed that they contribute to the effectiveness of phosphate uptake. However, rather more information is required before the role of these enzymes in the mineral nutrition of the host can become more than a matter of pure conjecture. An important question still to be answered is whether or not these phosphatases contribute significantly to phosphate availability in the soil. Obviously any hydrolysis by these enzymes can take place only when the substrate is in contact with them so it would depend upon such factors as the degree of exploitation of the soil by the mycorrhizal roots and their hyphae, hyphal density and the mobility of the substrate. Some information would also be needed as to what proportion of the total phosphate taken up is derived from organic sources before the ecological significance of the possession of such enzymes can be evaluated. Surprisingly the experiment of labelling the labile pool of phosphate in the soil has not been carried out with sheathing mycorrhizas.

The dual role of the sheath

The large and organized sheath is a conspicuous feature of ectomycorrhizas. In *Fagus*, it comprises some 35-40% of the dry mass of the mycorrhizas and is responsible for about half of their carbon dioxide emission. Harley (1975) argues that it must be an expensive structure to upkeep and so must have some selective value. It is a storage structure for fungal carbohydrates. Such a storage structure may be of selective advantage to fungi which produce a perennating mycelium and seasonally produce large reproductive structures such as basidiocarps. They store up carbohydrates when the host is most active in photosynthesis in the spring and summer and utilize this store to produce basidiocarps in the autumn as and after the leaves fall. Most fleshy agaric-type basidiocarps produced by mycorrhizal fungi and indeed the litter-inhabiting Basidiomycotina are ephemeral structures. The main flush of these in northern temperate climates starts towards the end of August. A heavy rainfall at that time and subsequent warm humid weather will be followed in about a week by the appearance of basidiocarps of various species of *Boletus*, *Lactarius* and *Russula*. The main season lasts from early September to late October when the litter is moist and the atmosphere humid. Most are frost sensitive and their disappearance is correlated with the first frosts of autumn. They cannot withstand drought. Spore production requires an ample supply of water and spore discharge involves the activity of turgid cells. They also rely upon the turgidity of their hyphae for support rather than upon any distinct mechanical tissues. In dry conditions they lose water very easily to the atmosphere and shrivel. Spore discharge ceases. The majority do not recover on return to more humid conditions. Thus the development and appearance of basidiocarps is restricted mainly to the autumn. This periodicity of reproduction is thus a reflection of a periodic climate. But the sheath is also a storage tissue for mineral nutrients which in a way benefits the host. The availability of these nutrients in the litter is also seasonal. After leaf-fall the litter is moist and microbial and microfaunal activity is high. Minerals will be made available for absorption by leaching and mineralization. However, in the dry summer both of these processes will be depressed. Efficient absorption by the fungi in times of plenty and storage for release to the host in times of scarcity has obvious merits.

Ectomycorrhizas thus appear to be adaptations shown by trees to seasonal climates, particularly the northern temperate one, where host and fungus grow and reproduce out of phase with each other. Such a system would be less well adapted to herbaceous plants where in particular new roots as well as shoots are produced each year from reserves in some special storage organs or seed. In the wet tropics the trees grow and fungi grow and reproduce throughout the year and in the dry tropics in the dry season everything is virtually dormant. Most trees in such climates have endomycorrhizas of the vesicular-arbuscular type.

Other beneficial effects on the host

It is becoming apparent that the beneficial effects on the host cannot be ascribed solely to the increased efficiency of the mycorrhizal roots in mineral nutrient absorption. The fungal symbiont also supplies the host with growth substances such as auxin, indole acetic acid, and probably cytokinins and gibberellins. Since these growth substances are also formed endogenously by the host, it may possess above normal levels of these and they must greatly influence its growth and development; in all probability they change the physiological and biochemical processes going on in the roots.

It has been known for a long time that cell-free extracts from cultures of mycorrhizal fungi induce the development of structural features, such as branching pattern, swollen appearance and lack of root hairs, associated with mycorrhiza in non-mycorrhizal roots. Such changes can also be brought about by the application to roots of above optimal levels of auxin. Many mycorrhizal fungi produce indole acetic acid and several other indole compounds in culture and it has been suggested that the mycorrhizal fungi produce extracellular auxin on infection and that it produces the characteristic morphology of ectomycorrhizas. Certainly high levels of auxin are characteristic of mycorrhizal roots and support for the idea that it is the fungus which induces the changes comes from the observation that, if the association terminates for any reason, any new roots assume the appearance of non-mycorrhizal ones.

High auxin levels induced by the fungs may thus help to explain, perhaps too naively, the increase in absorbing surface provided by the fungus. It may also go some way to explaining why mycorrhizas develop only in mineral deficient soils. High available minerals can terminate the association. The synthesis of auxin is influenced by the concentration of available nitrogen. An ample supply of easily available nitrogen inhibits extracellular auxin production by mycorrhizal fungi in culture. If the fungi cannot produce sufficient auxin, successful infection is not established. If the same applies to the established mycorrhizal association, high external nitrogen levels would mean an increase in endogenous nitrogen in the roots leading to lower levels of fungal produced auxin and eventual breakdown of the association.

More recent investigations have provided evidence that several ectomycorrhizal fungi also produce cytokinins and substances related to gibberellins. Their role in the symbiosis is as yet even less well determined. These and the high levels of auxin could help to explain the diversion of host photosynthate to mycorrhizal roots and its conversion to fungal carbohydrates. They might enhance the synthesis of particular RNA and enzymatic proteins such as those necessary to convert sucrose to mannitol. Such an increase in metabolic activity would create the necessary sink for host metabolites. Cytokinins have also been implicated in directing both short and long distance transport of organic nutrients in the phloem in comparable situations in biotrophic parasitic symbioses, such in 'green island' formation by leaf-infecting rust fungi. However, at the moment we can only speculate as to whether or not they actually do so in ectomycorrhizas. They could also help to explain why mycorrhizal roots live longer, and hence function longer than non-mycorrhizal ones.

There is some evidence that mycorrhizal roots are less susceptible to fungal root pathogens. The sheath may act as a mechanical barrier to such fungi or may produce antibiotics which deter them. Root pathogens would not longer be attracted by root exudates. Ross and Marx (1972) found that non-mycorrhizal roots of *Pinus clausa* infected by *Phytoph-*

thora cinnamomi exhibited a massive root necrosis and after two months 60% had died. On comparable mycorrhizal seedlings, 25% of the absorbing roots were mycorrhizal, thus reducing the amount of susceptible root tissue exposed to the pathogen. This resulted in nearly 70% of the test seedlings surviving. The cortical tissues of mycorrhizal roots were free of *P. cinnamomi,* verifying their resistance to attack by the pathogen. This protection may be achieved physically by the sheath acting as a purely mechanical barrier to infection or chemically by the production of antifungal compounds. These, such as diatretyne nitrile, a polyacetylenic antibiotic, may be produced by the fungus. This compound is produced by mycorrhiza formed between *Pinus echinata* and *Leucopaxillus cerealis* and has been shown to contribute to the resistance of roots to *P. cinnamomi.* Others, such as tannins, may be produced by the host in response to infection by the mycorrhizal fungus and subsequently act as a chemical deterrent against pathogens. Mycorrhizal roots may produce different exudates from non-mycorrhizal roots and may also support a different rhizosphere population. These may also affect the competitiveness of any would be pathogen in the rhizosphere.

LICHENS

The ectomycorrhizal root system with its distinctive sheath to all intents constitutes a dual organism similar in many ways to that found when an alga and a fungus combine to form a lichen. But in a lichen the physiological interactions of the alga or cyanobacterium and fungus result in a new distinct morphological entity which is a self-reproducing functional unit behaving as a single organism with no obvious similarities to either of its components. For convenience they are named and studied as if they were a single organism.

Fungi involved and structure

There are some 13,500 known species of lichens representing about one sixth of all the known fungi. The fungi involved are mainly Ascomycotina with a few Basidiomycotina. A very wide variety of Ascomycete fungi are involved. The majority are Discomycetes producing abundant apothecia. Others are

Pyrenomycetes or Loculoascomycetes producing perithecia and pseudothecia respectively. In addition there are some 14 Basidiomycete lichens, such as species of the genus *Cora*. None of the fungi involved have been given genetic names as none have been encountered free-living and none have sufficiently similar characters to any known free-living terrestrial genera to indicatd that they are co-generic with any of these. It appears that they cannot survive independently in nature. In lichens, one particular alga or cyanobacterium is associated with one particular fungus. Unlike the fungi, the majority of the algae and cyanobacteria belong to well-known free-living genera. Some 21 genera of algae are involved in lichen associations. Twenty of these are green Chlorophycae. Of these, members of the genus *Trebouxia* are most common and they are also the only ones which are rarely, if ever, encountered free-living. Other common genera of green algae include *Trentepohlia* and *Coccomyxa*. The remaining genus is *Heterococcus*, a member of the yellow-green Xanthophyceae. Twelve genera of cyanobacteria are involved. *Nostoc* and *Scytonema* are the most commonly encountered ones.

Lichens are usually separated morphologically into three major growth forms-crustose, foliose and fruticose (Fig. 2.2). Crustose lichens form crusts in such intimate contact with their substrate that often they cannot be readily separated from it. Foliose lichens (Fig. 2.2) are leafly dorsiventral structures often attached to their substrate by hair-like rhizinae and are easily peeled off from it. Some show remarkable parallels with the conventional dicotyledonous leaf. Fruticose lichens are erect, shrubby or pendent, hair-or tassel-like structures attached only at their base and usually circular in transverse section. Two major types can be recognized on internal structure (Fig. 2.3). In a few lichens the alga and autotrophs are distributed at random throughout the vegetative body or thallus. Such undifferentiated mixtures are called homoiomerous. The majority are heteromerous or stratified with the alga confined to a well-defined layer just below the upper surface. The alga is covered and protected by a thin, tough pseudoparenchymatous cortical layer composed of thick-walled, heavily gelatinized hyphae firmly cemented together. In the orange and

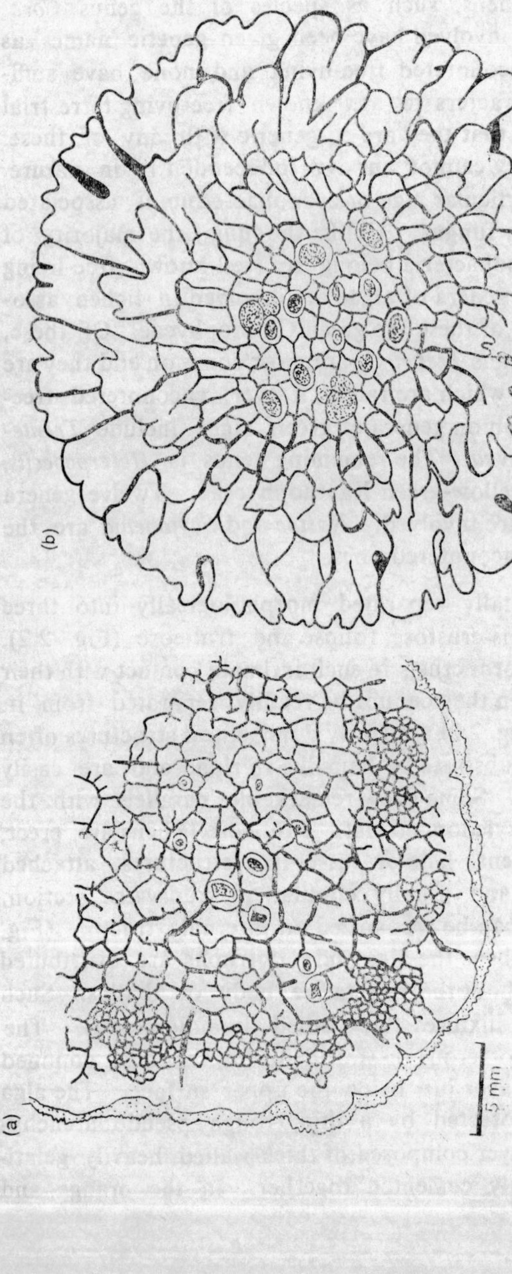
(b)
5mm
(a)
5 mm

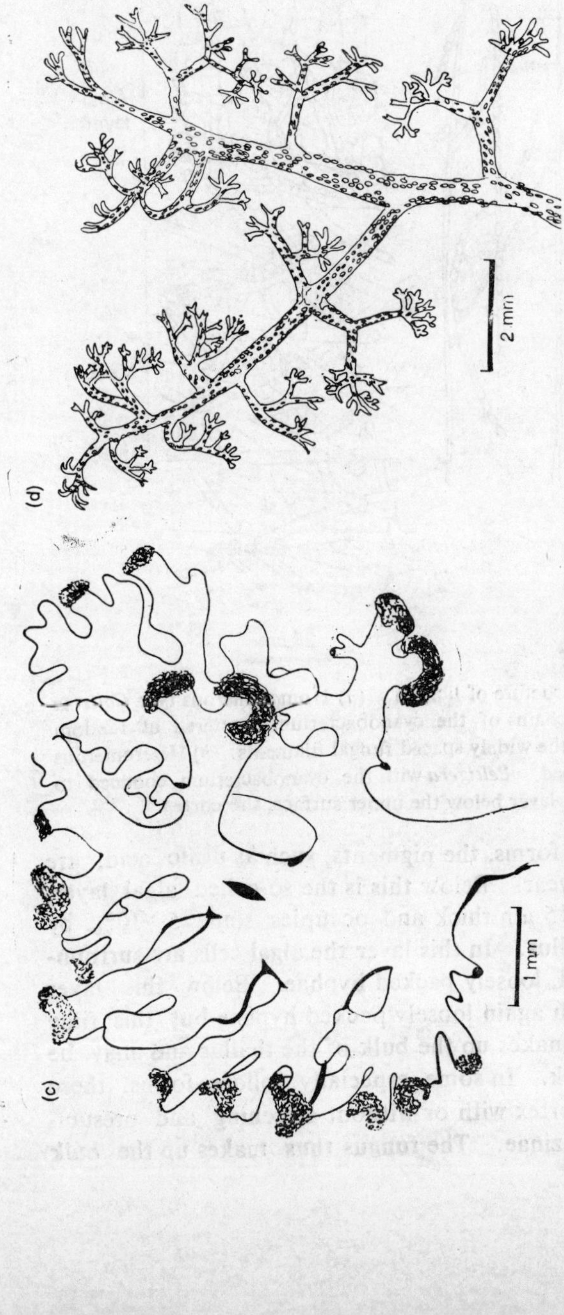

Fig. 2.2. Three major growth forms of lichens. (*a*) Crustose : part of a thallus of Ochrolechia tartarea dried out on a rock. (*b*), (*c*) Foliose (*b*) Xanthoria pariatina, with apothcia, growing closely applied to a rock ; (*c*) part of a thallus of Hypogymnia physodes with reflexed tips of branches bearing powdery soredia underneath. (*d*) Fruticose : part of an erect thallus of Cladonia rangiformis.

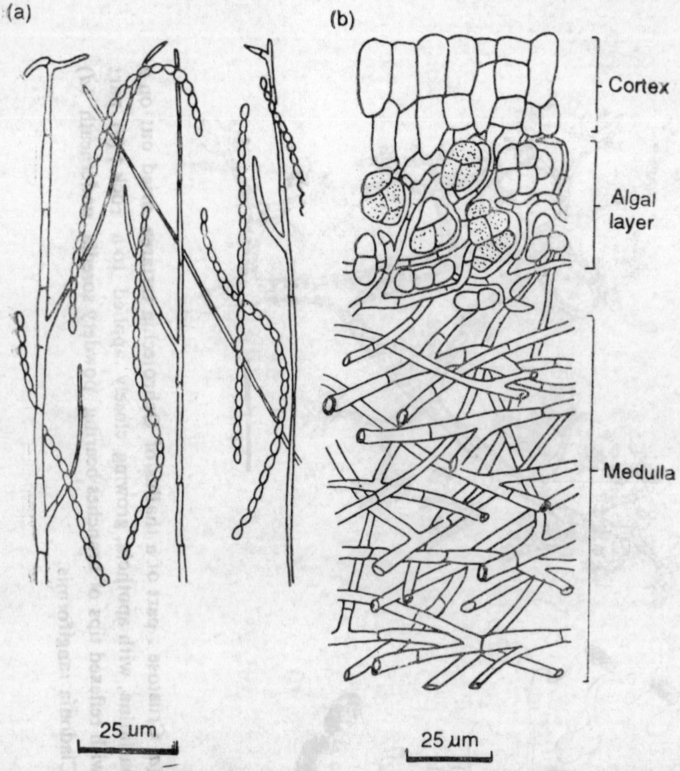

Fig. 2.3. Internal structure of lichens. (*a*) Homoiomerous type ***Collema*** with the chains of the cyanobacterium scattered at random through the widely spaced fungal filaments. (*b*) Heteromerous or stratified. ***Peltigera*** with the cyanobacterium confined to a distinct layer below the upper surface, the cortex.

yellow pigmented forms, the pigments, such as usnic acid, are deposited in this year. Below this is the so-called algal layer. It is usually 10—15 μm thick and occupies some 5—10% by volume of the thallus. In this layer the algal cells are surrounded by thin-walled, loosely packed hyphae. Below this layer is the medulla with again loosely packed hyphae but this time thick-walled. It makes up the bulk of the thallus and may be up to 0.5 mm thick. In some, especially foliose forms, there may be a lower cortex with or without attaching and presumably absorbing rhizinae. The fungus thus makes up the *bulk* of the thallus. As in ectomycorrhizal associations, both partners can reproduce separately, the fungi by ascospores or by basidiospores, and the algae by non-motile spores. Many of the green algae involved will produce zoospores in culture as their free-living relatives do but do not do so in the lichen. The spores of the fungi, at least, are actively discharged from the thallus and it is assumed that they unite with a suitable alga after dispersal to form a composite plant. This must be especially so in those, such as *Peltigera,* with no vegetative means of dispersal of both. Unlike ectomycorrhizal associations, the majority produce some combined reproductive structure, such as soredia (Fig. 2.4). These are separable clumps of algal cells closely enveloped by fungal hyphae. Individual soredia range from 25 to 100 μm diameter well within the range of wind dispersed propagules. Although most are dispersed by wind, some are dispersed by rain splash or animals. Aggregations of soredia occur on the surfaces of lichens as powdery areas. Their size, shape and position on the thallus are characteristic for each lichen species.

Physiology, carbon and mineral metabolism

The physiological relationships of the symbionts have been the object of a great deal of study since lichens were first discovered to consist of two organisms in the nineteenth century. As in ectomycorrhizal associations, the autotrophic host, in this case the alga or cyanobacterium, supplies the fungus with all its carbon requirements and with vitamins, such as biotin and thiamine. Movement of photosynthetically fixed carbon them the algal layer to the fungal medulla has been demonstrated in several lichens. In the foliose lichen *Peltigera polydactyla,* the thallus can be dissected horizontally into the algal layer plus the upper cortex and the fungal medulla. Smith and Drew (1965) exposed discs cut from the thallus of *P. polydactila* to ^{14}C labelled carbon dioxide for 4.1 in the light. The discs were then dissected and they showed that about 40% of the ^{14}C fixed by the cyanobacterial symbiont *Nostoc* had passed into the medulla, indicating a rapid movement of photosynthate to the fungus. In other lichens containing other algae this movement has been shown to occur but it may be slower.

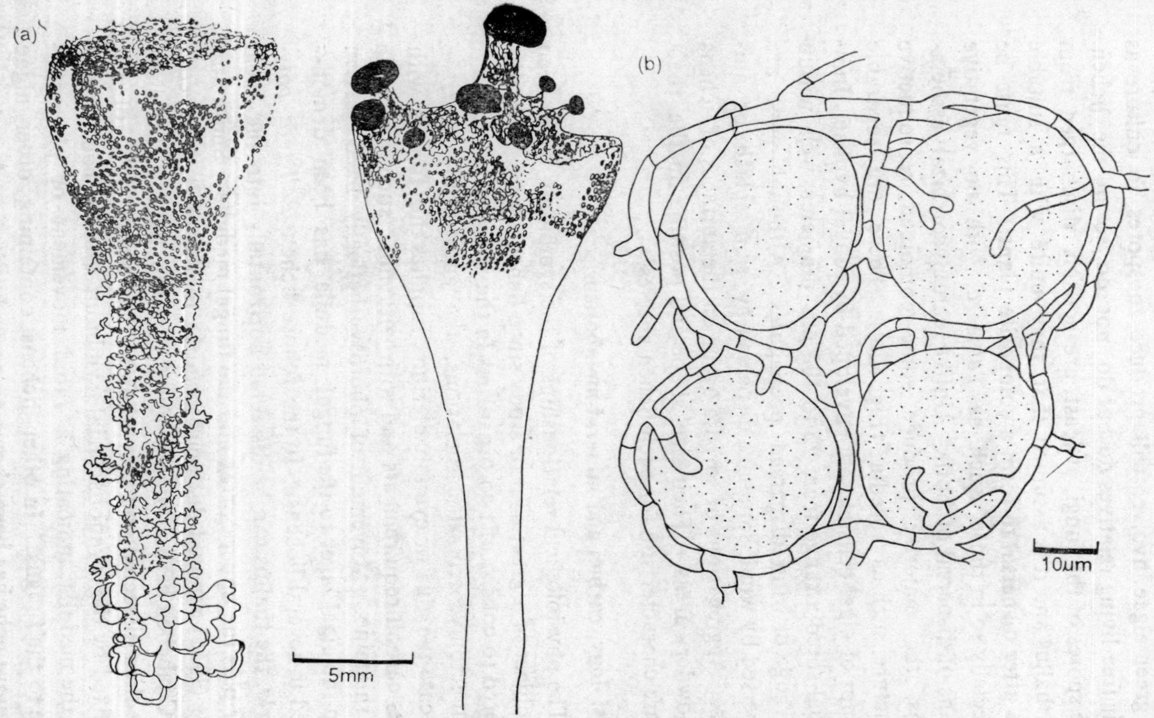

Fig. 2.4. Soredia (*a*) Two podetia of *Cladonia fimbriata*, one with apothecia around the rim and both covered in powdery soredia. These may be wind blown or rain from inside the cup-like apex. (*b*) A soredium, a separable clump of algal cells closely enveloped by fungal hyphae.

The algae of lichens can be cultured ; immediately after isolation into culture solutions, they release appreciable amounts of carbohydrates from their cells as they photosynthesize. Each alga usually releases a single, simple carbohydrate. In *Nostoc* and other cyanobacteria, it appears to be glucose. In the Chlorophyceae it is always a polyhydric alcohol. *Trebouxia* always releases ribitol and *Trentepohlia* erythritol. These are also the forms in which carbon moves from alga to fungus in the lichen. The carbohydrates, in whatever form they are released, become converted in the fungus into fungal carbohydrates, usually mannitol but sometimes into arabitol as well. *Nostoc* and lichen algae are unable to utilize these and as an in sheathing mycorrhizas there is again a one way flow of carbon from autotrop to heterotroph. It is not known how the fungus induces its partner to release carbohydrates. Lichens, but not its two components alone, produce a wide variety of chemicals called lichen substances or lichen acids. It has been suggested, with very little supporting evidence, that these may in some way alter the selective permeability of the cell membranes of the autotroph. Alternatively it may be a special feature of physical contact between the symbionts. A further possibility is that the fungus interferes with the metabolism of the alga or cyanobacterium so that the carbohydrate intended for cell wall or sheath synthesis is released instead. For example, in *Nostoc*, immediately after isolation into culture from *Peltigera*, 50% of the carbon fixed in a 3 h period was released as glucose. After two days in culture, the amount released had dropped to 12% and glucose was no longer released but polysaccharides were Free-living *Nostoc* and symbiotic ones grown in culture for some time have thick mucilaginous sheaths surrounding their filaments. This sheath is lacking or greatly reduced in *Nostoc* in the lichen. The loss in ability to release glucose after a period in culture is correlated with an increase in thickness of the mucilaginous sheath. Thus glucose intended for sheath synthesis may be diverted by the fungus and converted to mannitol. Hill (1972) has provided evidence that before the glucose is released the fixed carbon is converted to sugar phosphates and then a glucan. It is suggested that the glucan is produced outside the cyanobacterial plasmalemma as part of the wall or sheath and that it is hydrolysed to glucose by an extracellular fungal glucanase, the free glucose formed being converted to mannitol (Fig. 2.5). The green alga *Trebouxia* isolated from

Fig. 2.5. Redistribution of ^{14}C in *Peltigera polycactyla* in various fractions after pulse feeding with labelled CO_2 followed by incubation in buffer.

the lichen *Xanthoria aureola* behaves very similarly to *Nostoc*. Initially in culture it releases about 40% of the carbon fixed in photosynthesis but in this case as ribitol. Only about 2% is converted into insoluble materials such as storage products, cell walls, proteins and lipids. It also has a very thin wall. With time in culture, the cell wall becomes much thicker and some 50% of the carbon fixed is incorporated into insoluble materials and only 2% is released. However in this case ribitol is not a component of any algal polysaccharides and a different mechanism for its release must operate.

The fungi then can derive their carbon requirements from their algal partners but they must also do this. In culture they grow very slowly and they are poor competitors. Most can utilize only simple carbohydrates and as most lichens live in environments where these are certainly in very short supply or even non-existent, they, like the fungi of sheathing mycorrhizas, must be regarded as ecologically obligately dependent for their carbon on their autotrophic partner. They may also obtain other materials from them. Most lichen fungi are deficient for biotin and thiamine. It is assumed that these are made available by the autotrophic partners. Some lichens, containing cyanobacteria, such as *Nostoc*, are known to fix atmospheric nitrogen. Most of the nitrogen fixed passes to the fungal partner. Many lichens with green algae possess cephalodia, small outgrowths on the surface, which usually contain a cyanobacterium. These fix nitrogen and constantly release it to the thallus where it is utilized almost exclusively by the fungus. For example *Peltigera aphthosa*, which unlike other species of *Peltigera* has *Coccomyxa* as its algal component, has cephalodia containing *Nostoc*, normally found as the autotroph in other species of *Peltigera*. In this case only 3% of the nitrogen fixed by *Nostoc* is eventually used by *Coccomyxa*.

Lichens have very efficient mechanisms for the accumulation of substances from any watery solution which they are in with contact. In nature supplies of nutrients, especially mineral ions, are again likely to be both very low and very erratic. Such efficient absorption mechanisms would be advantageous. In laboratory experiments, the absorption of mineral ions, such as phosphates, and other nutrients, such as amino acids and sugars, is very rapid and metabolically dependent. For example, *Peltigera polydactyla* may take up the equivalent of dry mass of phosphate in 24 h and the equivalent of 25% of glucose. It is also clear that these are accumulated in periods of relative abundance and utilized at a slow rate in periods on prolonged scarcity in between. There is no direct evidence that the fungus absorbs mineral ions and passes them on to the autotroph but, because the fungi are such efficient absorbers in general, it is often assumed that they do so. In the lichen it is the algal layer with its fungal hyphae which is the most active in ion uptake but it is not known whether uptake by the autotroph is direct of via the fungus. The medulla does not play the role of the sheath in ectomycorrhizal associations.

There is in fact no experimental evidence that anything passes from the fungus to the autotroph. This is a major point of contrast between lichens and ectomycorrhizas. From the evidence presented for nitrogen fixation by cephalodia in green alga lichens, it could even be argued that the fungus actually reduces the amount of minerals reaching the alga.

DETRIMENTAL CONSEQUENCES OF EFFICIENT ABSORPTION MECHANISMS

Such efficient mechanisms for absorption would also help to account for the well-documented reports of relatively high accumulation by lichens of radioactive fall-out and their extreme sensitivity to atmospheric pollutants.

Accumulation of strontium and caesium

The so-called 'reindeer moss', *Cladonia rangiferina*, and 'Iceland moss', *Cetraria islandica*, are fruticose types which have covered much of the tundra. They have accumulated large quantities of radioactive strontium (^{90}Sr) and caesium (^{137}Cs) from atomic fall-outs. These lichens are eaten by reindeer in northern Sweden and Russia and by caribou in Canada and Alaska. They supplement their diet of sedges and willows with the lichens during the winter. If the winter is very severe, lichens may form up to 95% of their food. Their nutritive value is due to such storage carbohydrates as lichenin, a β—1, 3 and β—1, 4 linked glucose polymer and the less common isolichenin, with α—1—3 and α—1, 4, linked glucose. The lichens are eaten and the radioactivity built up in the animal and passed on in the food chain to especially the Eskimo and Lapp. Eskimos who consumed caribou have been found to contain over four times as much ^{90}Sr as the average populations of the northern temperate zone. This can be attributed to three factors : the rapid and efficient uptake by the lichens, their slow growth rate and great age so that they persist, and the reindeer and caribou grazing over very wide areas.

Sensitivity to sulphur dioxide

Their extreme sensitivity to atmospheric pollutants, such as sulphur dioxide, means that they are markedly absent from towns and cities and around industrial sites such as off refineries and brickworks. This sensitivity is no doubt enhanced by their efficient absorption mechanisms. The sulphur content rises markedly as the amount of sulphur dioxide in the atmosphere increases. As a group they show differential sensitivity and can thus be used as reliable biological indicators of pollution. There appears to be a direct relationship between the zonation of Lichens and the mean annual, but more especially the mean winter, sulphur dioxide levels. A better correlation exists between zonation and winter means because lichens are more sensitive to sulphur dioxide at high humidities. Relative humidities are on average higher in winter. The sulphur dioxide penetrates the lichen tissues more easily if they are moist. Lichens are also physiologically more active in the winter months.

Lichens vary to such a degree in their tolerance to sulphur dioxide that around a city or an oil refinery one finds concentric patterns of exclusion of particular species. The zones are related to the mean winter sulphur dioxide levels. All except *Lecanora conizaeoides* show these patterns but it occurs even in city centres. It is the most tolerant of all lichens to sulphur dioxide. This tolerance can be explained in part by the fact that it is virtually un-wettable in the natural state. This is probably due to heavy incrustations of non-wettable lichen substances. In this sense it avoids the pollutant by reducing the amount that it takes up dissolved in water. It is also a very poor competitors with other lichens; it seems probable that the elimination of its competitors by levels of sulphur dioxide which it can tolerate leads to its abundance in polluted areas. Although it was unknown before the middle of the last century, it is now extremely common and very abundant, in terms of density, on tree bark and wood throughout Britain where mean winter sulphur dioxide levels are in the range 55—150μm^{-3}. Hawksworth and Rose (1970) have devised a qualitative scale for the estimation of mean winter sulphur dioxide levels using epiphytic lichens growing on tree bark.

It is possible to oversupply nutrients to the symbionts so that the association breaks down. This is exactly the situation that occurs in ectomycorrhizal associations where, given high levels of mineral nutrients in the soil, the host is non-mycorrhizal. However, in both cases the fungi are incapable of an independent existence in nature.

GENERAL BIOLOGY AND POSSIBLE BENEFITS TO THE AUTOTROPH

The benefits which the autotroph derives from the association are much more difficult to assess. There is no unequivocal evidence that any specific compounds move from fungus to autotroph. Although for some algae the assocition may not be an obligatory one, once in an association, constraints are applied on the alga by the fungal partner. Green algae no longer reproduce sexually or produce their motile stages in the lichen. The alga makes up only 5—10% of the dry mass of the lichen yet must supply the whole with all its carbon requirements. It is often argued that it can produce only sufficient organic carbon photosynthetically to support very slow growth of itself and the fungus. The fungus would thus appear to be too large a sink for the alga to grow quickly and the alga equally too small a source for the fungus to grow quickly. Thus growth of both is restricted. There is, however, some evidence now that the algae photosynthesize more efficiently in the lichen than in culture at least and that the supply of photosynthate from the alga in a lichen is substantially greater than that needed to sustain the fungus (Table 2.3). The massive movement of photosynthate to the fungus and its conversion to the soluble mannitol may be the key. The cell contents of many lichen fungi must contain a near saturated solution of mannitol. Its function may not be so much for growth and metabolism as seen in other symbiotic systems but its high concentration may be part of the mechanism enabling lichens to withstand environmental extremes. For example, under conditions of moisture stress a low internal osmotic potential is obviously of great advantage.

The algae and cyanobacteria benefit to the extent that they are physically supported and effectively displayed to the light by the fungi, protected to some extent from environmental extremes and may have their habitat range extended. These

Table 2.3. Rates of growth and photosynthesis in lichens and lichen symbionts. (After Farrar, 1978.)

Maximum relative growth rates	
Fungi in culture	= ⊳ 0.33 d^{-1}
Algae in culture	= > 0.39 d^{-1}
Thalli in nature	= > 0.01 d^{-1}
Thalli in laboratory	= > 0.015 d^{-1}
Photosynthetic rates (mg $CO_2 g^{-1}$ algae h^{-1})	
Algae in culture	= 3—14
Algae in thallus	= 2—28
Growth and photosynthesis in Hypogymnic physodes	
Maximum growth rate	= 0.005 g g^{-1} dry wt in 24 h
Net carbon fixation	= 0.02 g $(CH_2O)g^{-1}$ dry mass (12 h light in 24 h)

benefits can best be considered in relation to the general biology of lichens.

Growth rates

Lichens are exceedingly slow growing plants. Foliose types usually grow the fastest but rarely grow more than 10 mm per year. Crustose types often grow less than 1 mm per year. Individual plants may be very old. Some have been estimated to be able to maintain themselves as individuals for 4500 years. Such age indicates a very well-balanced and stable association. Their slow growth rate may be attributed to a number of factors including the inherently slow growth rate of the fungi, their inability to conserve water and their restriction to habitats where the supply of minerals is very low. When dry they are exceedingly resistant to environmental extremes. These features enable them to grow where neither partner can alone. As such they tend to be pioneer colonizers of such inhospitable habitats as rock faces, even as far as 7300 m up Mt Everest, tree bark and man-made habitats such as stone walls, tombstones, roofing tiles, asbestos and concrete. Then resistance, their low demand for minerals and their highly efficient absorption mechanisms enable them to grow usually where no other organisms can. They are favoured by high humidities and low light intensities.

Water relations, photosynthesis and respiration

Lichens are in many ways a paradox. They are able to survive in drier habitats than either of their two components. Fungi as a group are somewhat tolerant of water stress while most algae are very susceptible to desiccation. Brock (1975) infers that the fungus confers desiccation resistance upon the lichen. The lichen autotrophs which he tested were able to photosynthesize down to a lower a_w within the lichen than when liberated from it. The lowest a_w permitting photosynthesis of the whole lichen ranged from 0.96 to 0.80 and the lowest of the autotrophs separated from these ranged from 0.995 to 0.90. He suggested that one function of the stored fungal carbohydrates, such as mannitol, is to act as osmotica to lower the osmotic potential so that water accumulation could occur in environments of low water availability. Because of the intimate association of the fungus and the autotroph some of the water becomes available to the autotroph, allowing it to photosynthesize at a lower a_w. This may be so but the polyols may also have other major roles.

Lichens are metabolically active only when moist. One reason for their low productivity is that they spend the major part of their time dry and thus metabolically inactive. The fungal medulla, although it does not appear to have an important role to play in mineral uptake and storage has two important functions which are beneficial to the autotroph. It provides a supportive skeleton to the whole over and around which the cells of the autotroph are displayed in a thin layer to the light. However, more relevant to the present discussion is that it also has a higher saturated water content than the remainder of the thallus. In *Peltigera polydactyla*, for example, the medulla contains 25% more water per unit dry mass than the algal layer and upper cortex. It thus has an important role to play in the water relations of the whole thallus. Since the medulla is either beneath or internal to the algal layer, it provides water for it when the thallus is drying out. Lichens have no physiological control over their water content. Water loss and uptake are purely physical phenomena. Loss is governed by the physical factors that affect transpiration.

Thus lichens can show large and rapid fluctuations of water content during the course of the day. They dry out rapidly in the sun with the water content falling as low as 5% of their dry mass and they become saturated immediately it rains as water is taken up in the extra hyphal capillaries and into the thick hyphal walls. The water content of the thallus has profound effects on the rates of respiration and photosynthesis and as such affects the carbon balance of the lichen. The saturated water content of lichen thalli lies between 100—300% of their dry mass. The maximum rate of photosynthesis occurs at between 65—95% of complete saturation, depending upon the species, and the rate decreases above and below this, declining particularly rapidly at water contents higher than those for maximum photosynthesis. The maximum rate of photosynthesis is also relatively low per unit area compared with leaves. The autotroph occupies only 3—10% by mass of the lichen yet has to supply the fungus with all its carbon. The chlorophyll content may be less than one quarter of that of leaves and the several layered and often pigmented fungal cortex absorbs far more of the incident light than does the epidermis of a leaf. The maximum rate of respiration is reached somewhere between 40—95% of complete saturation. The rate of respiration increases approximately linearly with water content up to this point but does not decrease in rate above or in saturated thalli. Respiration also continues at around 6—7% of the maximum rate even in air dry thalli in which photosynthesis has ceased. Thus it is clear that under many natural conditions there may be only a few hours each day, as in the early morning following wetting by dew and before they dry out in the sun, when there is a positive carbon balance. This is even more likely in extreme environments. Lange, Schulze and Koch (1970) have shown that in a typical day in the Negev Desert *Ramalina maciformis* fixes 1.32 mg CO_2 g^{-1} (dry mass) in the period between sunrise and drying out. It then remains dry until night fall when it is re-wetted by dew. It subsequently loses 0.78 mg CO_2 g^{-1} (dry mass) before dawn by basal respiration. Thus some 60% of the fixed carbon is lost within 24 h of fixation. Net production is therefore a relatively low percentage of gross primary production.

Obviously any means of delaying drying out, for however brief a period, would be beneficial to the carbon balance. The water storage capacity of the medulla should be evaluated in this light.

The decreased rate of photosynthesis which occurs below the optimum water content of the thallus can be attributed to two factors, reduced light transmission and eventual dehydration of the alga. The water-holding capacity of the medulla delays the latter. The fungal cortex acts as a light screen. Air-dry *Peltigera polydactyla* is quite grey but on wetting with water turns blue green in under a minute as a result of the cyanobacterial cells showing through the more translucent moist fungal cortex. When saturated, the cells of the fungal cortex are fully expanded and allow maximum transmission of light. With loss of water they contract so that there are more vertical walls per unit area and the whole becomes opaque as less light is transmitted. The cyanobacterial cells also contract so that there are more of them per unit area (Fig. 2.6).

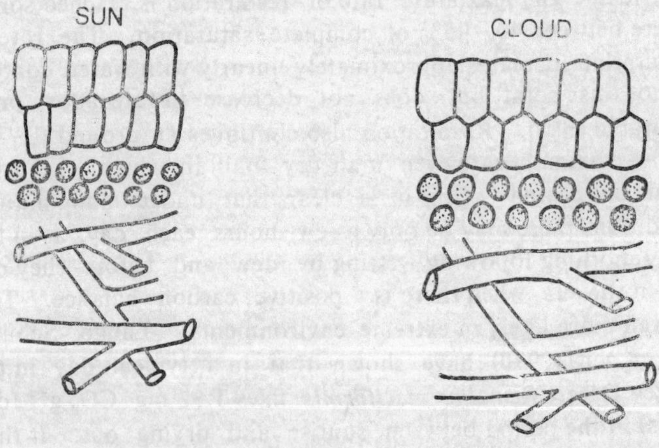

Fig. 2.6. Control of light transmission by the cortical cells. In the sun the cortex contracts as it dries out so that there are more cells per unit area and thus less light is transmitted through the thick walls. There are also more algal cells per unit area because they contract as well so each receives less radiant energy.

Thus a fixed number of cyanobacterial cells are exposed to less radiant energy than they are when the thallus is saturated.

Significance of fluxes in water content

Fluxes in water content, *i.e.*, alternating periods of wetting up and drying out, may be far more important to the lichen than its water content at any particular moment of time and may be absolutely critical in maintaining the lichen in a healthy state. Most attempts to maintain lichens in the laboratory have been successful only when the thalli were alternately wetted and dried. Slow alternating phases of wetting and drying also appear to be one of the factors that favour the synthesis of a lichen from its components. Constant saturation has a marked deleterious effect. Farrar (1976a) has shown in *Hypogymnia physodes* that the capacity to photosynthesize, the total polyol content and the ability to absorb phosphate all declined over 7 days at complete saturation. He suggested that this might be due to anoxia consequent on reduced gas exchange through the water saturated hyphal walls. He also concluded that lichens may actually require alternating periods of wetting and drying when he investigated the effects of various wetting and drying periods on *H. physodes*. It maintained high photosynthetic activity and high polyol content when so subjected but not when maintained continuously saturated. The anomaly of the situation is that subjection to wetting and drying cycles leads not only to restricted periods of photosynthesis but to high losses in the fixed carbon. A substantial proportion of the carbon fixed when photosynthesis is possible under such conditions is converted to polyols to maintain the pool of these at a high level. *H. physodes* commonly contains as much as 10 mg g^{-1} ribitol in the alga and 10 mg g^{-1} mannitol and 50 mg g^{-1} arabitol in the fungus on a dry mass basis. The physiological response of the lichen to such cycles of stress was characterized by both a depletion and turnover of the polyol pool. Levels of insoluble materials such as proteins remained unchanged. Farrar suggests that the polyols are used to buffer the latter against change. They could, for example, protect such macro-molecules by direct substitution for water molecules in their hydration shells

Sudden re-wetting by rain after a dry spell results in a burst of carbon dioxide emission. This lasts less than two minutes and is probably a purely physical effect. This is followed by a period of several hours of resaturation respiration which is well above the basal rate. The intensity of this respiration is less in lichens from dry habitats than in those from wet ones and it may be related to drought tolerance. It may be due to an expenditure of energy in repairing membranes damaged as a consequence of desiccation. There is also a period of one or two minutes when organic solutes are lost. There is a loss of polyols of both algal and fungal origin and sucrose. The amount of carbon lost by leakage may be as much as that lost by resaturation respiration. During the initial stages of resaturation the rate of photosynthesis is low and recovery is slow ; rapid cyclic wetting and drying could thus be disastrous to the lichen in that carbon losses could exceed gains.

The factors discussed above go some way to explaining the slow growth rates of lichens but this tolerance of marked fluctuations in their water content, or drought resistance, enables them to persist in such severe habitats as rock faces where no other plants, certainly no other fungi or algae, can, and thus extends the habitat range or ecological tolerance of either component. When dry, lichens are extremely resistant to high temperatures. Thallus temperatures of 53—69°C, some 20—40°C above ambient, have been recorded in lichens exposed to the sun on rocks. Such lichens experience higher temperatures than the vegetative phase of any other organisms except those living in hot springs. But such lichens are thermoduric—they survive but do not grow at these temperatures. High temperatures on rocks are invariably the result of high solar insolation. Yet many lichen algae are quite intolerant of strong sunlight. The contraction of the cortical cells helps to decrease the intensity of the exposure. This is also aided in many by deposition in the fungal cortex of coloured lichen substances, such as the deep orange anthraquinone parietin of *Xanthoria parietina.* Many lichens are normally deeply pigmented when growing in sunny areas and weakly or not pigmented in shade. This can be verified in the case of

X. parietina by comparing its colour on a north and a south facing rocky cliff. Parietin has maximum absorption at 436 nm filtering out part of the blue region of the spectrum. Such pigments thus also change the quality of the light reaching the alga.

Lichen substances

Lichens thus have a number of quite different properties from those of either of its components, especially in their ability to withstand desiccation and high temperatures. These series of peculiar characteristics are further illustrated by their production of lichen substances, often also called lichen acids but not all are acids. A wide variety of these are formed as granules or crystalline encrustations on the surfaces of hyphae. The commonest fall into two major groups—the depsides and depsidones (Fig. 2.7). These are usually formed from two simple phenylcarboxylic acids, mainly orsellinic acid, held by an ester linkage in the depside and with an additional diphenyl oxygen linkage in the depsidone. Neither the alga nor the fungus produces these substances in culture. Many monocyclic phenyl carboxylic acids, especially orsellinic acid (methyl-dihydroxy-benzoic acid), have been identified in pure cultures of lichen fungi. It is thought that the alga is responsible for converting these units to the dimers. Because they do not appear in cultures of isolated symbionts or in cultures of non-lichenized algae, cyanobacteria and fungi and because similar ones are produced by different lichens, it is often assumed that they have some role to play in the symbiotic state. They may have a varied protective role. Their function in increasing the opacity of the fungal cortex has already been mentioned. The general resistance of lichens to insect and microbial attack is well-known and it has been suggested that lichen substances may be responsible for this. Many have antibiotic properties which could protect them from bacterial and fungal attack. Most have a bitter tasts which would make them less palatable to slugs and other herbivores. Reindeers, however, appear to enjoy the very bitter 'reindeer moss'. Simple phenolies are toxic and it may be that depside and depsidone formation is a protective response of the algae to decrease their toxicity. A

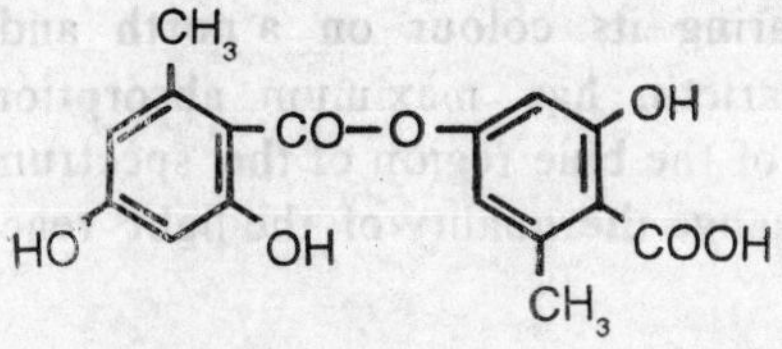

LECANORIC ACID-DEPSIDE (ester of phenylcarboxylic acids, often orsellinic acid, methyl–dihydroxy-benzoic acid)

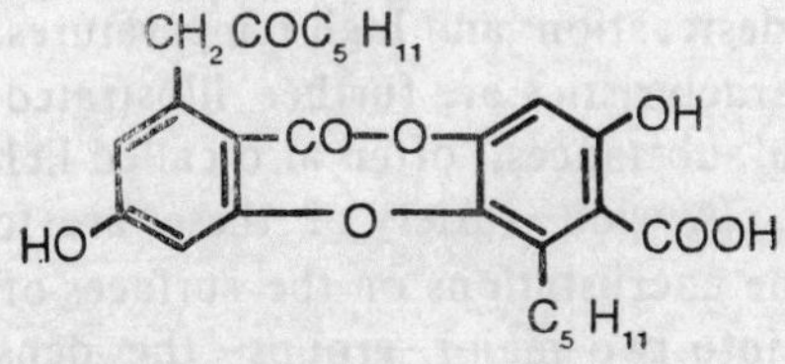

PHYSODIC ACID - DEPSIDONE (like depside but with additional diphenyl linkage)

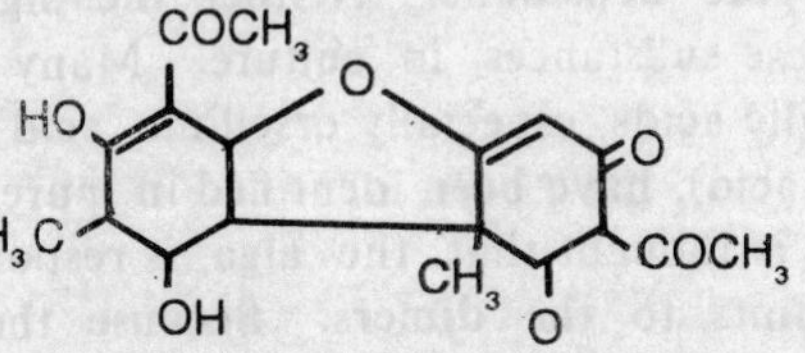

USNIC ACID-DIBENZOFURANE derivative

Fig. 2.7. Lichen substances.

further suggestion, already made, is that they may increase the selective permeability of the algal membranes and thus be important in the flow of carbohydrates to the fungus. They may also be important in two respects in the chemical weathering of substrates such as rocks. Any acids may bring ions into solution and some lichen substances are known to chelate these. Both of these processes may be important in maintaining an adequate supply of minerals when growing on a substrate very low in available mineral ions. Whatever their role, they are not produced by all lichens and so cannot be essential to the symbiotic association.

Lichen chimeras

The association has conferred ecological versatility on the two partners by changing a number of their physiological attributes, enhancing in particular their resilience to extreme environmental conditions. Both partners are also morphologically changed. Lichenized green algae have thinner cell walls than their free-living relatives and cyanobacteria lack a typical mucilaginous sheath in the lichen. The fungus shows tissue differentiation often with a pseudoparenchymatous cortex and an organized medulla, features not found in the vegetative phase of non-lichenized fungi. In ectomycorrhizal associations, it appears that the fungus induces both morphological and anatomical changes in the root system of the host thus determining its form but it is not known what controls the development and form of the sheath. Since in the lichen the fungus forms the bulk of the thallus, it is often assumed that it determines the form, the size and the rate of growth of the whole.

The processes by which lichens develop their complex form and high degree of internal organization remains at the moment as just another fascinating unsolved problem. Recent evidence suggests that the algal partner plays a vital role in determining both the pattern of development and the final form of the lichen in spite of the fact that it usually makes up no more than about 5% of the whole. This evidence comes from the discovery of lichen chimeras. Several physically united pairs of lichens containing two different algae but the same fungus have been found. Each member of the pair was known previously as an entirely species.

Many green algal lichens have developed a subsidiary association with a cyanobacterium. The latter develops in special, outgrowths of the surface, a millimetre or so in diameter, called cephalodia. This is so in *peltigera,* a genus easily recognizable on mirphological and anatomical features, in which there are some species with cyanobacteria and others with green algae and cephalodia. However in these chimeras or composite morphotypes the two algal fungus associations have developed more or less to the same extent. This has led marked divergence in morphology in the two parts. For example, chimeras have been found in which shrubby species of *Dendriscocaulon* with cyanobacteria have united with a leafy dorsivental species contianing green algae belonging to the genus *Sticta*. In all cases, clear continuity of fungal tissue between the parts has been established, eliminating the possibility that the association may be epiphytic or parasitic. The difference in morphology between the two components of these lichen chimeras plainly demonstrates the importance of the algae in controlling form. This is particularly obvious in *Dendriscocaulon-Sticta* chimeras where the radial arrangement of the internal tissues around a central axis of hyphae in the cylindrical branches of the *Dendriscocaulon* part containing cyanobacteria changes in the immediate vicinity of the green algal cell of the *Sticta* part to a layered dorsiventral structure. The growth pattern is also different. In the part with cyanobacteria, growth is confined to the tips of the branches whereas in the green algal part growth is essentially marginal. The mechanism by which the growth of the fungus is controlled to form the lichen thallus are not known.

Environmental factors appear to play an important part in determining which algae is selected by a particular fungus and by virtue of being associated with two algae the range of ecological tolerance of the fungus may be considerably increased. He studied the distribution of composite morphotypes between *Dendriscocaulon* and *Sticta filix* and normal ones in New Zealand. *S. filix* is widespread in many well-wooded areas in New Zealand. It requires a variable regime of relatively high illumination and fluctuating humidities whereas *Dendriscocaulon* prefers very sheltered and shaded habitats of unvarying high humidities and very low illumination. At one site consisting of two large northwest facing boulders, some 6 m tall and influenced by spray from a small water-fall, there were some 90 separate thalli of *Sticta* and *Dendriscocaulon* and almost one-third of these were composite ones (Fig. 2.8). On the upper parts of the boulder, which although partially shaded by the tree canopy received some 5 hours of sunshine per day through October to May, only

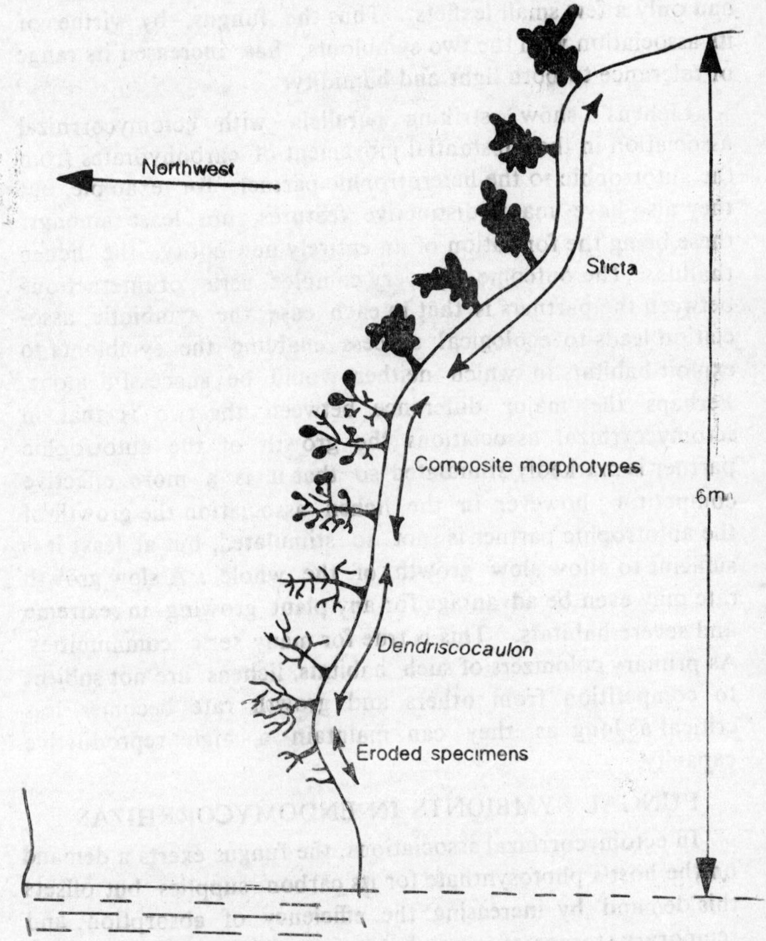

Fig. 2.8. Distribution of green algal morphotype (*Sticta filix*) and cyanobacterial morphotype (*Dendriscocoulon* sp.) with composite morphotypes in relation to aspect and associated environmental factors at a site in a forest in New Zealand. The fungus in the same in all States.

Sticta filix occurred. At the base of the boulders with were in permanent shade only *Dendriscocaulon* occurred. However in the middle zone only composite plants were present. Even so there was a gradation ; those at the top had many broad leaflets of the green algal morphotype and those at the bottom had only a few small leaflets. Thus the fungus, by virtue of its association with the two symbionts, has increased its range of tolerance to both light and humidity.

Lichens show striking parallels with ectomycorrhizal association in the substantial movement of carbohydrates from the autotrophic to the heterotrophic partner, for example, but they also have many distinctive features, not least amongst these being the formation of an entirely new entity, the lichen thallus. The outcome of a very complex series of interactions between the partners is that in each case the symbiotic association leads to ecological success enabling the symbionts to exploit habitats in which neither would be successful alone. Perhaps the major difference between the two is that in ectomycorrhizal associations the growth of the autotrophic partner is markedly stimulated so that it is a more effective competitor ; however, in the lichen association the growth of the autotrophic partner is not so stimulated, but at least it is sufficint to allow slow growth of the whole. A slow growth rate may even be advantage for any plant growing in extreme and severe habitats. This is true for many xeric communities. As primary colonizers of such habitats, lichens are not subject to competition from others and growth rate becomes less critical as long as they can maintain a high reproductive capacity.

FUNGAL SYMBIONTS IN ENDOMYCORRHIZAS

In ectomycorrhizal associations, the fungus exerts a demand on the host's photosynthate for its carbon supplies but offsets this demand by increasing the efficiency of absorption and temporary storage of mineral ions. A wide variety of endomycorrhizal associations exist and some of these, vesicular-arbuscular and ericoid mycorrhizas in particular, afford interesting parallels and also contrasts with ectomycorrhizal ones.

VESICULAR-ARBUSCULAR MYCORRHIZAS

Occurrence

Vesicular-arbuscular mycorrhizas are produced by aseptate mycelial fungi and are so-called of the two characteristic structures—vesicles and arbuscules—found in roots with this type of infection. They are by far the commonest of all mycorrhizas and are found in Bryophytes, Pteridophytes, Gymnosperms, excluding the Pinaceae which have sheathing mycorrhizas, and in virtually all families of Angiosperms. They are of general occurrence in the Gramineae, Palmae, Rosaceae and Leguminosae, which all include many crop plants. Indeed most crop plants, including herbs, shrubs, and some trees, possess the type of mycorrhiza. In the Angiosperms, apart from such families as the Ericaceae and the Orchidaceae which have other types of mycorrhiza, the Chenopodiaceae, the Cruciferae, the Cyperaceae and the Resedaceae are odd exceptions in that most species from these families appear to be either non-mycorrhizal or at least only very sparsely infected. However, the striking fact is that the majority of all plants that are mycorrhizal have this type of infection.

Associations resembling modern day vesicular-arbuscular mycorrhizas were present very early in the evolution of land plants. Kidston and Lang (1921) examined the 'thallophyte' flora of the petrified fossil genera *Rhynia* and *Asteroxylon* which are now believed to be some 370 million years old. Illustrations and slides made from Kidston and Lang's material show aseptate hyphae bearing structures with striking resemblance to vesicles and spores which are of the size and with a well complexity similar to those seen belonging to these fungi in present day soils. It is not known whether vesicular-arbuscular mycorrhizas existed before *Rhynia*. Earlier plants, such as *Cooksonia*, have not been examined for mycorrhiza because their fossil remains are mainly in the form of compression material.

Features of infection and fungi involved

In these types of mycorrhizas there is no sheath of fungus around the host root only a loose and very sparse network of rather thick-walled, irregular and aseptate hyphae. These are often aligned more or less parallel to the length of the root with numerous cross connections. Hyphae grow out from this network into the soil and may spread from one to several centimetres away from the root. These hyphae, depending on the genera of fungi involved, produce one or more of a number of spore types in the soil. These may be very thick-walled, brown to black, balloon-like chlamydospores, some 100-200 μm in diameter. They are produced singly or in cluster as the blown-out ends of hyphae (Fig. 2.9*a*). Loose aggregations of spores may also be formed within a poorly differentiated reproductive structure called a sporocarp (Fig. 2.9*b*) or thick-walled zygospores may be formed from the fusion of two hypae. Chlamydospores are the most common.

The hyphae which establish the infection penetrate the epidermal cells after forming an irregular disc-or lozenge-shaped appressorium on the surface and then grow intercellularly, often forming an intracellular coil in the first cell penetrated. Inside the root the parallel alignment of the aseptate hyphae is again most noticeable. The meristem of the root is not infected. The intercellular hyphae completely permeate the entire cortex but do not invade the endodermis or stele. Arbuscules are formed inside the perenchymatous cortical cells by the repeated dichotomous branching of the penetrating trunk hyphae. The ultimate branches may be less than 1 μm in diameter and are very difficult to resolve. They often virtually fill the host cell in which they are produced and resemble little trees—hence their name. The life span of an arbuscule is about four days. They then quickly disintegrate and it is often assumed that they are 'digested' or lysed by the host. It may be however that the arbuscules are merely ageing and undergoing autolysis just as fungal hyphae do in culture. During this process the host's nucleus enlarges considerably and the arbuscules release oil as droplets into the host cell. The hyphal branches in the arbuscule and the haphae between the cells often become separate at this stage. The hypha or hyphal branch producing the arbiscule penetrates the wall but only invaginates the host's plasmalemma so that the ultimate branches of the arbuscules are surrounded by it. Arbuscules are usually regarded as complex haustoria through which a two-way transfer of materials may occur. If they are haustoria, they are unusual in that they often lyse. This may be a consequence of a host defence mechanism comparable with the

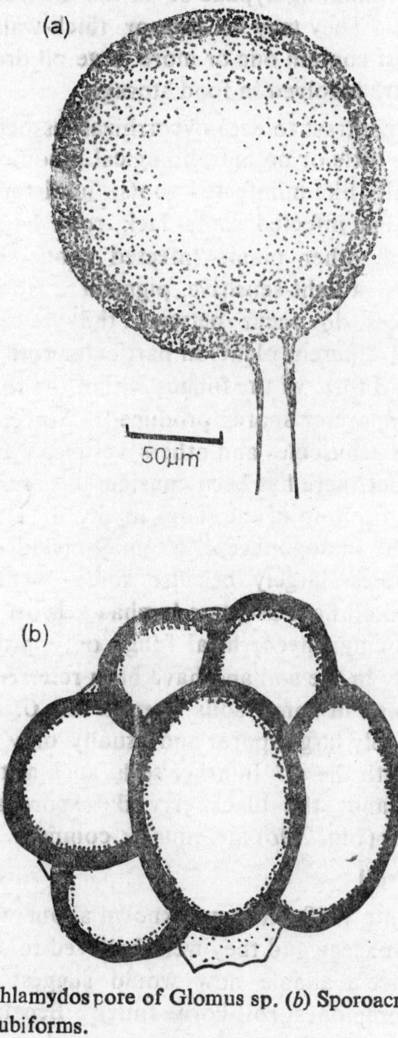

Fig. 2.9. (*a*) Chlamydos pore of Glomus sp. (*b*) Sporoacrp of Sclerocystis rubiforms.

laying down of callose arround the haustoria of some obligate biotrophic parasites. The hyphae may also form vesicles. Tney are readily discernible in roots, using suitable stains such as trypan blue in lactophenol, as they are usually in excess of 100 μm diameter. This is sufficiently large often to distort the cell or intercellular space in which they occur. Vesicles arise as swellings terminating hyphae or in an intercalary position along a hypha. They may be thin-or thick-walled, ovoid or spherical. Most contain one or more large oil droplets suggesting that they are involved in food storage.

In marked contrast to ectomycorrhizal associations, infected roots show little or no morphological modifications. Most look superficially like uninfected roots, undistorted and with root hairs. Some infected roots lack root hairs ; in maize, onion and some other plants infected roots are coloured a bright yellow by a light-sensitive pigment. However considerable differences do occur between the vesicular-arbuscular mycorrhizas of different plants in particular with regard to the distribution and form of the fungus within the root in addition to the type of spore or spores produced. Some, for instance, rarely produce arbuscules and others vesicles. In spite of this and the fact that there has been considerable controversy over the taxonomic position of the fungi involved, all are believed to belong to the Endogonaceae, a family included in the Zygomycete Mucorales largely because some members produce zygospores and all have aseptate hyphae. Most of the British vesicular-arbuscular mycorrhizal fungi only produce chlamydospores singly in the soil and have been referred to the genus *Glomus*, although in some soils species of *Gigaspora*, which produce relatively large spores and usually only arbuscules in roots, occur with these. In other soils, such as those of boulder clay woodlands, the blackberry-like sporocarp of *Sclerocystis rubiforms* (Fig. 2.9*b*) are equally common.

Spores in the soil

Until the late 1950s little was known about these members of the Endogonaceae and they were believed to be fairly rare but the evidence available now would suggest that they are some of the commonest soil-borne fungi. Because their spores are so large, 100–250 μm in diameter, they can easily be extracted from soil by decanting and wet sieving or by using a flotation technique. In the former technique about 250 g of soil are suspended in a litre of water, the heavier particles are allowed to settle out for a few seconds and the liquid decanted and washed through a series of size-graded sieves. The fraction passing through a 250 μm mesh sieve but retained by a 100 μm one contains the majority of the spores. A simple flotation technique is to add soil to a beaker some two thirds full of water, stir thoroughly to dislodge the spores from the soil crevices and allow to settle. The balloon-like spores will float to the surface and collect in the superficial scum. Most of these will adhere to the surface of a liter paper if it is partially submersed at a very acute angle to the surface and withdrawn. By such techniques, large numbers of spores have been recorded from soils, for example 1 – 8 g^{-1} under at Rothamsted. 20-70 g^{-1} under wheat in Ontario and over 400 g^{-1} under *Mercurialis perennis*. Dog's Mercury, in a Cambridgeshire boulder clay woodland. These numbers may be deceptive, High spore numbers are not necessarily indicative of high spore productivity. Because the spores have such thick melanized walls, they are very resistant to decay and may persist for very long period so accumulating and appearing abundant. A large number of spores encountered in the soil are devoid of contents and are no longer viable. Be this as it may, these large spores which appear to be poorly adapted for widespread dissemination are present in all soils whether they support pioneer or climax plant communities.

Spores extracted by such techniques have been used to infect plants and have produced typical vesicular-arbuscular mycorrhizas but so far these fungi have not been grown in pure culture. Chlamydospores will germinate on nutrient agar but the hyphae stop growing when the food supply in the spore is used up so they cannot be sub-cultured. They are thus obligate biotrophs but are unusual in displaying a surprising lack of host specificity, contrasting markedly with other obligate biotrophs such as the parasitic rusts and powdery mildews. Sprores produced from an infected maize plant have been used to produce typical vesicular-arbuscular mycorrhiza in strawberries, onions, red clover and soybeans.

BENEFITS TO HOST

Enhanced host growth and increased phosphate uptake

The inability to culture vesicular-arbuscular mycorrhizal fungi has made them more difficult to work with in terms of understanding their physiology and conducting experiments involving inoculating plants with these fungi. Isolates are maintained in 'pot cultures', *i.e.*, on the roots of living plants in pots. Plants are inoculated with single spores and grown in pots of sterile soil. The spores produced are collected by decanting and wet sieving the soil in the pots and used to inoculate others. Observations on the effect of infection on a wide range of host plants have been variable but in general enhanced plant growth occurs and this is associated with increased mineral uptake, especially phosphate. The extent of the enhanced growth depends on the level of infection in the root system and on the nutrient status of the soil. There is a much more marked stimulation of growth with low phosphate availability in the soil. Other factors may also be involved ; increased growth will not occur if other mineral nutrients are limiting in the soil. The degree of root hair development on the host roots may also affect the response. Plants with many and very long root hairs are less dependent upon vesicular-arbuscular fungi for their mineral supplies than are those with very few, short ones. This may be one explanation why members of families, such as the Resedaceae, are rarely, if ever, infected, their root hairs possessing sufficient surface area to exploit at large enough volume of soil to satisfy their mineral requirements.

There are a wealth of published reports which now provide quite indisputable evidence that infection of plant roots by vesicular-arbuscular mycorrhizal fungi increase phosphate uptake, especially under conditions of low phosphate availability. For example, Hayman and Mosse (1971) found that in some phosphate deficient soils they obtained up to a 20-fold increase in shoot mass of non-mycorrhizal onions merely by adding phosphate in the order of 1 ton superphosphate acre^{-1} (2.5 Mg 10^4 m^{-2}). Comparable increase in shoot mass were obtained by growing mycorrhizal onions in the same soil but without the added phosphate. It appears, as with sheathing micorrhizas, that the host obtained maximum benefit when the mineral nutrient regime is least favourable for growth. Most attention has been paid to the effect of vesicular-arbuscular mycorrhizas on crop plants such as apples, maize, onions and tomatoes but one of the earliest reports was of their effect on plants of the mixed rain forests of New Zealand. Much more recently attention has been paid to their role in other natural or semi-natural ecosystems, such as grasslands in the Pennines, tussock grassland in New Zealand, lowland tropical rain forest trees in Costa Rica and in man-made habitats such as plant communities establishing on coal tip spoils in Pennsylvania and Scotland.

Data from Khan (1975) using maize and wheat in Lahore, W. Pakistan may be used to illustrate some of the effects which these fungi are reported to have on crops grown in the field. Micorrhizal plants were obtained by growing seedlings on steam sterilized sand mixed with spores of *Glomus mosseae*. After 3 weeks mycorrhizal and non-mycorrhizal control plants were transplanted into field plots with both a very low phosphate content and indigenous spore population. Half of the plots containing mycorrhizal and control plants were treated with triple superphosphate at 280 kg 10^4 m^{-2}. Thus there were plots containing mycorrhizal plants without phosphate (M) and with phosphate (MP) and similarly non-mycorrhizal without phosphate (NM) and with phosphate (NMP) but with time the non-mycorrhizal plants gradually become infected from the indigenous spores in the soil. At harvest, after 90 days, the mycorrhizal plants without phosphate had grown the most as indicated by the fact that they had the greatest dry mass (Table 2.4). They had also taken up the most phosphate. Furthermore they produced the heaviest grains and more grains per cob. The non-mycorrhizal plants fared badly all around. The trends were the same in wheat except that the dry mass of 1000 grains was similar in all treatments. However mycorrhizal plants without phosphate produced more ears per plant and ear bearing stems per square metre, thus giving a greater yield. A number of workers have found that, in some soils, non-mycorrhizal plants grown with added phosphate were smaller than those with mycorrhiza in the same soil but without added phosphate. This is true of Khan's plants. In such soils, the added phosphate rapidly becomes unavailable. It is 'fixed' so that, even if added at high levels, it is still less available to non-mycorrhizal

Table 2.4. Effects of mycorrhizal infection with and without added phosphate on maize and wheat.

		Treatment			
Post	*Character*	*NM*	*NMP*	*M*	*MP*
Maize	No. of grains per cob	31	279	354	321
	100 grain mass (g)	2.4	19.8	23.7	20.9
	Total dry mass (g)	119	175	264	244
	% P	0.18	0 48	0.61	0.54
Wheat	Ear bearing stems m^{-2}	20 0	26	32	28
	Ears per plant	3	5	8	4
	No. of grains per ear	36	60	72	70
	1000 grain mass (g)	38.5	69.0	38.0	37.5
	Calculated yield (gm^{-2})	85.8	215 3	275.5	235.2

NM non-mycorrhizal without added phosphate
NMP non-mycorrhizal with added phosphate
M mycorrhizal without added phosphate
MP mycorrhizal with added phosphate
Phosphate added is superphosphate at 280 kg 10^2 m^{-2}
Seedlings (N and NM) planted out into field plots in or 10 after 21 days
Figures at harvest 90 days later.

plants than the small amount initially available in the unamended soil is to mycorrhizal plants. Only about 20-30% of any added phosphate is ever available to the standing crop. The rest becomes slowly available over the next 10-15 years. The mycorrhizal plants in plots with added phosphate performed on the whole less well than mycorrhizal plants in plots without added phosphate. With a high soil phosphate content, infection is often suppressed. Khan noted this in his maize plants. The incidence of infection was much lower than in mycorrhizal plants without phosphate and fewer arbuscules and vesicles were formed. This is also reflected in the numbers of spores produced. Initially there were 550-600 spores kg^{-1} soil. At harvest in plots with mycorrhizal maize without added phosphate the number had risen to about 2800 spores kg^{-1} soil whereas in plots with mycorrhizal maize and with added phosphate there were only about 1800 spores kg^{-1} soil. In addition the external hyphae attached to the roots did not extend so far into the soil so that they exploited a smaller volume of soil. Others have noticed in such circumstances that the growth of the fungus along the surface of the roots may be quite extensive even although actual infection is less. This might lead to competition for phosphate between fungus and host in the rhizosphere. Uptake by the host may thus be reduced and its growth rate decreased.

Possible use as 'biological fertilizers'

Because virtually all crop plants possess vesicular-arbuscular mycorrhizas and because they benefit so much from these in their mineral nutrition, the research effort which has been put into these is quite enormous. Much emphasis is being placed on their use as 'biological fertilizers'. Techniques as to how this might be achieved vary. Attempts at field inoculation with the most effective fungus have been made by placing suitable infected roots at sowing depth, decanting and wet sieving spores from plots where a particular crop has grown well and adding these to fields and pre-inoculating plants before planting out. The ultimate might be to incorporate spores into pellets around seeds. The results of Azcon—G. de Augwlar, Azcon and Barea (1979) illustrate some of the possibilities (Table 2.5 and Fig. 2.10). They grew alfalfa,

Table 2.5. Effects of Glomus and Rhizobium on the yield of Medicago sativa growing under normal cultivation in an arable field soil.

Inoculation treatment	*'Shoot dry mass g^{-1}
Uninoculated controls	15.2+1.0
Rhizobium	15.9+1.2
Glomus	22.6+1.1
Rhizobium+Glomus	32.1+1.8

*Mean of four root cotes.

Medicago sativa, in an arable field soil with very low available phosphate in Granada, Spain and investigated the effects on yield of prior inoculation with *Glomus*, the symbiotic nitrogen-fixing bacterium *Rhizobium* and the two together. In terms of shoot dry mass, an important criterion of yield as a folder plant. *Glomus* improved yield but *Rhizobium* did not. But the synergistic effect of both was most obvious. It was argued

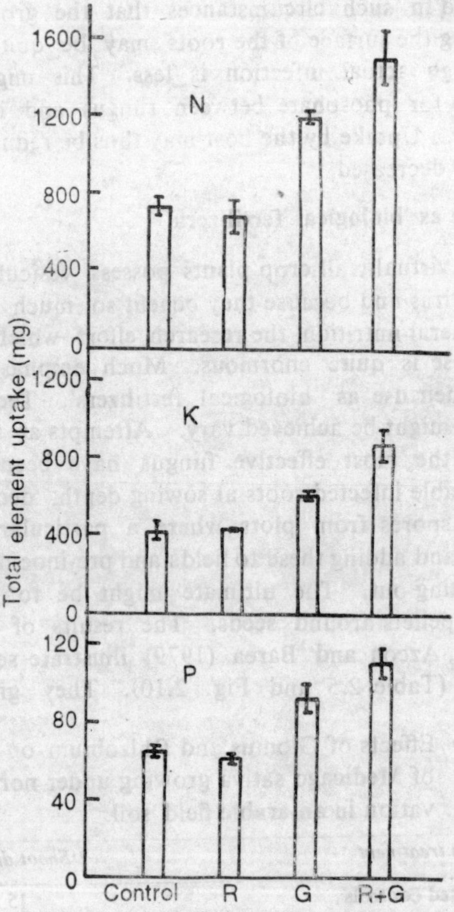

Fig. 2.10. Effects of *Glomus* (G), *Rhizobium* (R) and both (R+G) on nitrogen phosphorus and potassium uptake by *Medicago sativa* in an arable field under normal cultivation. The total N, P and K uptake by the plants was calculated from the data of shoot dry masses and the percentage of the element. The figures are the mean of four replicates. Standard errors are given.

that *Rhizobium*-inoculated plants were growth limited by phosphate so that yield was no greater than controls. *Glomus* increased the phosphate uptake of plants so that inoculation with both greatly improved the yield. This argument is borne out by analysis of the shoots for nitrogen, phosphorus and potassium. *Glomus* improved the uptake of all three but inoculation by *Rhizobium* did not unless *Glomus* was present.

Vesicular-arbuscular mycorrhizas may be of equal importance in natural communities in assisting plants to obtain sufficient phosphates and in the recycling of phosphate. In such communities the available phosphate levels are often exceedingly low. In the New Zealand forests in which Baylis worked it was 4-8 p.p.m., as compared with what he considered as a minimum agricultural standard level of 28 p.p.m. All the trees in these forests possess vesicular-arbuscular mycorrhizas and infection by appropriate mycorrhizal fungi was essential if seedlings were to become established and grow. Agricultural and horticultural crops are grown in soils usually with much higher phosphate level and mycorrhizal infection may be of less significance but even so all the available evidence indicates that mineral nutrient uptake by mycorrhizal roots is the norm rather than the exception. There is also a need for recycling because there are no natural processes, such as there are for nitrogen, by which phosphorus can be increased after its depletion. Phosphorus from plant remains, unless rapidly reabsorbed, is quickly converted to various insoluble organic compounds, phytates, in the soil and, although as such it is less likely to be leached than are nitrogenous compounds, it is unavailable to plants. The role of the mycorrhiza can thus be seen as ensuring efficient and rapid transfer of phosphate from the available pool in the soil into the root, thus decreasing the time period in which it might be subjected to conversion to organic forms and to leaching.

The main interest in vesicular-arbuscular mycorrhizas has centred around the enhanced phosphorus nutrition of the host. Their effect is thus seen to be much more marked in phosphate deficient soils. Evidence is also available that they may increase the nitrogen, potassium, calcium, magnesium and iron uptake. There are two possible ways, or a combination of two possible ways, in which this increased uptake can occur. The fungus could increase, the effective absorbing area of the root so that a much large volume of soil is exploited for its labile phosphate or it could make soluble insoluble or organic forms of phosphate which are not available to the uninfected plant. They may do both but there is not much evidence for the latter and so are exactly comparable with the sheathing mycorrhizas.

Source of phosphate taken up

If various phosphate are added to the phosphate deficient soils it can be shown that mycorrhizal and non-mycorrhizal plants of the same species grow equally well on available phosphate such as calcium monophosphate ($CaHPO_4$) and that mycorrhizal plants grow better on tricalcium phosphate ($Ca_3(PO_4)_2$) which is only slowly available. However, this may be because they have a large surface area to absorb what little phosphate is available rather than making it more soluble and hence more available. If the labile phosphate pool in the soil is labelled with ^{32}P and the non-labile pool is left unlabelled the specific activity (c.p.m. mol P^{-1}) of phosphorus compounds in plants grown in the soil is the same in both mycorrhizal and non-mycorrhizal ones, indicating that they both use the same pool of phosphate. If the mycorrhizal plants used some of the non-labile insoluble soil phosphates, the specific activity of the absorbed phosphate would be less than that of the non-mycorrhizal plants.

Phosphate ions have a very slow diffusion coefficient in soils and there is likely to be a marked depletion of phosphate in solution immediately around the root. This depletion zone usually extends 1-2 mm around the root up to the tips of the root hairs. The hyphae of the mycorrhizal fungus may grow well outside the depletion zone and considerably extend the volume of soil explored. In mycorrhizal onions, hyphae have been demonstrated to absorb at 70 mm from the root surface. The enhanced phosphate uptake by the mycorrhizal plants thus depends on a more complete exploration of the soil by hyphae outside the depletion zone of the root hairs. There is also some evidence that there is more complete exhaustion of the soil explored by the hyphae including the root hair depletion zone. It appears that non-mycorrhizal plants of different species differ in their ability to utilize the low levels of phosphates which occur in phosphate deficient soils. They may not take up phosphates until it reaches a certain miminum or threshold level. This level is different for different species. Mycorrhizal plants either have no threshold value or a very much lower one and can thus more completely exhaust a soil of its labile phosphate.

Iner-plant transfer of nutrients via fungus

In many natural plant communities endomycorrhizal infection arises when uninfected roots make contact with hyphae spreading from infected roots. Since any one fungus may be capable of infecting a number of different plant species, a series of plants may become connected via the hyphae of their common mycorrhizal fungus. This could lead to inter-plant transfer of carbon and mineral nutrients. Recent autoradiographic studies by Francis and Read (1984) have shown that such inter-connections provide a direct pathway for the transfer of carbon compounds between different plants. They used seedlings of *Plantago lanceolata* as 'donors' of both infection and ^{14}C labelled carbon. After their establishment, uninfected seedings of *Festuca ovina* were planted amongst them and they became mycorrhizal. The 'donor' plants were then allowed to photosynthesize in ^{14}C labelled CO_2 and the 'receiver' plant of *F. ovina* were kept in full light, half light or the dark. After two days, autoradiography showed that the label moved from 'donor' to 'receiver' roots only if they were connected to one another by hyphae. At that stage uninfected 'receiver' roots contained no label. Quantitative analysis of the radioactivity in the plants confirmed the transfer of assimilate between the two plants but the quantity of label moved was strongly influenced by the light regime. The levels of activity were six times higher in the *F. ovina* plants kept in the dark than in those fully illuminated. Thus the movement appears to be governed by source-sink relationships. The ecological and physiological significance of this may be far-reaching. In numerous plant communities many young seedlings may spend part of their early life in poorly lit environments and even in full shade. To such seedlings only mycorrhizal infection would be vital on the one hand in providing a necessary enhanced mineral nutrient supply but the carbon drain imposed by the fungus, to provide such enhancement, could be critically disadvantageous on the other hand with regard to growth. Direct transfer of assimilate from more fully illuminated neighbouring and connected mature plants may sustain seedling infection. Furthermore during the critical establishment phase the whole seedling may live for a period at least as a partial heterotroph depending upon its neighbours to which it is connected by its bridging fungus.

Benefits to fungus

Virtually all the evidence indicating that the fungus benefits from this association is at the moment based on assumptions. It is assumed that since these fungi have never been grown in culture they are like all other biotrophs in being dependent on their hosts for their carbon supplies. It is also assumed that they are like all other Zygomycotina in being unable to utilize polymers such as cellulose and are thus 'sugar fungi' requiring simple carbon compounds. Again, as with sheathing mycorrhizal fungi, because of the scarcity of such carbohydrates in the soil, they must obtain these from their host. Potted plants have been allowed to photosynthesize in ^{14}C labelled CO_2 and the label has appeared in spores formed around the roots. This indicates, but does not necessarily prove, direct translocation of photosynthate to spores. The carbon metabolism may be quite different from that of sheathing mycorrhizal fungi but again there is no clear indication of this. There are generally more soluble sugars present in mycorrhizal than in non-mycorrhizal roots. Something like 75% of the labelled photosynthate reaching mycorrhizal roots is in the form of soluble sugars. Although there is diversion to mycorrhizal roots they do not, unlike sheathing mycorrhiza, contain higher levels of trehalose and mannitol but some glycogen and lipids are present. These may be storage compounds but very little carbohydrate appears to be stored. This may be because storage occurs in spores and vesicles and these are scattered.

There is no obvious bulk storage tissue, such as a sheath. The absence of polyhydric alcohols such as mannitol is to be expected as they have not been recorded from other aseptate fungi.

Mosse and Phillips (1971) grew *Trifolium parviflorum* aseptically on agar slopes and inoculated them with spores of *Glomus mosseae*. The addition of calcium phytate to the agar medium greatly stimulated the growth of the hyphae out from the root. Growth of the hyphae was virtually as good when inositol was added and they argued that the stimulation caused by the addition of calcium phytate was due to the fungus using its inositol component as a carbon source. If this is so the fungus has a potential carbon source in the soil for which there may be very little competition.

In this mycorrhizal system, the host depends on the fungus for soil-derived nutrients and the host has appeared to adopt the strategy that it is less costly in carbon and more flexible in operation that the absorbing surface should be an extensive hyphal system rather than a much branched root system of a much larger biomass.

ERICOID MYCORRHIZAS

Occurrence, structure and fungi involved

Ericoid mycorrhizas are another type of endomycorrhiza but here the fungi involved are septate ones. Ericaceous plants, such as those of *Erica* (heather), *Calluna* (ling) and *Vaccinium* (bilberry), are mostly calcifuges associated with acid, peaty, mor-humus soils of very low mineral status. These soils are particularly low in exchangeable nitrogen and phosphorus, most being bound as insoluble inorganic or organic compounds. They are plants which you would expect to benefit from mycorrhizal infection.

They have a dense root system ending in very fine, branched absorbing rootlets, often unfortunately called 'hair roots'. These rootlets have a very narrow central stele, surrounded by a cortex of 1-3 layers, often only one, of parenchymatous cells.

There are not root hairs. The rootlets are covered by a very sparse, loose weft of dark brown, septate hyphae. From this weft branches penetrate the cortical cells forming compact intracellular coils invaginating and enclosed by the host's plasmalemma. They do not penetrate the stele or the apical meristem. After a time, some 3-4 weeks, the cell contents of the host begin to degenerate with the plasmalemma losing its integrity. The intracellular hyphal coils then collapse, although they may persist intact for some time in the degenerate host cell. They lose their filamentous structure and the coils become an amorphous mass of fungal tissue (Fig. 2.11). This process

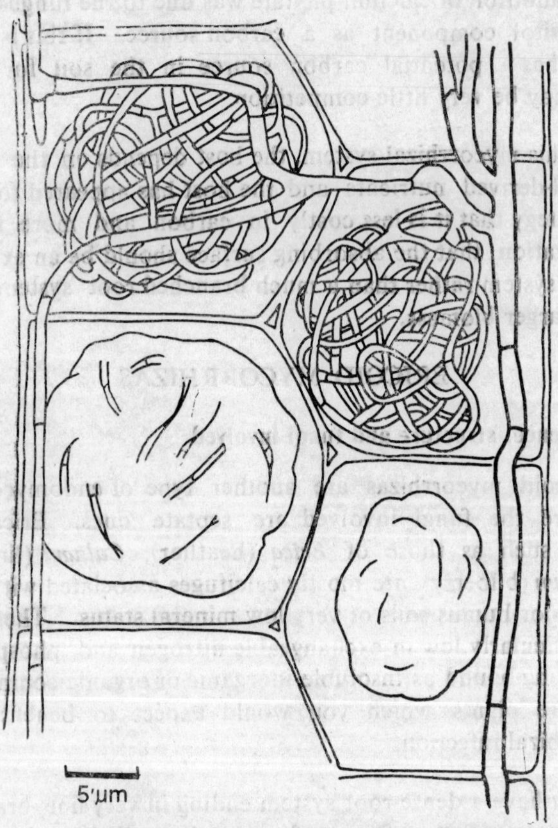

Fig. 2.11. Ericoid mycorrhiza, showing four cells of a rootlet with a loose weft of septate hyphae and hyphal coils, some undergoing lysis, in the cortical cells.

may be compared with the degeneration of arbuscules in vesicular-arbuscular mycorrhizas but is much more comparable with similar phases in orchidaceous mycorrhizas.

A number of sterile, slow-growing, dark, septate fungi have been obtained in culture from root segments and have been shown to form mycorrhiza with various ericaceous plants. Most work on these has been done by Read and his associates. They first obtained their own isolates by thoroughly washing hair roots in 20-25 changes of sterile distilled water. They then macerated the roots and plated out onto distilled water agar separate cortical cells containing hyphal coils. They isolated only those fungi which were seen to grow out from the hyphal coils within the cells. These fungi were all slow growing, dark and septate. Most proved to be sterile but one isolate produced ascocarps, apothecia, and they described these as belonging to a new species *Pezizebla ericae*. Most of their work has been done with this Ascomycete which does form typical mycorrhiza with several ericaceous genera. Other isolates may yet be shown to belong to this species but other fungi may also be involved. For example, a group of Australian workers have produced evidence that *Clavaria vermicularis* may be mycorrhizal with *Azalea indica* and various rhododendrons and ericas. Basidiocraps of *C. vermicularis* are almost always found in close association with ericaceous plants. Their attempts to grow the fungus from basidiocarps failed so that they could not synthesize mycorrhiza directly. They used a fluorescent antibody technique and claimed that the mycorrhizal fungus within the root cells was serologically the same as the *Clavaria* growing in the neighbourhood of the plants. In this country *Clavaria argillacea* is always associated with *Calluna vulgaris* and may be mycorrhizal with it.

Benefits to host and fungus

It has become clear from the work of Read and his associates that ericoid mycorrhizas behave like sheathing and vesicular mycorrhizas. They improve the mineral nutrition, especially the nitrogen and phosphorus uptake, and growth of their hosts while the fungus is dependent upon its autotrophic host for its carbon supply. It is important to appreciate that the nitrogen supply to plant roots or mycorrhizas may be in the form of ammonium ions, nitrate ions both and that organic nitrogen in the soil may or may not be available to the absorbing systems. Nitrate and ammonium ions have different mobilities in soils. Ericaceous plants, such as *Calluna* and *Erica*, typically grow in acid soils rich in organic matter in which nitrification rates are very low and the available mineral nitrogen source is ammonium, in very low concentrations. Ammonium ions, in contrast to nitrate ions, are relatively immobile in soil although they are not as firmly bound as phosphate ions. In acid peaty soils only a small fraction, less than 0.5% of the total nitrogen, is present as ammonium and not all of this is exchangeable and thus available to plant roots. Most of the nitrogen, over 7%, is present in the organic form, often as unidentified complex compounds and again unavailable to roots. The story is very similar to that for phosphorus. It has been argued, without substantive proof, on many occassions that mycorrhizal plants can make use of these compounds. The ability to utilize these would be very important to ericacceous plants growing in such acid moorland soils.

Source of nitrogenous compounds utilized by mycorrhizal plants

Mycorrhizal ericaceous plants grown in such soils are larger and healthier and their nitrogen, and phosphorus, content are significantly higher than in non-mycorrhizal ones. For example, *Vaccinium* seedlings grown on sterile moorland soil for six months after inoculation with a mycorrhizal fungus contained over twice as much nitrogen on a dry mass basis as did not-mycorrhizal seedlings (Table 2.6). Stribley and Read (1974) investigated this further and labelled moorland soil with ^{15}N by long incubation with ^{15}N ammonium sulphate. After removing excess ammonium ions, they found that the labelling pattern of the nitrogen in the soil was complex but basically there were two fractions. One was very small with high labelling. This was the exchangeable or labile ammonium fraction. It was highly labelled as the pool was

Table 2.6. Nitrogen content and dry mass yield of mycorrhizal and non-mycorrhizal seedlings of *Vaccinium macrocarpon*.

Growth stage	*Total nitrogen content (mg per plant)*	*Total dry mass yield per plant (mg)*
Seedlings at time of condition	0 11	18.50
Mycorrhizal	0 88	124.60
Non-mycorrhizal after three months	0 62	121.0
Mycorrhizal	1.82	235.0
Non mycorrhizal after six months	0 81	184.0

small. The other was relatively large with low labelling. This was the organic unavailable nitrogen. After sterilizing this soil by gamma irradiation, they grew mycorrhizal and non-mycorrhizal seedlings of cranberry, *Vaccinium macrocarpon*, in it in pots for six months before analysing their shoots for their nitrogen content. Mycorrhizal plants were heavier, larger and healthier. They contained more nitrogen and a greater concentration of nitrogen on a dry mass basis. But what they called ^{15}N-excess (labelled atoms %N) of mycorrhizal plants was lower than that of the shoots of non-mycorrhizal plants—that is to say that the non-mycorrhizal plants contained a high proportion of ^{15}N which suggests that they mainly utilized the exchangeable ammoniun-N. Mycorrhizal plants, although they contained more total nitrogen, had a lower ^{15}N enrichment and had thus absorbed more unlabelled nitrogen (Table 2.7). This implies that they had a source

Table 2.7. Nitrogen content dry mass and ^{15}N-excess of shoots of mycorrhizal and non-mycorrhizal plants of *vaccinium macrocarpon* after six months growth in ^{15}N-labelled soil.

	N content % oven dry mass	*Yield mg oven dry mass*	*Total N mg plant⁻¹*	*^{15}N-excess (atom %)*
Mycorrhizal	1.20	30.32	0.36	15.38
Non-mycorrhizal	0.98	20.97	0.21	20.03

of nitrogen which was unavailable to non-mycorrhizal plants and Stribley and Read argue that it seems likely that the mycorrhizal fungus is able to assimilate nitrogen from organic compounds in the soil and conduct these directly to the host plant. This seems a fair argument as they have also shown that the fungus does not cause net mineralization in the soil. They grew the fungus in sterile soil for six months and found that it did not bring about an increase in exchangeable ammonium nor did it stimulate the growth of other species such as the grass *Festuca ovina* grown in that soil.

Not all the organic nitrogen in soils is in the form of complex compounds. Free amino acids are present in low concentrations in organic soils but, because they are not in equilibrium with a large source of supply, it is generally believed that they cannot be an important source of nitrogen for plants. They are released from two sources—from humified organic materials and as leachates from freshly fallen leaf litter. Stribley and Read (1980) investigated the relationship between mycorrhizal infection and the capacity to utilize simple organic nitrogen source, such as amino acids, and complex ones, such as humic and fulvic acid. They grew mycorrhizal and non-mycorrhizal seedlings of *Vaccinium macrocarpon* in sterile sand which was saturated with 125 cm^3 of filter-sterilised mineral nutrient solution at pH 5.8 and which contained as the nitrogen source one of the amino compounds glycine, glutamic acid, glutamine, asparite acid or alanine at a nitrogen concentration of 20.5 mg dm^{-3}. Control plants received either ammonium nitrogen at the same concentration or no nitrogen. They were grown in a controlled environment room for 8 weeks. The shoots were then removed and dried. The results are given in Table 2.8. Mycorrhizal plants of *V. macrocarpon* utilised all the amino acids except aspartic as readily as they utilized ammonium. Even so growth on aspartic acid was still some 5 times greater than in the controls without nitrogen. The non-mycorrhizal plants had restricted ability to utilize these nitrogen sources, growth being better on ammonium than on any of the amino acids. They also tested the ability of the mycorrhizal fungus to degrade humic and fulvic acids extracted from soils under ericaceous plants ; the fungus was grown in liquid culture containing 0.1% (w/v) of the acids as sole

Table 2.8. The effects of different nitrogen sources on shoot dry mass (mg pot^{-1}) of mycorrhizal (M) and non-mycorrhizal (NM) *Vaccinium macro-carpon* seedlings.

Nitrogen source (20.5 mg dm−³)	*M*	*NM*
Ammonium	43.4	41.4
Glycine	43.9*	11.6
Alanine	47.6*	22.5
Aspartic acid	29.5*	7.0
Glutamic acid	39.4*	8.2
Glutamine	44.6*	18.2
No nitrogen	6.0*	8.3

*Significant difference in dry mass between M and MM at P 0.05.

nitrogen sources and after 40 days at 20°C both the percentage loss of the acids (*i.e.* amounts utilized) and the dry mass of the mycelium produced were measured. The fungus had very little capacity to utilize these acids as nitrogen sources. Stribley and Read suggest that it is the capacity of the fungus utilize simple organic compounds, such as amino acids, in the soil which explains the enhanced nitrogen uptake by mycorrhizal plants.

Read and his associates have done much less work on phosphate uptake but have shown that the fungus shown that the fungus shows acid phosphatase activity and that it can use phytates at least in the form of the sodium salt of myo-inositol hexaphosphate. They grew their isolate on a glucose-mineral medium amended either with inorganic or organic phosphorus. For unit phosphorus in the medium, the yield of the fungus, in terms of dry mass mycelium, was about the same. This would help to explain the enhanced phosphorus content of mycorrhizal plants in such acid nutrient poor soils but, as in sheathing mycorrhizas, this aspect of their nutrition requires further investigation.

Nutrition of fungi

There is a direct evidence that the fungi of sheathing mycorrhizas absorb carbohydrates from their hosts. It is assumed that the fungi of vesicular-arbuscular mycorrhizas also do this. The fungi of ericoid mycorrhizas are again mainly intracellular and as with vesicular-arbuscular mycorrhizas it is impossible to separate fungus from host. This makes any assessment difficult. When ^{14}C labelled carbon dioxide is supplied to photosynthesising shoots, sucrose becomes the most strongly labelled sugar in both mycorrhizal and non-mycorrhizal roots of *Vaccinium*. But mycorrhizal roots contain both trehalose and mannitol which are not present in non-mycorrhizal roots as they are fungal compounds and both of these are labelled. Thus it appears that, in their carbon nutrition, ericaceous mycorrhizal fungi are more like those of sheathing mycorrhizas than those of vesicular-arbuscular ones. Mycorrhizal roots also show a striking increase in a mannose polymer. This is a cell wall component of the fungus. Such a component is also present in sheathing mycorrhizas and in many rust-infected tissues. Although isolates of *Pezizella ericae* can utilize such polymers as pectins and carboxymethyl-cellulose, but not native cellulose, in culture, their relative slow growth rates are indicative of low competitive saprotrophic ability, which would suggest that this fungus too is an ecologically obligate biotroph.

ECTENDOMYCORRHIZAS

A number of general in the Ericaceae and members of some other families in the Ericales possess mycorrhizas which are different from those discussed above. They are intermediate in form between ecto-and endomycorrhizas and have been called ectendomycorrhizas. For example, in the Ericaceal *Arbutus* and *Arctostaphylos* possess such mycorrhizas. The root system of *Arbutus* is differentiated into long and short roots. The short roots are swollen and are invested with a hyphal sheath. There is no Hartig net as in sheathing mycorrhizas but intracellular coils develop in the outer cortical cells and these are eventually lysed. Virtually nothing is known of the fungi involved or the physiology of these arbutoid mycorrhizas.

Monotropa hypopitys

Mycorrhizas somewhat resembling these are found in the Monotropaceae, a family in the Ericales. A good example is

Monotrowa hypopitys, the yellow Bird's nest or Pinesap. It is a non-green or achlorophyllous herb and its short, fleshy roots are invested by a sheath of hyphae with some of these forming a network between the epidermal cells comparable with the Hartig net of sheathing mycorrhizas. But such mycorrhizas must be physiologically different as the non-photosynthetic *Monotropa* cannot possibly supply carbohydrates for its fungus. The hyphae of the sheath and net however also form structures resembling simple haustoria, usually only one per cell, within the outermost cells of the host root. These so-called haustoria are not true ones but consist of fungal intrusions from the sheath into the outer cortical cells. They cause the host cell wall to invaginate, inducing the formation of structures reminiscent of transfer cells. Duddridge and Read (1982) have called these 'fungal pegs' and suggest that they facilitate enhanced nutrient supply during periods of peak demand. They could be involved in the rapid transfer of carbohydrates from fungus to plant in the period of expansion of the inflorescence. Such cellular structures are common in healthy plants and in some other symbiotic associations. They have been recorded recently in the ectomycorrhiza of *Pisonia grandis* which have a poorly developed Hartig net.

Bjorkman (1960) isolated a fungus, a *Boletus*, a genus well-known to be mycorrhizal with conifers, from mycorrhizal roots of *Monotropa* growing amongst pines and found that it also formed typical mycorrhiza with pine seedlings. He injected ^{14}C glucose into the phloem of mature pine trees with *Monotropa* growing amongs them and showed that after 5 days low but significant levels of radioactivity occurred in young *Monotropa* plants 1—2 metres away but not in other herbs close by, indicating that labelled compounds had been translocated from *Pinus* to *Monotropa*. Support for this idea also comes from physically separating *Monotropa* from the roots of its associated trees with metal plates driven into the soil. Its subsequent growth is very poor. It thus appears that photosynthesis by the pine provides carbon compounds not only for its symbiotic fungus, as one would expect but also for *Monotropa*, so there is a three tier system involving a chlorophyllous symbiont and a non-chlorophyllous one and a fungus as a bridge. The remarkable feature is the dual role of the fungus, extracting carbohydrates from the one and contributing carbohydrates to the other. This state of affairs may be relatively more common than is generally credited. The movement of nutrients between the pine and *Monotropa* may not be all one way. ^{32}P labelled inorganic phosphate injected into *Monotropa* plants has been detected in associated pine roots within 2 h of injection. It is not at all easy to see what advantage the fungus gets out of this sort of association.

Sequestration of heavy metals by ericoid mycorrhizas

Ericaceous plants are normally mycorrhizal and are very succesful colonizers of natural and man-made environments which contain very high levels of metallic elements. Members of the family colonize soils of high mineral composition such as those on serpentine and even natural heathlands have potentially toxic levels of metals such as aluminium, because of its availability increasing at low pH. *Calluna vulgaris* is frequently the dominant and only colonizer of coal tip spoils and *Vaccinium macrocarpon* is prominent on soils contaminated by metal smelting. Enhanced uptake under these circumstances would be physiologically damaging rather than beneficial.

Bradley, Burt and Read (1982) investigated the relationship between mycorrhizal infection and resistance to metal toxicity in *Calluna vulgaris*, *Vaccinium macrocarpon* and *Rhododendron ponticum*. Plants were grown from seed and some inoculated with mycorrhizal fungi. Mycorrhizal and non-mycorrhizal plants were grown in acid washed sand containing a dilute nutrient solution supplemented with copper over the range 0—75 mg dm^{-3} and zinc over the range 0—150 mg dm^{-3} for 12 weeks. The growth of non-mycorrhizal plants was severely inhibited in all copper treatments and all but the lowest level of zinc. Mycorrhizal infection provided considerable resistance to the effects of the metals. Some growth occurred in all concentrations of copper and zinc. Mycorrhizal plants showed larger increases in shoot:root ratios in the metal treatments compared to control plants. The shoot : root ratios were higher levels of metals, reflecting a greater diminution of root relative to shoot growth with increasing concentration of metals. Compared with non-mycorrhizal plants, mycorrhizal plants had a lower concentration of metals in their shoots. The concentration of metals in the roots of mycorrhizal plants was considerably higher than that in their shoots which suggests that metals are being complexed in the roots. They propose that hyphal complexes of the fungus in the cortical cells of the root provide absorptive surfaces which facilitate the exclusion of metals from the shoot and avoids metal toxicity. The surface for heavy metal absorption is present in mycorrhizal plants but non-mycorrhizal ones can rely only on root extension to provide such a surface ; in the presence of heavy metals this is reduced by restriction of root growth. Thus as in plants, such as *Agrostis capillaris* (*tenuis*), heavy metal tolerance depends upon sequestration of the metal in the roots, the normal site to complexing being the wall, *i.e.* the wall provides a series of catchment surfaces upon which metallic ions can be fixed. Resistance is thus based on an exclusion mechanism and stress is avoided rather than tolerated. If such a mechanism is to operate, it is clear that the fungus itself must be resistant to heavy metal contamination. The mycorrhizal fungi were cultured on the same levels of copper and zinc. They grew without any significant reduction up to 50 mg dm^{-3} of copper and 100 mg dm^{-3} of zinc. Thus in Ericaceous plants in addition to increased nutrient uptake, a further benefit of the mycorrhizal association is the exclusion of undesirable elements.

Mycorrhiza of the Gentianaceae

A number of other flowering plant families have mycorrhizas in the form of hyphal coils. Two good examples are the Gentianaceae and Orchidaceae. Little is known about the mycorrhiza of the former family. Gay, Grubb and Hudson (1982) investigated the mycorrhiza of *Blackstonia perfoliata*, *Centaurium erythraea* and *Gentianella amarella*, three turf compatible biennial members of the Gentianaceae found in chalk grasslands. All three plants are usually mycorrhizal. Their seedlings become infected within two weeks of germination. The cortical cells of the roots become full of irregular coils of aseptate hyphae. The coils are eventually lysed. Vesicles can occasionally be seen attached to these coils (Fig. 2.12). This

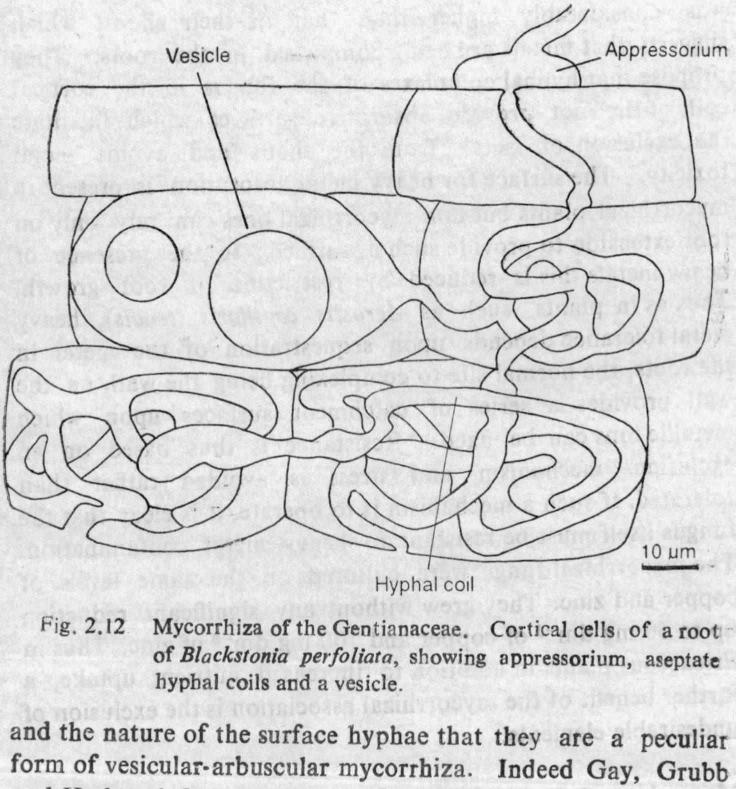

Fig. 2.12 Mycorrhiza of the Gentianaceae. Cortical cells of a root of *Blackstonia perfoliata*, showing appressorium, aseptate hyphal coils and a vesicle.

and the nature of the surface hyphae that they are a peculiar form of vesicular-arbuscular mycorrhiza. Indeed Gay, Grubb and Hudson infected seedlings of these three species by growing them in posts of sterile soil in which plants of *Sanguisorba minor* and *Leontodon hispidus* infected with typical vesicular-arbuscular mycorrhizal fungi were growing. Such a view has been confirmed by Jaquelinet-Jeanmougin and Glaninazzi-Pearson (1983). They grew various species of *Glomus Gigaspora margarita* in roots of onions or raspberries in pots of γ irradiated soil and used chopped roots from these to inoculate seedling of *Gentiana lutea*. All produced typical intracellular coils in the roots of the gentian. The turf compatible biennials have two periods of maximum growth and these are associated with maxima in the accumulation of mineral nutrients. These are in the mid-summers of their first and second years. These periods also coincide with the periods of greatest density of mycorrhizal infection. At these times the hyphal coils are intact and thus presumably active whereas they become lysed in the autumn periods. The assumption is that they are behaving as typical vesicular-arbuscular mycorrhizas in enhancing mineral uptake while being dependent upon their hosts for their carbon supplies.

Orchidaceous mycorrhizas

Orchidaceous mycorrhizas, although very similar to ericoid ones, are unusual in number of respects, especially in their carbon nutrition. At the onset it must be appreciated that most of the work on these has been done with heterotrophic orchid seedlings rather than with the autorophic adult plant. For most green orchids the association with a particular fungus is obligatory in nature if their seeds are to germinate and become successfully established. For the non-green or achlorophyllous orchids, this association is obligatory for all their lives.

Orchids produce vast numbers of tiny seeds, often millions per capsule. Each seed weighs only 0.3-14.0 μg. The embryo consists of 10-100 cells and there is no, or virtually no, food store. The embryo is merely enclosed in a loosed-fitting, thick-walled, net-like testa which is essentially a dispersal structure (Fig. 2.13*a*). The majority of seeds are incapable of successful germination unless assisted by an external source of carbohydrate. For this in nature, they relay upon fungal infection. Any orchid until it is photosynthetic relies upon its fungus for its carbon supply. It may be some considerable time after germination before the first photosynthetic leaf is produced. In many orchids, such as species of *Doctylorchis*, the first green leaf may arise as early as in the second year of growth. In *Spiranthes spiralis*, it does not arise until the eleventh year and in others it may take even longer before any photosynthesis occurs. So all have a prolonged and some a very prolonged

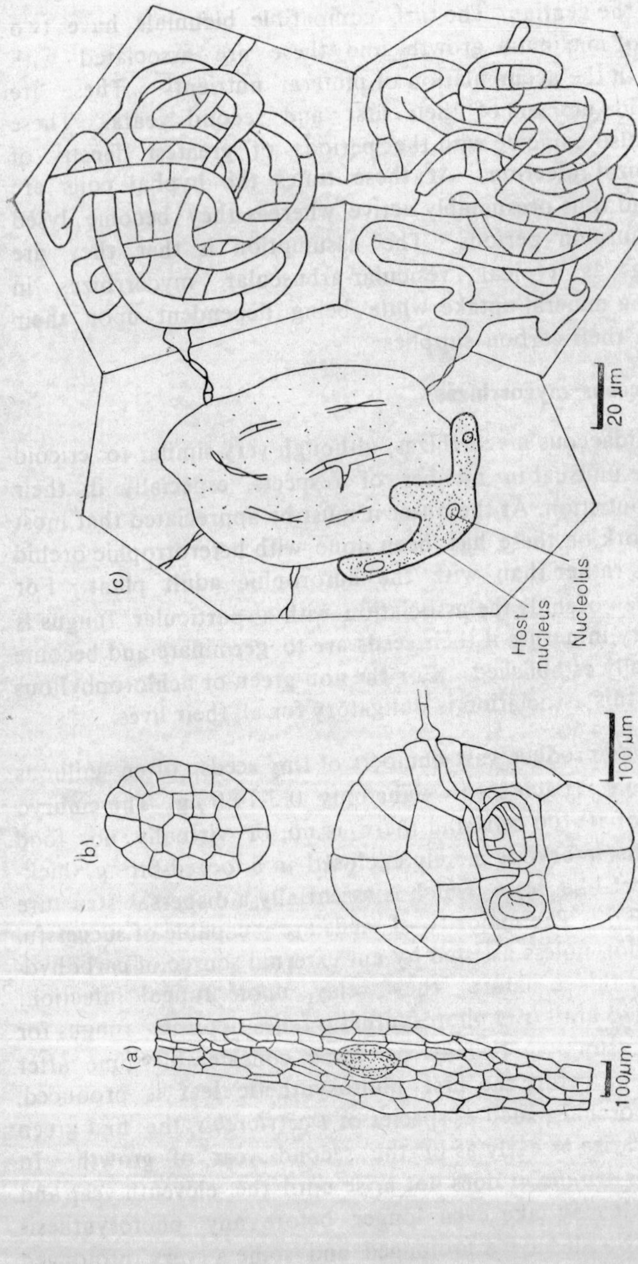

Fig. 2.13. **(*a*) An orchid seed (*b*) Embryo swelling and initial stages of infection (*c*) Three cortical cells from the root of *Dactylorchis purpurella*, two with infact hyphal coils and one with coils undergoing lysis.**

phase in the seedling stage when they are non-green, subterranean and obligatorily mycorrhizal.

Fungi involved

By far the commonest of the fungal associates are members of the Basidiomycete genus *Rhizoctonia*, a genus recognized on hyphal characteristics. Many of these have been demonstrated also to produce ill-defined crust-like basidiocarps in such genera as *Corticium, Ceratobasidium* and *Tulasnella*. Unlike the Basidiomycotina of sheathing mycorrhizas, these fungi are not members of the Agaricales but of the Aphyllophorales. The contrast markedly with the Agaricales. The majority are known to be capable of existing as free-living saprotrophs in the soil. They are relatively fast-growing and can compete successfully with other soil saprotrophs. Furthermore they can utilize cellulose so that they are self-supporting for carbon compounds. Many have been demonstrated to be aggressive necrotrophic parasites of other hosts. They cause root diseases of cereals, tomatoes, lettuce and other crop plants and 'damping-off deseases of seedlings, such as cress and swede. Other fungi which have been less frequently isolated as orchid mycorrhizal fungi include such Agarics as *Armillaria mellea*, better known as a devastating root parasite of trees. It is known to form mycorrhizal associations with the achlorophyllous orchids *Gastrodia elatu* and *Galeola septentrionalis*.

Features of infection

Initially the fungus enters the embryo via the suspensor and this is followed by an immediate marked stimulation of growth of the embryo. The central region of the embryo initially becomes completely colonized (Fig. 2.13*b*). The protocorm, or seedling axis, then develops. It differentiates a conducting strand, an apical group of leaves and roots from the axis near the leaf bases. The fungus then becomes restricted to the outer parenchymatous cells of the axis with hyphae passing out into the soil. As the young roots develop, they apparently become infected by the fungus in the soil and not from that in the protocorm.

The fungus enters only the parenchymatous cortical cells of the root and it becomes restricted to these cells where it forms dense intracellular coils or tangles of separate hyphae (Fig. 2.13*c*). As in the protocorm, with time these hyphae swell, lose their contents, collapse and eventually degenerate into a structureless, disorganised mass as in ericaceous mycorrhizas.

Carbon nutrition of fungi and heterotrophic growth of the orchid

There is good evidence that the fungi concerned utilize complex carbon sources in the soil and make the products of the hydrolysis available to the young orchid. If seeds are placed on moist filter paper, they do not germinate, but if the filter paper is inoculated with an appropriate *Rhizoctonia*, the seeds will grow. If pads of filter paper on which the fungus has been growing are washed, the washings contain hydrolytic products of cellulose such as glucose and cellobiose as well as the fungal carbohydrates trehalose and mannitol. From this it can be argued that the fungi either release, by hydrolysis, sugars in excess of their needs from polymers, such as cellulose, so that the sugars are available in the external medium for direct absorption by the orchid, *i.e.*, the orchid is behaving as a commensal, or the fungi take these up and translocate them in their hyphae to the orchid, *i.e.*, there is a transfer across a living interface. The former is most unlikely as there would be too much competition from microorganisms in the soil for these free sugars for the orchid to obtain sufficient. Also Hadley and Williamson (1971) have shown that not all protocorms in a population of *Dactylorchis purpurella* become successfully infected. Uneffected individuals in the presence of the fungus do not show the same increase in growth rate as do infected ones. In addition to these two possibilities the host must gain something in the way of nutrients from the periodic degeneration of the hyphae. This must release the contents of the hyphae into the host cell. One view is that hyphal collapse and degeneration is caused by the host. Such a view has led some to consider the orchid as being a necrotrophic parasite of the fungus. It host mediated, this action by the orchid is more likely to be a defence mechanism against complete parasitic invasion. There is some sort of balance in this association between the aggressiveness of the fungus and the host's defence mechanisms. This is certainly a much more precarious balance than is evident in other mycorrhizal associations. This can be seen in germination experiments with the fungus present. Three types of seedlings can be observed. There are seedlings which are growing vigorously—that is where a balanced association has become successfully established. There are seedlings which have germinated but which have subsequently been killed. In this case, the fungus has become, after initial successful establishment, wholly parasitic and incapable of being controlled by the host. Finally, there are seedlings which are no longer growing. In these, after successful establishment of the infection, degeneration of the fungus has been of effective that the fungus has been eliminated.

In this association both biotrophic transfer across orchid-fungus membranes and any nectrotrophic activity by the host may be involved in the movement of carbon sources from fungus to orchid. The extent of the movement would indicate that the former is the major pathway. Translocation to the developing orchid seedling by its fungus has been demonstrated by Smith (1966-1967). She used what is known as the double dish technique and inoculated the inner dish which contained full mineral nutrients and cellulose as a sole carbon source with the *Rhizoctonia* fungus of *Dactylorchis purpurella*. In the outer dish she used a carbon free medium, silica gel, and sowed this with seeds of *D. purpurella*. The fungus grew out from the inner dish and spread across the outer and stimulated the seeds to grow. Over such a distance translocation of the products of cellulolysis in the fungal hyphae to the orchid seed could be the only feasible explanation of this stimulation. She then followed the uptake and incorporation of ^{14}C labelled glucose by the fungus growing out from the infected seedlings. Uninfected tissues of *D. purpurella* contain glucose, fructose and sucrose and the hyphae of *Rhizoctonia* glucose, trehalose, little mannitol but no surose. After feeding the hyphae of

Rhizoctonia with labelled glucose, the label appeared in trehalose in the hyphae. Carbohydrates are usually translocated as trehalose in fungi. In infected parts of the seedling, the label appeared in trehalose and sucrose, *i.e.*, in both host and fungal sugar, so that there must be a transfer between the fungal and host cells. In a time course experiment after a short feeding period with ^{14}C glucose trehalose contained the highest proportion of the label in the infected host. This proportion decline with time and was accompanied by a rise in sucrose (Fig. 2.14). This suggest that trehalose was being

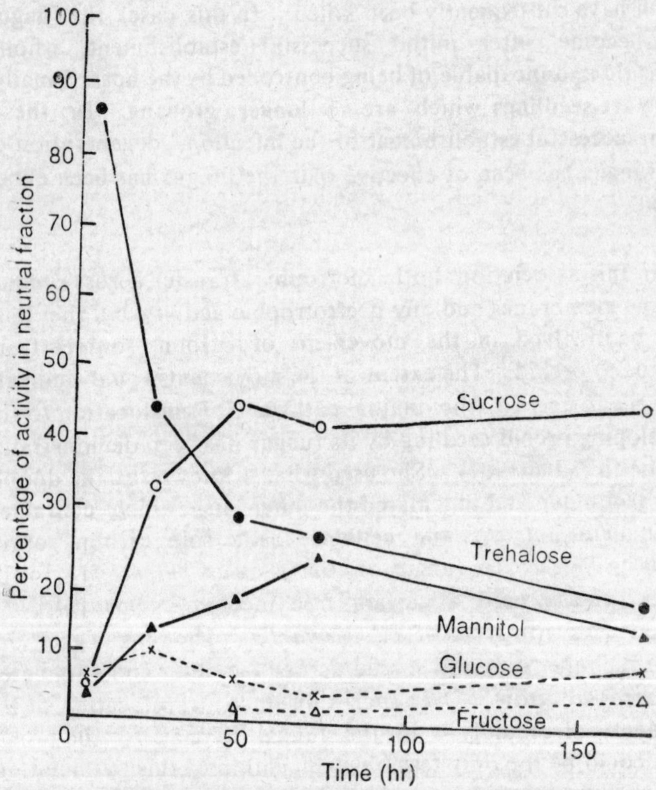

Fig. 2.14. Distribution of ^{14}C activity in a neutral extracted of mycorrhizal seedlings of *Dactylorchis purpurella* after the fungus, *Rhizoctonia* sp., had been fed with ^{14}C glucose. (After Smith. 1967)

utilized by the seedling and thus maintaining a concentration gradient in the system. What is not clear is whether the trehalose is transferred directly or whether it is externally hydrolysed and the glucose formed rapidly absorbed by the orchid tissue.

In similar experiments, Smith showed that ^{32}P labelled orthophosphate supplied to the fungus could later be detected in the infected seedlings. Thus in the heterotrophic seedling phase there is a net movement of carbon and minerals from fungus to host-orchid. At this stage this association must be regarded as a parasitic symbiosis with the orchid parasitizing the fungus. The ultimate source of carbon supplied to the orchid may be cellulose from dead plant remains in the soil or simple carbon compounds from another seed plant on which the fungus is a necrotrophic parasite in a second symbiotic relationship. As in the *Monotropa-Pinus* association, the fungus acts as a bridge.

Relationship between the adult green orchid and the fungus

In adult orchids only the absorbing roots ever become infected. Root tubers or photosynthetic roots never become infected not does infection spread to the stem and leaves. These parts are thought to produce a phytoalexin, orchinol, as a defence reaction on infection by mycorrhizal and other fungi. The roots may produce orchinol on infection but in insufficient quantity to inhibit the fungus. An odd feature of this associtation is that infection of the seedling does not initiate a permanent state of infection in the adult. The overwintering storage root tubers are free on infection and, in orchids such as *Dactylorchis fuchsii*, new roots are infected from the soil as the arise each year in the spring. As adults, some orchids are completely free of the fungus.

At the present very little is known about the relationship between the adult green orchid and the fungus. Reciprocal movement of carbon compounds to the fungus from the autotrophic orchid has not been demonstrated although suitable experiments to test the possibility have been carried out ;

however, at the moment it can only be concluded that this association is not a mutualistic symbiotic. There is much which still needs to be explained in this association, especially the significance of the re-infection of the new roots of the adult green orchid every growing season. It could be argued this is inevitable because of the aggressiveness of the fungi as root parasites but the possibility of there being a reciprocal exchange of carbon from the orchid to the fungus at this stage cannot be definitely ruled out or that the fungus does not still supply the orchid with minerals. *Dactylorchis fuchsii* and *Listera ovata* often grow-side-by-side in the same woodland. The former has three relatively short roots arising from the leaf bases above the tuber. The latter has very nemerous and very long roots. The roots of the former are virtually always very heavily infected and those of the latter only spasmodically infected. It would well be that *D. fuchsii* is still relying on its fungus for mineral supply. For some orchids at least one can see the possible advantages of maintaining the association. Some, for instance, by reverting to the association and becoming heterotrophic or partially heterotrophic can persist over one or more years without bearing leaves or sufficient leaves to maintain themselve fully. This would be advantageous in particularly adverse seasons or under conditions of very heavy shade or when their leaves are removed by grazing.

3

Symbiotics Relationship with Insects

A very wide range of fungi parasitize insects although perhaps the two best known genera are *Cordyceps* and *Entomophthora*. Such entomogenous fungi attack all stages of insects-eggs, larvae, pupae and adults. Some of these are not obviously antagonistic. This is especially true of members of the Laboulbeniales, a very little known group of the Ascomycotina which are obligate ectoparasites of Coleoptera in particular, and the Trichomycetes, a group with close affinities to the Zygomycotina, which inhabit the midgut, hindgut and rectum of many insects and other Arthropods. The Trichomycetes obtain all their nutrients from the aqueous fluids of the digestive tract in which they are bathed. They are attached to the lining of the digestive tract by a distinct device – a cellular or non-cellular hold fast and the thallus consists of a simple, cylindrical filament or a sparingly branched one, with or without septa. There is no penetration of the host tissues and perhaps these fungi may best be regarded as endocommensals (Fig. 3.1*a*).

In contrast, members of the genus *Coelomomyces*, one of the Mastigomycotina, as the name suggests, develop in the coelomic cavity, especially of the larvae of mosquitoes. A branched coenocytic hyphal system is produced attached a one end to the host's tissues. All members of the genus are

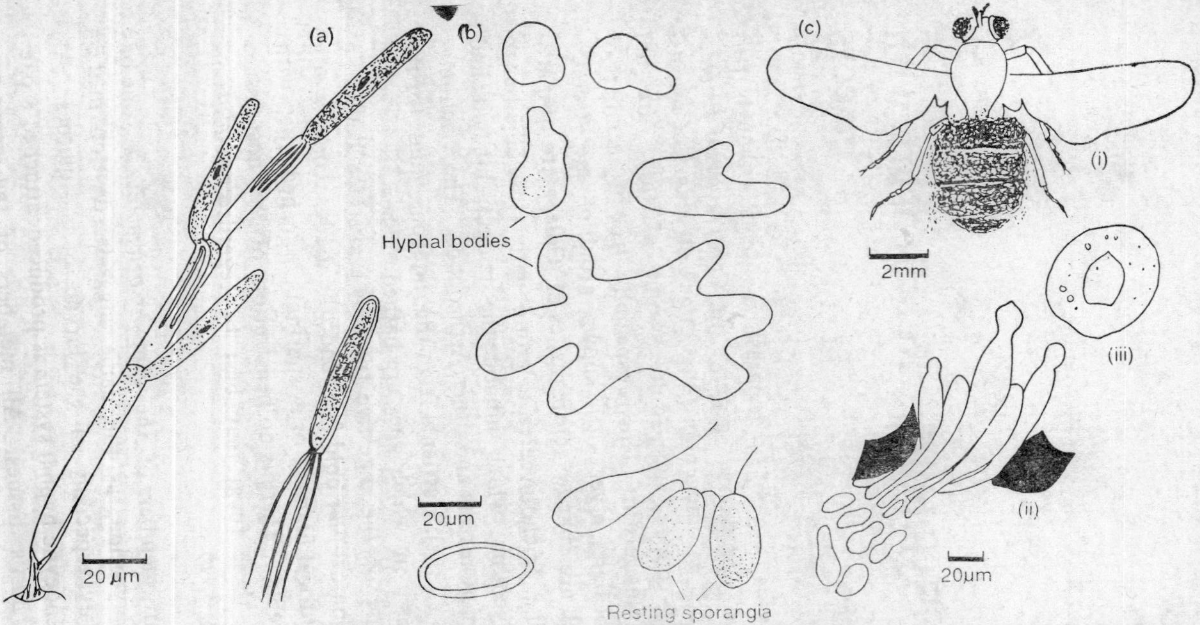

Fig. 3.1. (*a*) ***Harpella melusinae***. a Trichomycete, an unbranched vegetative thallus with three conidia developing and a released conidium, (*b*) ***Coelomomyces pentangulatus***, Hyphal bodies, vegetative thallus, developing resting sporangia and a released resting sporangium. (*c*) ***Entomophthora muscae***. (*i*) Dead fly with distended abdomen and bands of conidiophores between the segments. (*ii*) Hyphal bodies giving rise to unbranched conidiophores. (*iii*) A discharged conidium.

obligate parasites utilizing the adipose tissue and killing larvae before they pupate. Hyphal bodies, rounded off fragments of hyphae, are also produced and these circulate in the haemolymph (Fig. 3.1*b*). Similar bodies are found in the Entomophthorales. These are members of the Zygomycotina, which parasitize Hemiptera and Diptera although some are saprotrophic in soil and dung. On infection either a coenocytic mycelium develops or more usually this segments into rounded hyphal bodies which are produced in such profusion that the whole insect becomes more or less filled with them. The various tissues of the insect's body disintegrate at different rates-first the abdominal contents, then tissues of the head, thorax and legs and finally the trachea and tracheoles. Eventually the hyphal bodies develop conidiophores which protrude to the exterior and produce multinucleate conidia which are violently discharged. In houseflies infected with *Entomophthora muscae* the infected dying flies crawl towards the light-the top of a grass stem or a window pane. Their abdomens become distended and white bands of conidiophores project between the segments of the exoskeleton. The dead flies soon become surrounded by a white haio of discharged conidia. Some of these conidia will adhere to other flies as they crawl over them. Others, which have become air-borne, will impact upon flies in flight. The cycle of infection thus continues. At death, the flies virtually become mummified. They are no more than a sclerotial mass of hyphal bodies contained only by their own chitinous exoskeletons (Fig. 3.1*c*).

Another genus of fungi which similarly fills insects with a mass of mycelium is *Cordyceps*, one of the Ascomycotina. Most species of *Cordyceps* are parasitic on insects, especially Lepidoptera, but some are found on spiders and others on the subterranean ascocarps of *Elaphomyces*. *Cordyceps militaris* produces orange coloured, club-shaped stromata, 1·4 cm tall, bearing numerous ascocarps, perithecia. These develop in the autumn and arise from buried mummified pupae of Lepidoptera. On infection from ascospores, cylindrical hyphal bodies develop in the haemocoel of the pupa. They multiply by budding. On death, mycelial growth occurs and the pupa becomes converted into a sclerotium. Its tissues are extremely resistant to decay due to the production of an antimicrobial nucleotide cordycepin by the fungus. Thus it does not rot before the fungus has produced its reproductive structures.

Mutualistic Associations between Fungi and Insects

Not all associations between fungi and insects are so one-sided. The ability of many insects to exploit potential food resources to their best advantage, or perhaps even at all, may be dependent upon a reciprocally beneficial association with fungi. Some insect species are always found to be associated with a particular fungus. The insect may feed on the fungus or the fungus may digest or partly digest food for the insect. In many cases complete interdependence exists. In ectosymbiotic associations the fungus lives outside the insect but it becomes the food supply of the insect or it renders food digestible to the insect or it creates a favourable environment for the insect. It may, for example, protect the insect. The fungus is disseminated by the insect which may even have developed special pouches, mycetangia or mycangia, for keeping and carrying the fungus.

ECTOSYMBIOTIC ASSOCIATIONS

Ambrosia Fungi and wood-boring beetles

Wood, although a favourable habitat for many insects, is a most unpromising food source for them. It is resistant to degradation by normal animal enzymes. It is deficient in B vitamins and sterols which insects require. For some the role of the fungi is to grow on the wood and thus provide the insect with something nutritious and palatable to gnaw on. Ambrosia beetles, either Scolytidae or Platypodidae, are the most successful of wood-inhabiting beetles. They show a specialization of the wood-boring habit by boring deep into the wood, the Platypodidae even into the heartwood. Although wood borers, they are not wood feeders. They introduce into their tunnels ambrosia fungi which grow on and line the walls of the entire tunnel system and serve as the chief source of food for all stages of the beetles. The fungi are constantly associated with the beetles and are transmitted by them frequently in specialized organs, mycetangia. The role of the fungi here is seen in their ability to convert a virtually indigestible substrate into a digestible form resulting in the conservation of time and energy for the bettle. These ambrosia beetles should be distinguished from bark beetles which feed on living tissues especially the phloem. Many of these are constantly associated with a specific fungal flora in their brood chambers at the interface of wood and bark. The fungi are transmitted from tree to tree, not in specialized mycetangia, but usually accidentally by their slimy spores adhering to the integuments of the beetles. Although the fungi may have a nutritional value to the beetles, especially the young larvae, they are not essential. Some such associations have proved disastrous to the host tree, for example, the one between elm bark beetles and *Ceratocystis ulmi*, whereas others cause blue-stain of coniferous sapwood.

In northern temperate forests the Scolytidae are best known from the numerous species of bark beetles ; the ambrosia beeles are comparatively uncommon. In tropical forests the position is reversed and Platypodidae are more common. In the world as a whole species of ambrosia beetles far outnumber species of bark beetles. Although the adults of the two groups differ greatly in their appearance, their larval forms are very similar. They prefer dead, dying or weakened trees as breeding material, such as trees weakened by drought, age, fungal attack, fire, competition or defoliation and as pests are referred to as secondary insects. This preference is because the moisture content of the tree is important in fungal-establishment. As with wood decay fungi, the moisture content of healthy trees is too high. Most attack a very wide range of hosts. They destroy a great deal of timber. The tunnel to their brood chambers are called 'shot holes' and extend deep into the wood. In addition they may be surrounded by streaks of fungal stains.

The adult female Scolytid beetles which do all the boring are very odd. They fly to and attack the wood of a new host when their guts are emply and do not feed while excavating the entrance tunnel to their new gallery systems. Once this is achieved they mate, sow their ambrosial fungi and wait for the spores to become established. They then feed on the fungi, resume boring and lay eggs in the newly constructed brood chambers. If the fungi do not develop in the entrance tunnel, they eventually starve. In any case considerable resorption of the flight muscles occurs during this initial period of boring.

Mycetangia are usually found only in female Scolytids but are present in the males, which do all the boring, in the Platypodidae. The location and structure of the mycetangia vary enormously. In *Xyleborus*, they may be a pair of sacs alongside the brain opening to the inside of the mouth, membranous pouches at the base of the mandibles or sclerotized pouches in the base of the elytra. More usually they consist of depressions, flask-shaped invaginations or pouches on the body surface, such as the prothorax, where oil, secreted from specialized hypodermal cells or glandular hairs, accumulates (Fig. 3.2*a*). The fungi are sown as the tunnels are constructed when the output of the secretions is increased, possibly for lubrication during boring, and slowly washes out oidia or yeast-like fungal cells. These germinate on the walls using the oil as an initial carbon source. The fungi were called ambrosia by Schmidberger (1836). This was at the time when spontaneous generation was still acceptable and as the material on which the beetles fed appeared to have no earthly source, he called it ambrosia, the food of the gods. But it was Hartig (1844) who recognized it as fungal and he called it *Monilia candida*. It is now apparent that there are more than one ambrosial fungus and that there are specific associations between beetles and fungi. They are conidial states of Ascomycotina. Their taxonomy is still in a state of confusion. Almost all superficially resemble species of *Monilia*, a genus including a heterogenous assemblage of fungi bearing their conidia in chains. The majority have now been placed in the genus *Ambrosiella*. They produce a greyish brown to dark brown mycelium bearing branched or unbranched conidiophores which produce acropetal chains of thin-walled, colourless blastoconidia or a thickwalled, solitary, terminal conidium (Fig. 3.2*b*).

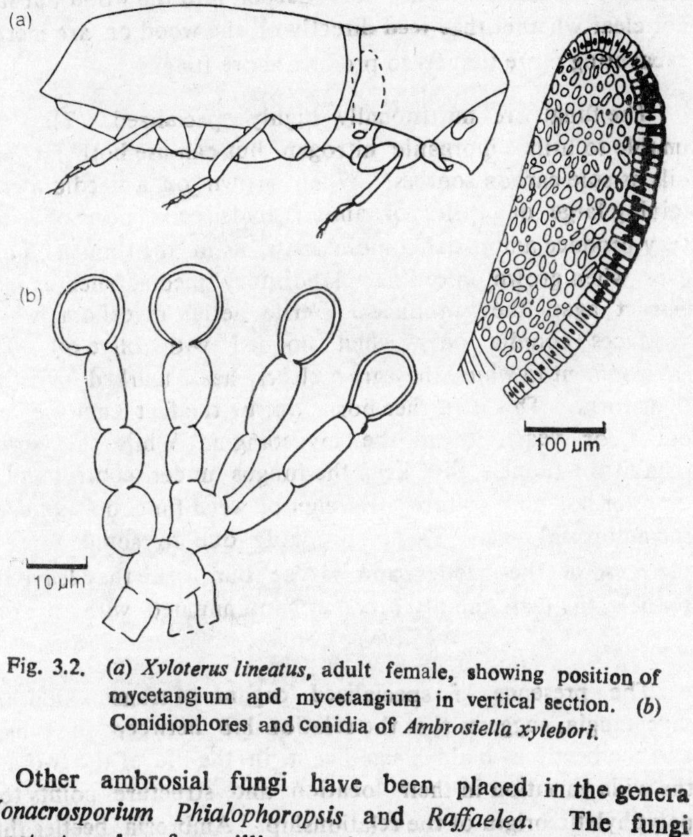

Fig. 3.2. (*a*) *Xyloterus lineatus*, adult female, showing position of mycetangium and mycetangium in vertical section. (*b*) Conidiophores and conidia of *Ambrosiella xylebori*.

Other ambrosial fungi have been placed in the genera *Monacrosporium Phialophoropsis* and *Raffaelea*. The fungi penetrate only a few millimetres into the wood, especially the parenchyma around the tunnels. They stain the wood a dark brown and form a hyphal mat on the tunnel walls from which the closely packed conid iophores project into the tunnels. The beetles eat these. In many species of *Xyleborus* the head capsules of the larvae are poorly chitinized and their mouthparts are not sufficiently strong to be used to gnaw wood ; they rely solely on the fungi as food. Very often there is insufficient fungal growth to satiate all the beetles in a particular brood. The last adults to develop in a brood are often much smaller due to a deficiency of food. Platypodid larvae feed exclusively on the fungi during the first three instars but the last two instars have a well-chitinized head capsule and much stronger mandibles. They bore deeper into the wood but it is not clear whether they feed directly on the wood or are merely excavating more tunnels to produce more fungus.

The fungi are nutritionally highly specialized. They are unable to utilize inorganic nitrogen but can use both fats and oils as sole carbon sources. When grown on a medium containing large quantities of amino acids or in pure olive oil, they produce ambrosial conidial mats as in the tunnels ; however, when grown on ordinary laboratory media, such as malt extract agar, they produce a sterile aerial mycelium which produces conidia only when flooded with olive oil. The environment in which they grow clearly has a marked effect on their form. This is further borne out by the fact that they are oidial or yeast-like in the mycetangia. While the larvae inhabit the tunnels, they keep the fungus under control and in more or less pure culture. Foreign or weed fungi do not spoil the ambrosial mats. This is probably due to some form of secretion of the beetles and larvae but, once they leave the tunnels, the mats rapidly become contaminated with all sorts of fungi.

The presence of specialized organs of transmission, the mycetangia, suggests that the relationship between the fungus and the beetle is highly significant in the life of the two and the wide variation in their location and structure points to a polyphyletic origin of the relationship. Ambrosia beetles thus live in an obligatory symbiosis with special domesticated ambrosial fungi. Many details of this intriguing symbiosis need to be worked out. The beetles appear to gain the main benefits in that the fungi provide them with a rich and easily available source of food, vitamins and perhaps sterols, such as ergosterol, which is necessary for the development of the adult from the larva. Although the fungi are disseminated, protected from desiccation within the mycetangia, and inoculated directly onto a suitable, moist substrate by the beetles, they are stricdy controlled by them. Readily available sources of carbon and nitrogen are scarce in wood and the quantity of nutrients in the form of fats and oils, as carbon sources, and urea and uric acid, as nitrogen sources, provided by the beetles may determine the extent to which the fungi develop. Constant grazing by the beetles also keeps them in check. This may be bondage by preference as the fungi are never found free-living.

Plant galls and fungi

Some insect plant galls may also contain fungi but, although there is a wide variety of interrelationships between insects and fungi, they are usually all casual ones. In galls caused by Itonid midges, the fungus grows on the gall tissue and forms a thick layer of hyphae on the inside of the gall. It appears unlikely that the fungus acts as a direct source of food for the larvae but it may assist the insect indirectly by partially breaking down the gall tissue so that the larvae can digest it.

A single fungus is usually associated with each midge species and its spores are probably implanted when the female lays her eggs. The stem galls of *Rubus fruticosus* caused by the midge *Lasioptera rubt* are called ambrosia galls. The midge larvae, instead of feeding directly on the gall tissues, feed on the hyphae of the fungus lining the gall in a manner reminiscent of the wood-boring ambrosis beetles. However, in another ambrosia gall, that caused by the midge *Asphondylia sarothanni*, on the buds and pods of broom, *Cytisus scoparius*, the larvae drive their nourishment from the gall tissues and not from the fungus. As long as the larvae are actively feeding the gall continues to grow and the fungus feeds saprotropically on the waste products of the larvae. But as soon as the larvae stop eating prior to pupating, the gall ceases to grow and the fungus fills the cavity with its hyphae so that the insect pupates in a dense mass of mycelium. The fungus then parasitizes the gall tissues. The fungi concerned have been very little studied.

The Sirex/Amylostoreum association

Wood wasps, especially species of *Sirex*, also attack weakened or suppressed trees or freshly felled logs as do many ambrosia beetles. They are also associated with a fungus but the association is a very different one. There is specialized transmission of the fungus but not strict culture. The female wasp does not bore out brood chambers but inserts her eggs several centimetres into the wood and it is the larvae which tunnel. The adults are on the wing all their lives. All females carry *Amylostereum*, a genus of the Basidiomycotina, which causes a white rot of wood. The females possess a pair of small invaginated intersegmental sacs protruding into the body and connecting by ducts to the anterior end of the ovipositor (Fig. 3.3*a,b*). These mycetangia contain large amounts of mucus, secreted by a pair of glands adjacent to the mycetangia, and the fungal mycelium within fragments into oidia or arthrospores consisting of 1-4 short cells with clamp connections at the septa (Fig. 3.3*c.d*). The female larvae, from the second instar onwards, also carry the fungus in hypodermal organs situated on both sides of the body in deep skin folds between the first and second abdominal segments. During egg-laying the mycetangia contract and the oidia are squeezed out and deposited in the wood with the eggs. They germinate and grow in the wood and rapidly break down both the cellulose and lignin. One effect of this is that it enables the larvae to bore more freely in the wood. Unlike the ambrosia beetle larvae, they do not feed solely on the fungus as its hyphae are too sparse, nor do they actually ingest wood but they may be able to digest the decaying wood. It seems probable that both the fungus and the decaying wood are utilized by the larvae. The hyphae may be broken down externally by larval enzymes secreted in saliva discharged into their cupped mandibles. Larva are able to live and grow for at least three months on pure cultures of the fungus.

Benefits to the fungus and wasp

The fungus clearly benefits from the association not only in being effectively dispersed but by being inoculated, free of competition from other wood decay fungi, into a suitable host without having to penetrate any protective tissues, such as bark. With such an efficient means of dispersal, the production of basidiocarp becomes almost superfluous. They are very

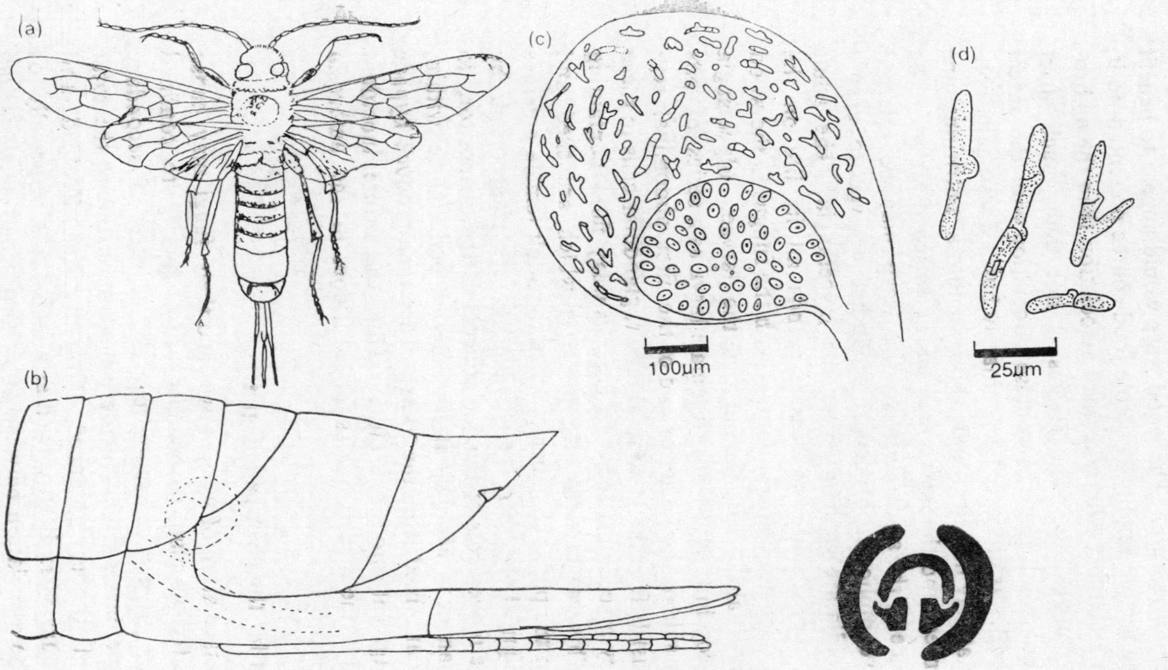

Fig. 3.3. *Sirex noctilio* (*a*) Adult female, (*b*) Posterior end of adult showing the ovipositor, position of the mycetangia and the ovipositor in section, (*c*) Vertical section through mycetangium and mucus gland, (*d*) Anthrospores from mycetangium.

rarely found in nature although isolates of the fungi will produce them in culture. The wood wasp would appear to benefit by having a more readily available food source presented to it. This is a much less strictly defined association than the ambrosia beetle/fungus one. The fungus is not under such strict cultural control. There is no close association with the insect in the wood. It grows indiscriminately but it does again convert a relatively non-nutritious substrate into a nutritious one for the insect and the larvae feed on the softened wood. The importance of the association to the wasp may be inferred from the complicated procedure which has evolved to ensure that the oidia are established in the mycetangia. The adult wasp, unlike the ambrosia beetle, spends all its life on the wing. It does not come in contact with the fungus in the wood and must obtain it from the pupa via the larvae. Mycetangia first develop in the fully-grown female larva as a series of pits on either side of the posterior end of the first abdominal segment. Just before the last larval moult, glandular cells in these temporary mycetangia secrete a copious waxy fluid containing oidia. This wax hardens into platelets which are shed and remain in the pupal chamber. When the adult female emerges from the pupal chamber, the wax platlets are transported, by retractive movements of the components of the ovipositor, to the permanent mycetangia at its anterior end. Once within, the oidia are freed from the wax. They then produce hyphae which fragment into more oidia. Thus the future inoculum multiples in the mycetangia as indeed it does in the ambrosia beetles.

Problems for the plant pathologist

The association of wood wasps and their fungi may cause problems for the plant pathologist. *Sirex* species are confined to conifers. They are usually regarded as pests of secondary importance, losses from which are small in relation to the total volume of timber produced. Only in New Zealand and Australia has *Sirex* been viewed as a major pest, especially of *Pinus*. The association between *Sirex noctilio* and *Amylostereum areolatum* has caused serious losses in *Pinus radiata* there. Improved silvicultural practice and the introduction of parasites of *S. noctolio* now keep the *Sirex* population in New Zealand at a level at which it is no longer a serious forestry problem. In Australia it is still a problem and it is generally accepted that about a 10% absolute loss in volume of timber produced may be caused under average climatic conditions. In prolonged drought losses may be far more severe.

A. areolatum is a relatively weak wound parasite and most trees resist attack by it alone but, in a common attack with *S. noctilio*, it is able to kill pines. Mass attacks by the wood wasp provide a large number of the inoculation points which are more effective in enabling the fungus to establish and spread than in a single inoculation from a natural wound. *S. noctilio* is attracted to physiologically stressed trees such as weakened, suppressed, over-crowded, fire damaged or drought striken ones. Injured trees exuding resin are particularly attractive. Volatiles in resin act as specific attractants; even so a high resin content is important in restricting growth of the fungus and it may also depress the survival rate of the larva. Resistance of *P. radiata* to attack by *S. noctilio* is related to the obility of the tree to obtain adequate water from the soil and the moisture content of the wood. The optimum moisture content for egg-laying is 40-70% on a dry mass basis. Dominant trees a high moisture content tend to be rejected by *S. noctilium* and the survival rate in such trees is often low. *A. aerolatum*, when deposited with the eggs, spreads slowly in wood with a moisture content of 70% saturation or more. As it grows it reduces the moisture content of the wood locally, with the result that the eggs hatch and the larvae bore into relatively dry wood. The moisture content of heavily infected wood averages about 35% of the dry mass. Eggs and young larvae become desiccated in wood below 20% but fairly mature larvae can survive in very dry wood.

The spread of *A. aerolatum* in wood of *P. radiata* is restricted by the formation of phenolics, such as the stilbene pinosylvin, and resins in the sapwood near oviposition holes as a reaction of the tree to attack. The killing of trees after attack by *A. aerolatum* was formerly attributed to the pathogen effects of the fungus. It was suggested that it cuts off the supply of sap to the crown. Although the fungus eventually dries the wood so much as to render it non-conducting, the restriction of sap flow is quite gradual and cannot explain the very marked physiological changes that occur in the tree within the first two weeks after attack by *S. noctilio*. The mucus, a protein mucopolysaccharide complex, secreted by *Sirex* and injected into the tree with its eggs and the fungus is phytotoxic. Mucus injections alone will induce the rapid early physiological changes typical of a *Sirex* attack. These include increased respiration in the stem, accumulation of starch in the leaves, decrease of starch in the bark of the stem, implying that translocation of photosynthate is inhibited and also some impairment of normal water relations of the needles causing desiccation, distortion and collapse of the translocating cells of the phloem. This is followed by yellowing, wilting and abscission of many needles. Such symptoms are reminiscent of premature senescence. This conditions the tree making it more susceptible to the fungus. It can then invade the sapwood and kill the tree.

Neither the fungus nor the mucus alone is capable of killing the tree but the combination of the two is lethal. The tree, having been conditioned by the mucus, is subsequently killed by the fungus invading the sapwood and cutting off the sap flow to the crown, death being preceded by a sudden increases in leaf water deficit. The mucus can be seen as a 'conditioning agent' which brings about a reduction in the effectiveness of the tree's natural resistance mechanism.

The Septobasidium/Aspidiotus association

The rather different mutualistic symbiotic relationship exists between scale insects and certain fungi. All species of the Basidiomycete genus *Septobasidium* are invariably found living in association with scale insects, such as species of *Aspidiotus*. In this association the fungus provides individual or communal prefabricated housing accommodation as well as protection from physical and biotic environmental factors, in particular protection from birds and wasps, for the scale insect. In return for being dispersed and being provided with the opportunity of deriving nutrients from insect secretions and excretions, the fungus also parasitizes some insects by sending out specialized, often coiled hyphal branches into the haemocoel of the insects (Fig. 3.4*b*). The insects can live unassociated with the fungus but probably never do so. The fungus is also obligately associated with scale insects in nature. This is a very delicately balanced mutualistic sombiosis as the scale insects may be killed by the fungus in which case the fungus also dies.

Septobasidium is a member of the Auriculariales and is found as an epiphyte on the bark of trees, especially on the lower sides of branches, in the trophics. Individual colonies form flat, radially ridged growth almost circular in outline. They may reach several centimetres in diameter and, being perennial, show concentric rings of growth. They often resemble a thick crustose lichen. Within, under the roof as it were, are a number of vaulted chambers and interconnecting tunnels or passages, opening to the outside but partially closed by a fungal flap. Uninfected female scale insects inhabit the tunnels and are free to move about but most chambers contain a single sedentary infected female (Fig 3.4*a*). Only the uninfected ones are fertile. They project their abdomens to the outside through one of the openings and are fertilized by wandering males on the outside. They then remain closeted within the colony and reproduce.

Basidiospores, which may themselves bud, are formed over the surface of the colony and they adhere to the young scale insects as they emerge from the parent colony (Fig. 3.4*c*). After they settle, the spores germinate and surround the insect with hyphae, some of which penetrate the setal and enter the haemocoel where they eventually form numerous hyphal coils. Inside the haemocoel the hyphae are composed of chains of wide spindle-shaped segments connected together by very fine hyphae, the whole resembling a string of cocktail sausages (Fig. 3.4*b*). The hyphal coils, each consisting of 4-8 symmetrical turns, like a coil of a hot water heater, arise on these hyphae connected end to end by fine hyphae or in clusters of

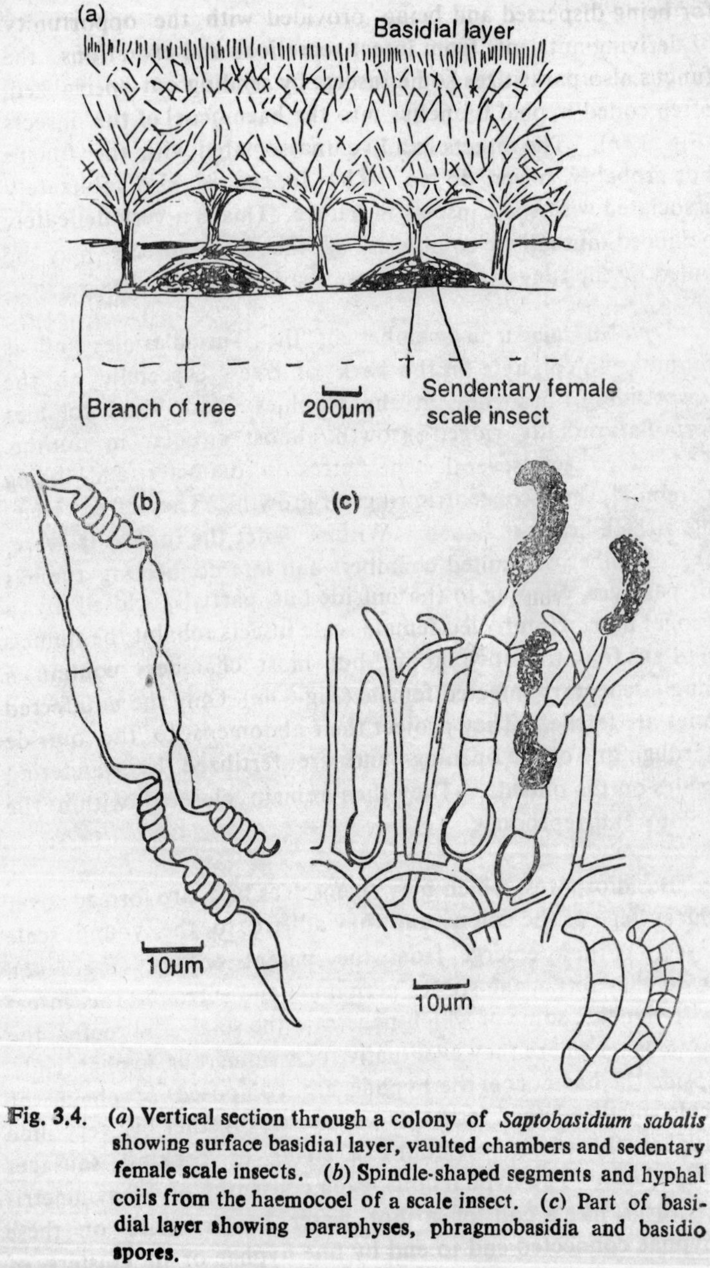

Fig. 3.4. (*a*) Vertical section through a colony of *Saptobasidium sabalis* showing surface basidial layer, vaulted chambers and sedentary female scale insects. (*b*) Spindle-shaped segments and hyphal coils from the haemocoel of a scale insect. (*c*) Part of basidial layer showing paraphyses, phragmobasidia and basidio spores.

two to three as lateral branches. The scales are sucking insects and the stylets of these females tap the phloem of the tree. Christensen (1965), in his rather splendid account of these, sees the insect at this stage merely as a pumping station transferring sap from the tree to the fungus. The insect converts surose from the phloem to its blood sugar trehalose which the fungus uses directly. A new colony soon becomes established around such a female.

In addition to shelter, gained at the expense of paratism and sterility of some individuals, the fungus affords protection from predator birds and parasitic wasps. The scale insects are hidden from the view of birds and the particular parasitic wasps have only short ovipositors so that they can reach and lay eggs only in the superficial scale insect and not those in the vaulted chambers and tunnels. The thick mat of hyphae above the chambers also to some extent insulates the scale insects from temperature extremes and prevents their desiccation. In return, through its parasitism the fungus obtains its organic nutrients, especially its carbon source, from the insect and is also dispersed.

The Attine ants and their fungi

The fungus-growing ants and termites provide further examples of mutualistic symbioses between insects and fungi. The Attine ants culture a saprotrophic fungus in their nests or so-called 'fungus gardens' and use the fungus as their primary and probably sole source of food. The Attini are a tribe of New World Myrmicine ants which, although presently distributed between latitudes 40°N and 44°S of the equator, are best represented in the tropics, especially in Central and South America, the southern and eastern United States of America and the Caribbean. They were probably a significant part of the fauna of the Americas even before the time of the American Indians. The two most common genera, *Atta* and *Acromyrmex* are leaf-cutting ants and use the fungi in their subterranean gardens. *Atta texana*, for example, because of its relentless leaf cutting, cause considerable damage to *Citrus* crops in eastern Texas and southern Louisiana. Because of their destructive activities, they are regarded as plague insects. There are several others genera of Attine ants such as *Trachymyrmex* and *Sericomyrmex*. Some of these use insect excrement and carcases, often of ants, and a variety of plant debris on which to cultivate their fungus. Others supplement these with leaves and flowers.

The *Atta* colony and their 'fungus gardens'

In *Atta* there are three castes—workers, females and males. The workers are polymorphic. They range considerably in size and show a gradual and continual change in size from smallest to largest. The largest are the 'soldiers', whose prime function is to guard the colony. The medium sized workers forage for leaves, cut out discs and carry them back to their nests on their backs and hence have earned themselves the name of 'proposal' ants. In foraging they make distinct well-worn tracks up to 300 mm wide through the vegetation. They also construct the nests which are underground ; to build these they excavate. The excavation craters of a single *Atta* colony may cover some $8m^2$ of ground and be about 1 m or so deep. An average sized colony has some 2000 chambers, each 200-300 mm in diameter. There would be some 1-2.5 million ants weighting about 10 kg in the colony. For every 2000 chambers, they excavate about 80,000 kg of soil and during the life of the colony about six times this mass of fresh leaf material is used in the gardens. Thus, in addition to their destructive activities, they are of some ecological value in creating fertile, organic-rich pockets in soil and in mixing the soil. They are of importance especially in soils where few roots penetrate to any depth and where there are few burrowing animals.

At the nest the leaf-cutting ants pass the discs over to the smaller workers. These clean the discs, scrape them and cut them into 1-2 mm pieces. They then pinch them to a pulp with their mandibles. Saliva and faecal droplets are added and they are deposited in the fungal gardens. Actively growing tufts of mycelium from other parts of the garden are planted on the surface of the newly added leaves. They are, in the 'mushroom growers' parlance 'spawned'. The garden is a speckled, brown, sponge-like or honeycombed mass of withered pieces of leaves, loosely connected by a weft of white mycelium. Exhausted or spent gardens may be deposited in certain chambers, in specially constructed larger ones or cast out by the medium-sized workers several metres from the colony.

These activities of the ants produce a flourishing growth of one fungus although the substrate is suitable for many. This pure culture is maintained in spite of the possibilities of contamination from the surr ounding soil or from spores carried in by the ants on themselv es or on leaves. When the ants are removed from or leave their gardens, as in the ambrosia beetle tunnels, the fungus is quickly replaced by alien fungi and bacteria. Only when the association is maintained is a healthy monoculture garden maintained.

The garden is the abode of the queen and her brood. Before swarming, a virgin queen tucks a small pellet from the fungus garden into a pocket, the infr abuccal cavity, below her mouthparts. After her mating flight, she settles, breaks off her wings and starts to excavate in order to establish a new colony. She digs a small tunnel, 150-250 mm deep, and near the end creates a chamber and plants the pellet in it. She continually adds faecal droplets and broken infertile eggs to it but also goes out and forages for fresh leaves. It may be several months before she lays fertile eggs. The offspring then attend to the garden and a new colony is formed.

The fungi concerned are Basidiomycete Agaricales which do not normally produce basidiocarps if the nest is attended, probably because the mycelium is grazed too much. A number of generic names have been given to these fungi. Most appear to be members of the white-spored genus *Lepiota* or of the macroscopically rather similar genus *Leucocoprinus*. A number of isolates have been cultured. They can be easily grown on ordinary laboratory media and are typically mycelial. When grown on acid media, rich in amino acids, and exposed to 0.5% carbon dioxide, as they are in the nests, they develop into more valvety growth producing clusters of hyphae with very swollen tips rich in cytoplasm. Such clusters are called staphyla or bromatia (Fig. 3.5). These are particularly abundant in the nests. The ants crop these. The hyphal tips are fed to the larvae and they are also the stable food of the adults.

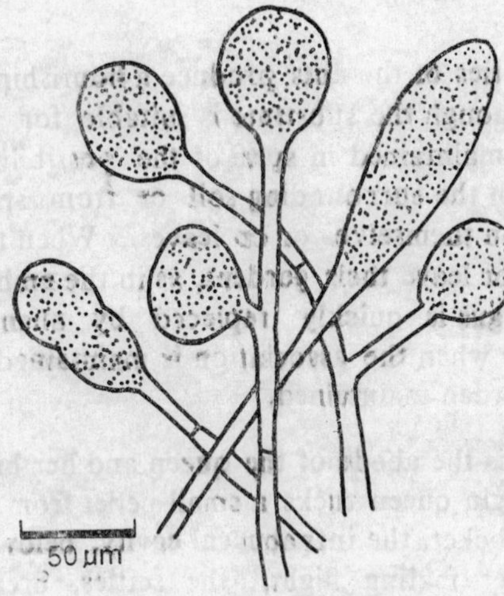

Fig. 3.5. A cluster of staphyla.

Neither the ant nor the fungus is found in nature without the other. The association is an obligate symbiosis for both. This is probably the most advanced of all the insect/fungus mutualistic associations. One intriguing aspect of this relationship which is not understood is how the ant maintains the growth, form and purity of the fungus in its garden as each species of ant cultures only a single species of fungus. One theory was that the ants' salivary and anal secretions, which the ants regularly apply to the gardens, contain antibiotics which prevent the growth of alien fungi. Although an appealing theory, it is an untenable one. Extracts from ants' fungal gardens and cultures of the fungus have no antibiotic activity.

The biochemical basis of the symbiosis

The biochemical basis of the symbiosis in the *Atta colombica tonsipes*/fungus association has been extensively investigated by Martin and Martin (1970) and they have clarified some of the principles and mechanisms underlying this symbiosis. The fungus grows very poorly on culture media containing polypeptide nitrogen and they initially suggested that the fungus lacks the necessary complement of proteolytic enzymes to make full use of protein nitrogen. Fresh leaf material would thus seem to be a singularly inappropriate substrate for the fungus as most of the nitrogen in leaves is in the form of protein. The application of faecal material to the leaves compensates for this. The two major nitrogenous excretory products of the ant are allantion and allantoic acid which would be available for the fungus to use as amino nitrogen. But, in addition, fecal material contains significant quantities of ammonia and some 21 amino acids. Glutamic acid, histidine, arginine, proline, lysine and leucine make up 80% by mass of the total amino acids excreted. These supplements would clearly be beneficial to the growth of the fungus but the benefit would be short lasting. They would be rapidly absorbed and utilized as they are excreted in very small volume in relation to the quantity of leaves used in the nests. In addition, Martin and Martin found the although the ants' salivary excretions showed no protease activity, the faecal droplets had significant activity. They contain three proteolytic enzymes, serine proteinase and two metallo-peptidases, which are highly stable and possess broad catalytic activity. The faecal material thus serves the additional function of catalysing the hydrolysis of leaf protein, thus further facilitating fungal growth. At first Martin and Martin though that these enzymes were the ants' contribute to the symbiosis. But because of their stability and general resistance to denaturation and deactivation and the fact that the enzymes are only slowly degarded by other proteolytic enzymes. Martin and Martin suspected that these enzymes were microbial in origin and have shown that they are derived from the cytoplasmic contents of the hyphae upon which the ants feed. Thus the ants do not contribute biochemically to the association but they are vehicles. They simply acquire, accumulate and move fungal enzymes from a region in the nest where the fungus is in a state of rapid growth and the enzymes are in ample supply to a site of inoculation, the newly added leaves, where the enzymes are in short supply. This permits substantial hydrolysis of protein providing a nitrogen source for the initiation of rapid growth by the fungus and simultaneously shortening the lag phase, during which time metabolic preparations for rapid growth are occurring. Any reduction in the lag phase would also shorten the time period in which the substrate would be susceptible to colonization by competitor saprotrophs. They were unable to detect any trace of digestive proteolytic enzymes produced by the ants. Thus the chief biochemical adaptation of the Attine ant to the fungus-growing habit has been the cessation of the production of any digestive proteases, since they would degrade any fungal proteases as they passed through their guts.

Much of the ants' culturing and maintenance activities can be considered as enhancing the competitive ability of the fungus so that it emerges in the nest as the dominant micro-organism excluding competitors. The scraping and pulping will liberate cell contents so that they are immediately available to be absorbed. Rapid and extensive colonization is enhanced by planting mycelial inocula. This gives a better competitive advantage over contaminants starting from spores. In many ways, in this association, the ants are dominated by the fungus. They forage for substrate for it, prepare the substrate for inoculation, inoculate and fertilize it, extract and transplant enzymes from one part to another, weed the gardens and remove old expended material. The fungus is cellulolytic and it indirectly contributes its cellulose degrading ability to the ants. It converts cellulose into fungal carbohydrates, especially glycogen and trehalose, which the ants can metabolize and so, by eating the fungus, the ants can tap the virtually inexhaustible cellulose supply in its environment. This they certainly could not do without the fungus. The fungus provides the ants with an excellent, rich and complete diet. More than 50% of the dry mass of the fungus is easily soluble. Carbohydrates make up 27%, free amino acids about 5%, portein 13% and lipids 0.2%. The lipid fraction also contains the vital, ergosterol, which the ant larvae may require for normal development. The lack of digestive proteases makes the amino acids particularly important to the ants.

One can attribute the success of the Attine ant in the New World tropics to the ecological soundness of its leaf-cutting, fungus-growing strategy. Much of this success is due to the ability of the ant fungi to grow on almost any kind of leaf material. By constructing their nests in the soil rather than above ground, the ants ensure that the fungi have a suitably most environment in which to grow and so may be found in a wide diversity of habitats, semidesert, grassland, tropical rain forest and even in suburban gardens where the ants plunder the leaves on man's crop and garden plants to build their fungus gardens.

Termites and their fungi

Enzymes of fungal origin have also been found in the digestive fluids of fungus-growing termites. Termites are best known as destroyers of wood. The lower termites possess symbiotic protozoa and bacteria which digest cellulose in the paunch, an enlarged region of the hindgut. But the higher termites, the Termitidae, which make up about 75% of the existing termite species, lack these and afford an interesting parallel with the Attine ants in that they maintain fungus gardens or 'combs' in their nests. Whereas species of *Microtermes* construct a series of chambers containing fungus combs below ground in a diffuse pattern, species of *Macrotermes* build steeple-like mounds up to 10 m tall, especially in tropical savannahs. The workers collect and eat all sorts of fallen plant debris but especially wood and they use their faecal pellets to make combs in chambers within the mound. Undigested or just chewed wood is not used. Each comb becomes covered by mycelium and is sponge-like in appearance but corky in texture. Numerous small spheres or nodules, 0.5-2.0 mm diameter, develop over the surface. In *Macrotermes natalensis*, these nodules are the conidia of the fungus. *Termitomyces*, a pinkspored member of the Agaricales. Several species of *Termitomyces* are known and all have an obligate association with termites (Fig. 3.6). Basidiocarps are produced early in the rainy season but only after the colony has abandoned its nest. All possess a stalk with a very distinctive long and tapering root-like base or pseudorhiza. The bsidio-

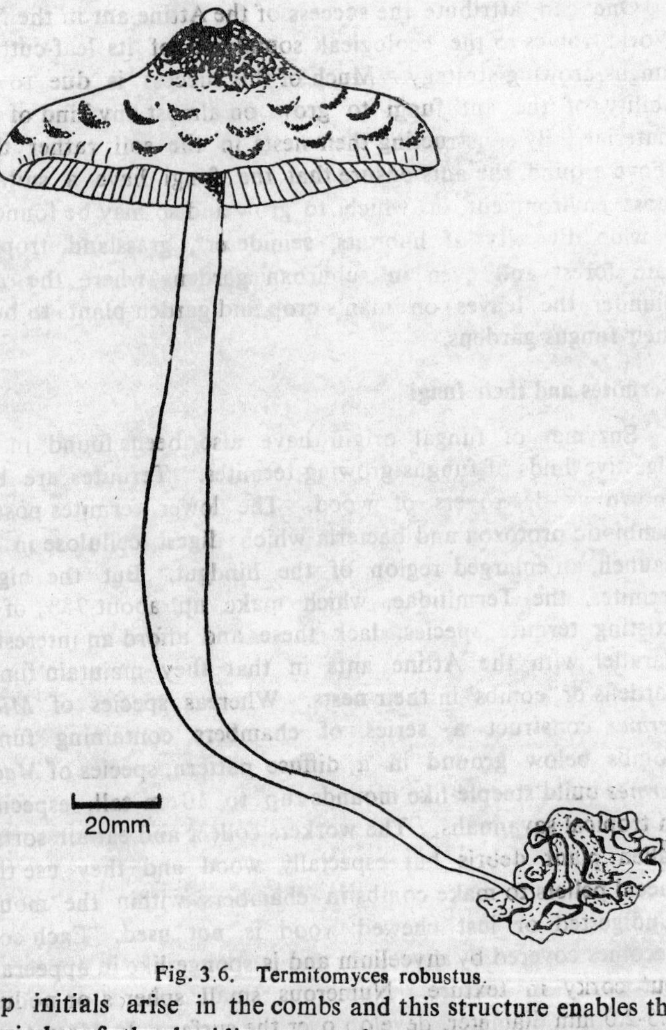

Fig. 3.6. Termitomyces robustus.

carp initials arise in the combs and this structure enables the remainder of the basidiocarp to be pushed up and expand above the soil on top of a nest or around the base of a mound. The basidiocarps are esteemed as delicacies by man in all parts of the world wherever they are found. For example, *T. robustus*, which grows in large numbers in contact with termites' nests on forest soils in Nigeria, is called 'ewe' by the Yoruba. The word means 'to expand' and is given because of the way in which the globose cap expands and flattens into a plate-like disc, some 200 mm diameter. The rains, which occur at that time, are called 'eji ewe'. 'Eji' means and hence they are the '*T. robustus* rains'. This and five or six other species are collected to eat and can be purchased at the roadside or local markets. Other species of *Termitomyces* are smaller except *T. titanicus*, common in Zambia. It currently holds the record for size in the fungi, with caps up to 1 m diameter.

Most termites culture one fungus in their nests but the same species may associate with different species of *Termitomyces* in different areas. Other fungi known to associate with termites include the Gasteromycete *Podaxis pistillaris* which is found in the nests of some species of *Trinervitermes* and a number of species of the Ascomycete genus *Xylaria* (Fig. 3.7). *X. nigripes* is usually associated with *Odontotermes* sp. and *X. furcata* may also associate with *Macrotermes natalensis*. Like *Termitomyces* and the ant fungi, they produce their reproductive structures only when the nests are abandoned.

Numerous suggestions have been made as to why the termites grow their fungi. It has been argued that the fungi by their metabolic activities, in what are virtually miniature compost heaps, produce a controlled high humidity and an equable temperature regime in the mounds and thus help to maintain a favourable micro-climate and that this would be of an obvious advantage in particular to termites inhabiting hot, dry savannahs. The insides of the mounds of *M. natalensis* are certainly maintained at 98-99% RH even in the annual very hot dry season. The mounds are architecturally very complex with walls up to 600 mm thick, with an elaborate air duct system in their walls. It is more likely that it is the design and structure which maintain the internal microclimate.

The acquisition of fungal enzymes

The fungus comb with its mycelium and nodules is an important part of the termites' diet. The combs are eaten away from below when they reach a certain stage of maturity

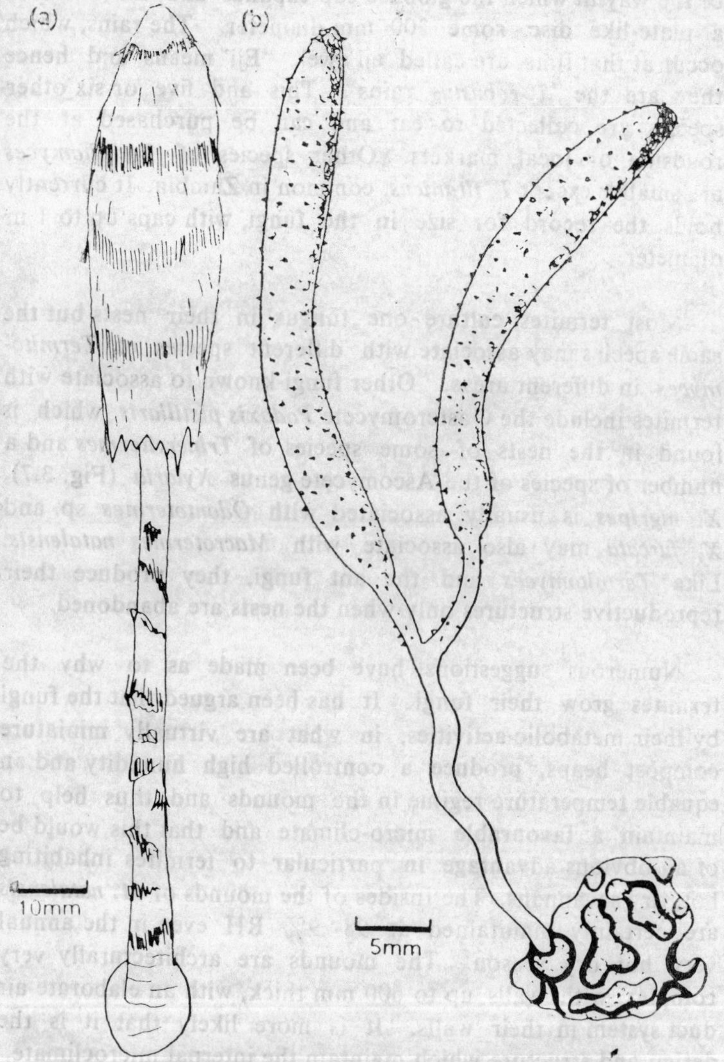

Fig. 3.7. Other termite fungi, (a) ***Podaxis pistillaris***, a Gasteromycete with a stalked basidiocarp, (b) ***Xylaria termitum***, ascocarps arising from an old piece of comb.

and are built up at the upper surface or are replaced by a new comb in the space beneath. Martin and Martin (1978, 1979) have investigated the *M. natalensis/Termitomyces* association. In this termite, ingested plant cellulosic wall material is digested in the gut. Any ingested material is finely chewed so that the cellulase enzymes can get at the cellulose, especially in the lignified walls. But the termites starve if fed only cellulosic materials such as wood with no access to the fungus.

Cellulose is a β-1,4 glucan and more than one enzyme is involved in its hydrolysis. Cellulose is first cleaved by an enzyme designated C_1 into linear glucan chains so that these can be hydrolysed by a second enzyme C_x to low molecular weight compounds such as glucose and cellobiose. The cellobiose is then hydrolysed by a β-glucosidase. The C_1/C_x story is a gross oversimplification and perhaps outdated. C_1 is now regarded as an enzyme complex consisting of an exo-β-glucanase, which removes cellobiose units from the end of the glucan chains, and an endo-β-glucanase with very special properties acting within the chains, but acting only on areas where the chains are highly ordered and strictly parallel, by disrupting H bonds between the chains, Cg^- is an endo-β-glucanase. The point that should be stressed is that the prior action of the C_1 component is required for ultimate hydrolysis by C_x.

Most celluloysis occurs in the midgut rather than the paunch of these termites and the entire set of enzymes required for cellulolysis are present there but the salivary glands and midgut epithelium only produce the C_x component and a β-glucosidase but not the vital C_1 component. Martin and Martin argue that the fungus is the source of the C_1. The termites feed on the nodules which contain high C_1 activity and they acquire the C_1 when they digest the fungus, Cellulose digestion in the midgut can thus occur only if the acquired fungal enzymes are present—thus the nutritional dependence of the termites on their fungus garden.

Such evidence as this has led Martin (1979) to put forward the very plausible hypothesis that many of the enzymes present in the guts of wood-and detritus-feeding Arthropods may be fungal enzymes acquired while feeding. The source of these may be the fungal tissue itself or simply the substrate into which the enzymes have been secreted. It also seems likely that the gut enzymes of many litter feeding invertebrates may be derived from fungal tissues consumed along with the detritus in their diet. He derived this hypothesis from the fact that wood- and detritus-feeding invertebrates routinely encounter fungal tissues during their feeding activities, that fungal tissues have been identified in the gut contents of many species and that the enzymes present in the guts of such species are common fungal enzymes. This acquisition of external sources of digestive enzymes can be seen as windfall of mycophagy and as an alternative to the maintenance of a culture of extra-cellular endosymbionts in the gut to produce the necessary enzymes. As a group, the termites have adopted both strategies.

The faecal pellets which the termites use to culture the fungus contain proportionally more lignin than the ingested wood because much, but not all, of the cellulose has been digested as it passes through the gut. It is assumed that the fungus is breaking down predominantly lignin in the comb material. It appears that the termites do not acquire fungal ligninolytic enzymes. As lignin degradation is an oxidation rather than an hydrolysis these enzymes in any case may not function in the gut because of the low oxygen tension. In addition some of the phenolic degradation products may be harmful or even poisonous to the termite.

ENDOSYMBIOTIC ASSOCIATIONS

Many insects, especially Coleoptera and Homoptera, possess intracellular endosymbiotic fungi. These are housed in specialized cells, mycetocytes, aggregated into special organs, mycetomes. The fungi are all yeasts or yeast-like and referred to such as *Candida* and *Torulopsis* ; they appear to provide insects with essential nutrients which are lacking from their diet. The fungus probably benefits by being provided with a suitable environment, free of competition, for its development and by being dispersed by special transmission procedures without the fungus having to undergo a free-living phase. Many insects of the wood-eating group have become nutritional specialists. *Sitodrepa panicea*, the drug store beetle, feeds on carbohydrate-rich substances such as grain and in particular flour. It houses a yeast in very much enlarged cells, lining four voluminous and lobed blind sacs situated at the anterior end of the mid-gut (Fig. 3.8). It is generally believed that the beetles rely upon

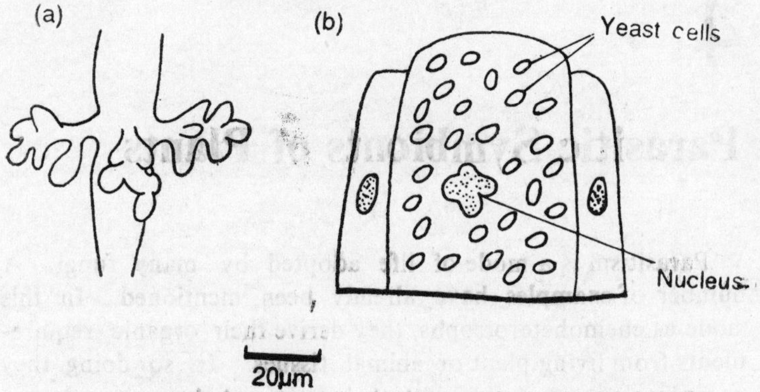

Fig. 3.8. *Sitodrepa panicea*, the drug store beetle. (*a*) Lobed, blind mycetomes at the anterior of the mid-gut. (*b*) A specialized cell, a mycetocyte, containing yeast cells from the lining of the mycetome.

the yeast to provide essential amino acids, vitamins, especially those of the B group, and sterols which are deficient in their diet. Larvae can be grown free of the yeast. They show a marked retardation of growth. Few even reach the adult stage. They require B vitamins, sterols and all the essential amino acids whereas normal larvae require only arginine, leucine and thiconine. Transmission of the fungus is ensured by other fungus-containing organs, intersegmental tubules which connect up with the ovipositor sheath and fungus-filled sacs at the base of the ovipositor. As the eggs are being deposited they are smeared with fungal cells from these. The larvae become infected as they hatch by eating the contaminated egg cases.

4

Parasitic Symbionts of Plants

Parasitism is a mode of life adopted by many fungi. A number of examples have already been mentioned. In this mode as chemoheterotrophs, they derive their organic requirements from living plant or animal tissues. In so doing they come into an association with their host and share a common life. A symbiosis is thus established. In such situations the fungus is usually antagonistic to its host. It benefits and the host is harmed. The fungus is referred to as a parasite and the relationship is called a parasitic symbiosis.

Modes of nutrition

The degree of dependence of the fungus on the association with its host varies enormously. In some it is absolutely necessary. These fungi need to be associated with the living cells of the host in order to grow and reproduce under natural conditions. They are obligate parasites in the ecological sense. They are normally unable to grow as saprotrophs in nature because they lack the ability to compete successfully in mixed saprotrophic populations. Many cannot even be grown successfully or at all in culture. In others the association is not an absolute necessity. They can grow as saprotrophs away from their hosts. They are facultative parasites.

Parasitic fungi can also be characterized by their modes of nutrition. Some can derive their nutrients only from living cells of the host. They are called biotrophs—'living feeders'. If the host or the part of it occupied by the fungus dies, the fungus also dies or goes into a resting state. Away from their hosts, biotrophs are encountered only as dispersal or dormant spores. Others derive their requirements from dead cells which they have killed. As parasites they continually create adequate substrate by killing host cells in advance of penetration. Since they derive their nutrients from dead host cells they are called necrotrophs—'dead feeders'. In this respect they differ from saprotrophs only in that they kill host cells. This is also reflected in the fact that many necrotrophs are facultative parasites which can live as saprotrophs on the non-living organic components in their environment. Indeed the mode of nutrition of a fungus very much depends upon the conditions in which it finds itself. A single fungus at different times and under different circumstances may show all three modes.

Obligate biotroph

Obligate plant parasites occur in three orders each from different sub-divisions of the fungi. These are the Peronosporales or downy mildews in the Mastigomycotina, the Erysiphales or powdery mildews in the Ascomycotina and the Uredinales or rusts in the Besidiomycotina. From these three orders, only a few of the rusts have as yet been cultured in the absence of living host cells. They are nutritionally very exacting and are all biotrophs.

Infection

These arrive on the surfaces of susceptible hosts in the form of wind-blown spores. These germinate and penetrate either via natural openings such as stomata or directly through the cuticle and cell wall. The germ tubes of the urediospores of many rusts extend until they encounter stomata, each forms an appressorium and enters via the pore. The germ tubes of the conidia of *Erysiphe* similarly produce an appressorium by the tip swelling to form an ovoid structure which increases the area of contact with an attachment to the host surface. From the underside of the appressorium a very narrow hyphal infection peg penetrates the cuticle and cell wall (Fig. 4.1.).

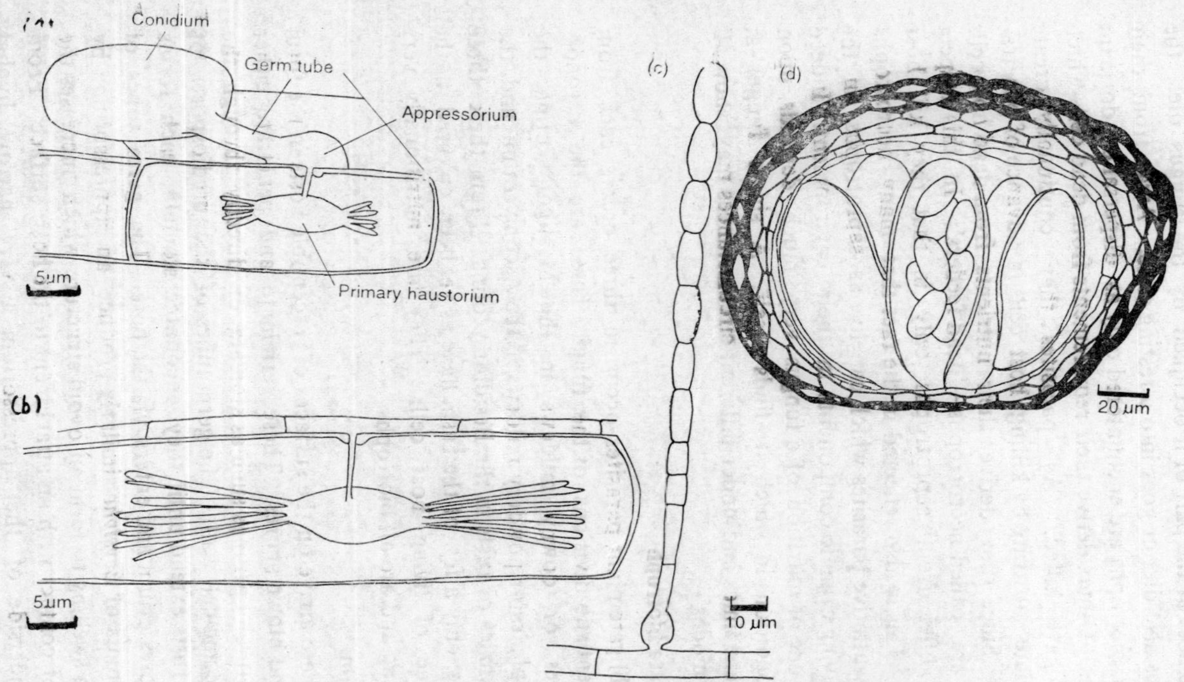

Fig 4.1. Life cycle of *Erysiphe graminis*, a powdery mildew of cereals and grasses. (*a*) Germinating conidium on a leaf, showing appressorium and primary haustorium developing (*b*) Section of an epidermal cell with a mature haustorium. (*c*) Chain of conidia developing from a swollen flask-shaped cell from the mycelium on the leaf (*d*) Vertical section of an ascocarp developing on older leaf,

Penetration in this case is partly enzymatic and partly mechanical. In *Erysiphe* differential staining techniques have revealed some dissolution of both the cuticle and cellulose of the wall immediately around infection pegs suggesting that hydrolytic enzymes are involved in penetration. In others, especially several facultative necrotrophs, such as species *Botrytis*, penetration may be purely mechanical. The infection peg of such fungi can penetrate thin metal films applied to leaf surfaces. Appressorial formation further illustrates the versatility of fungal hyphae in terms of penetrating and permeating tissues. The firmly attached appressorium prevents lift off from the surface allowing the infection peg to apply considerable mechanical pressure as it bores through the cuticle and wall.

Haustoria

After infection a swelling or vesicle usually arises from the tip of the infection peg and in most biotrophs hyphal branches arise from this and grow between the host cells (Fig. 4.2). They are intercellular and do not penetrate the host cells. The cells are penetrated by specialized hyphal branches of determinate growth called haustoria. The walls are penetrated by very narrow hyphal branches which balloon out to form haustoria of a variety of shapes (Fig. 4.3). These invaginate but do not penetrate the plasmalemma. The host-fungus interface is thus fungal cell membrane, fungal cell wall and host plasmalemma. Very often, in addition, there may be a distinct layer of amorphous material, the extrahaustorial matrix between fungal wall and host plasmalemma (Fig. 4.4). The base of the neck of the haustorium may also be surrounded by a collar-like sheath of host-derived callose. Haustoria are nucleated and rich in organelles such as mitochondria, ribosomes and vesicles, indicating that they are sites of high metabolic activity. The presumption is that they are the organs responsible for the absorption of major nutrients from the host cells. They most certainly provide a surface for close contact between the host and fungus and increase the surface area available for any such absorption. Nevertheless data to support this view are not easy to find. There is ample

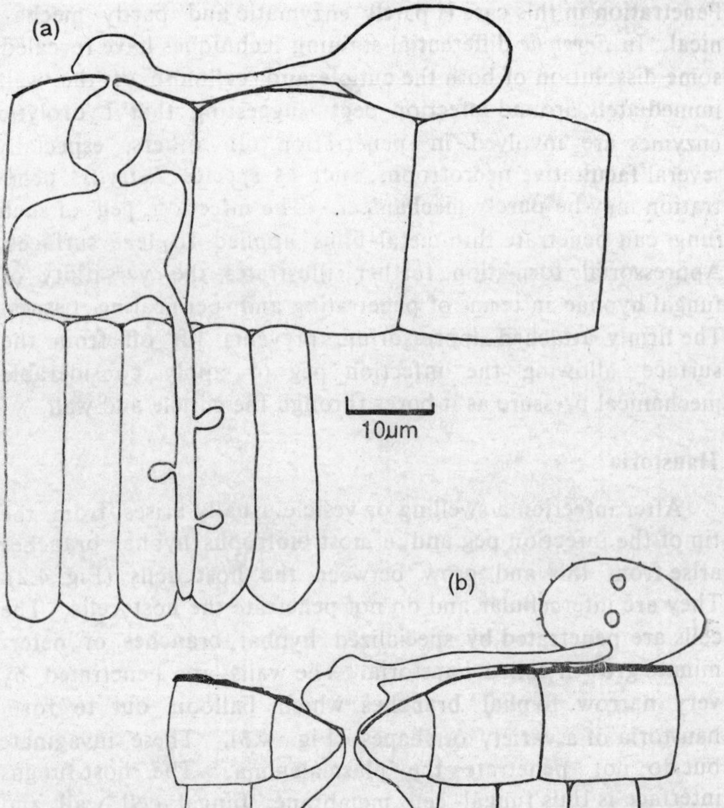

Fig. 4.2. Penetration structures in biotrophs (*a*) Bremia lactucae (*b*) Puccinia graminis.

evidence that substances do move from host to fungus but the very extensive intercellular hyphal system could be responsible for this. Some evidence for the nutritional functions of haustoria come from studies on wheat leaves infected

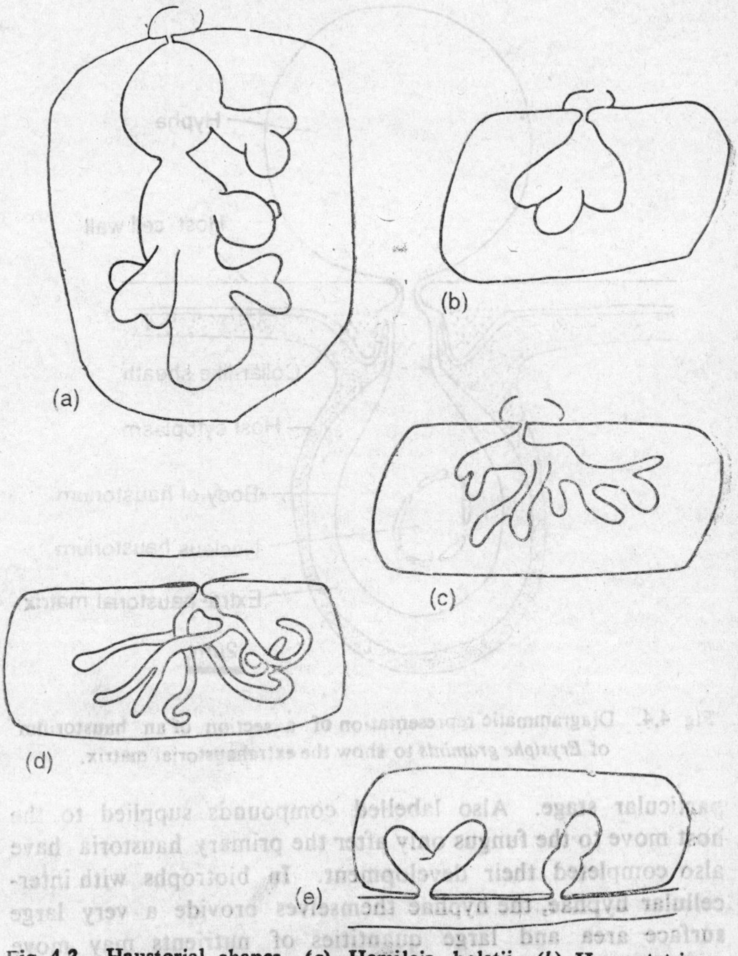

Fig. 4.3. **Haustorial shapes. (*a*) Hemileia holstii. (*b*) H. vastatrix. (*c*) Puccinia purpurea. (*d*) Peronospora calotheca. (*e*) Peronospora parasitica.**

with *Erysiphe graminis*. In the Erysiphales, unlike in the other two orders of obligate biotrophs, all the hyphae are outside on the surface of the host and only haustoria penetrate the cells and then only the epidermal cells. In infected wheat leaves, the surface hyphae start to grow only after the first haustorium arising from a germinating conidium has developed to a

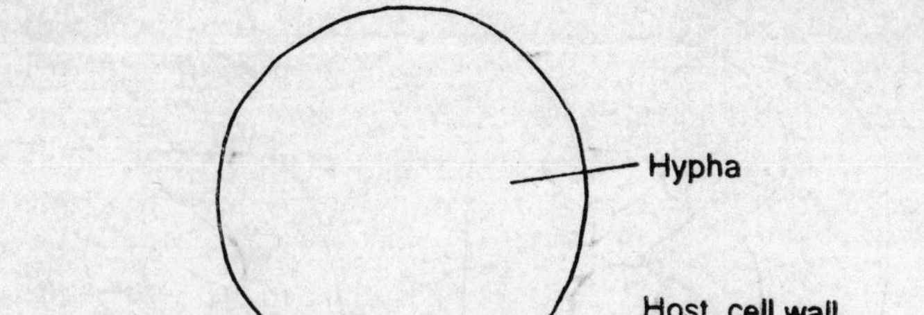

Fig 4.4. Diagrammatic representation of a section of an haustorium of *Erysiphe graminis* to show the extrahaustorial matrix.

particular stage. Also labelled compounds supplied to the host move to the fungus only after the primary haustoria have also completed their development. In biotrophs with intercellular hyphae, the hyphae themselves provide a very large surface area and large quantities of nutrients may move directly into these through the host cell walls as they do, for instance, in sheathing mycorrhizal associations.

Features of biotrophy

As the fungi are dependent upon living host tissues to complete their development, they usully cause minimal tissue damage, initially at least. There is little or no immediate cell death and they do not produce large quantities of extracellular lytic enzymes of toxins. They do produce these but their production is controlled. There are a number of possibilities. It might be that their production is switched on when they are required and switched off when they are not. For example, wall-degrading enzymes are produced only when wall penetration occurs. It could also be that they are only ever produced in the smallest of quantities. If they were also hyphal wall-bound they would not hydrolyse the walls beyond the sites of infection. Similarly, toxins might be produced in such small quantities that they increase membrane permeability to a degree without bringing about disruption.

The success of these fungi as parasites largely depends upon the extent to which they extract nutrients for their own growth before the host is sufficiently damaged to decrease the synthesis of these nutrients. Infection of part of the host creates a sink into which a variety of host-produced nutrients are rapidly removed. This induces the translocation of host metabolites into infected regions. It may well be that one function of haustoria is to introduce metabolites into the host which lead to such and other specific modifications of the host's metabolism in favour of the biotroph.

Perhaps the most marked feature of plants infected by obligate biotrophic fungi is a 2-4-fold increase in the respiration rate. By far the greater part of this is due to stimulation of the host's metabolism by the biotroph. The fungus itself directly contributes very little to this increase. In host tissues infected by rusts and powdery mildews, there is not only an increase in respiration rate but also a shift away from the major pathway of glycolysis and the tricarboxylic acid cycle to one involving the pentose phosphate pathway. This would not only provide more pentoses for the synthesis of nucleic acids, which in turn would affect protein synthesis, but also more NADPH which could be used in the synthesis of sugar alcohols and lipids which are all-important in the metabolism of the fungi.

Overall there is a tendency for a general increase in the biosynthetic processes of the host but the biosynthesis is of materials which the fungus requires. For instance there is increased synthesis of RNA. It has been suggested that once rust spores have germinated they may not be capable of further net protein synthesis and that for a successful infection to be established the biotroph must induce the host to produce proteins which it can utilize. In the early stages of infection of wheat leaves by rusts, the rate of RNA synthesis may be about three times that in healthy leaves suggesting enhanced protein synthesis.

Labelled amino acids are also incorporated about three times as rapidly into infected tissues as they are into healthy tissues and are rapidly built up into fungal proteins. Infection sites thus act as sinks for amino acids but the successful establishment and development of biotrophs within the host tissue also entails that there should be an adequate supply of carbohydrates from the host. Most biotrophs not only induce the diversion of host photosynthate towards infection sites but also ropidly convert it into substances which the host cannot utilize. This not only ensures a concentration gradient providing a continuous supply for the fungus but it also ensures that they are not re-utilized by the host. As in sheathing mycrorrhizal associations, host photosynthate, especially sucrose, is rapidly converted to trehalose, glocogen, arabitol or mannitol. Starch may also be preferentially deposited in host chloroplasts in and around infection sites. It disappears as the fungus beging to reproduce. Both its deposition and disappearance may be fungus-induced. This diversion and conversion of host nutrients by the fungus is the major cause of reduced growth and yield of cereals infected by powdery mildews and rusts. Infected leaves no longer become effective exporters of photosynthate. Healthy leaves of wheat export over 80% of the photosynthate produced whereas rust-infected leaves export less than 5% and in very severe infection may even import photosynthate. In cereals infected by *Erysiphe graminis*, an early attack on the first formed leaves, which usually export to the roots, restricts root development and as a consequence reduces the number of fertile stems per plant. A later attack on the upper leaves which nomally export to the inflorescences or ears causes poor filling and shrivelling of the grain. In some years, *E. graminis* reduces yield of barley in

Great Britain by 10-25%. An infection by biotrophs progresses the rate of photosynthesis in infected leaves usually declines; the decline is often accompanied by a decrease in chlorophyll content although this may occur later than the decline in photosynthesis. Chloroplasts in infected leaves are also usually smaller and often show signs of membrane damage. The loss of chlorophyll results in chlorosis and premature senescence.

In infections caused by powdery mildews, the chlorophyll content of the leaves initially falls, resulting in chlorosis, but the pigments are then resynthesized in areas adjacent to infections. In many rust infections. tissues adjacent to lesions remain green and healthy, producing 'green islands' in otherwise chlorotic leaves. The retention of chlorophyll in these islands is another indication of the ability of the biotrophic fungus to control the metabolism of the host. The photosynthate produced in these islands may help to maintain the fungus until after it has reproduced but the 'green islands' may also be associated with the diversion of photosynthate. Cytokinins spotted onto uninfected senescing leaves often produce 'green islands' and the host also translocates towards these. Increased levels of growth substances are characteristic of biotrophic infections. High levels of auxin (IAA) are found in many rust-infected tissues. In rust-infected wheat plants ten days after infection there is a 24-fold increase in IAA content. It is not known whether the fungus produces IAA in the infected host or whether it stimulates the host to produce it. Another alternative would be for the fungus to interfere with the process by which IAA is degraded in the host by, for instance, bringing about a decrease in the activity of IAA oxidase. The precise role of these growth substances in influencing the metabolism of the host to favour and maintain the biotrophs is not clear but a considerable volume of circumstantial evidence indicates that their role is an all important one, especially in mediating the increases in metabolism and the diversion of nutrients.

Infection by biotrophs while causing minimal tissue damage, often, but not invariably, leads to morphological disturbances in the host. These vary from uniformly elongated internodes to markedly curled and distorted stems, as in the white blister rust of crucifers, to markedly contorted leaves, as in peach leaf curl, slightly swollen to grossly and grotesquely enlarged roots, as in club root, and the stimulation of the growth of dormant buds to produce masses of spindly twigs, as in witches' brooms. Such growth responses would be expected from the known levels of growth substances in infected tissues and are a direct result of these levels but are probably purely incidental. Higher levels of growth substances also lead to morphological changes in sheathing mycorrhizal associations. The branching pattern of the host root system is changed on infection

In obligate biotrophic associations a relatively well-balanced relationship between the two partners is established initially since the fungi are physiologically highly specialized in being dependent upon living host tissues to complete their development. This is also reflected in the fact that most have limited host ranges. For example, the morphological species *Puccinia graminis*, the black stem rust of cereals and other grasses, can be divided into six pathogenic entities—formae speciales *tritici*, *secalis*, *avenae*, *agrostidis*, *poae* and *phlei-pratensis* according to their pathogenic adaptation to particular genera such as *Triticum* and *Secale*. They are named according to the host principally parasitized and crosses between them were usually infertile. Where man has developed genetically stable varieties or cultivars of the host as in wheat, it is possible to subdivide each forma specialis. For example, not all the strains within *P. graminis* f. sp. *tritici* are capable of infecting equally all the known cultivars of *Triticum*. By using a number of different cultivars of wheat with different degrees of resistance to the fungus as a so-called differential host series and assessing quantitatively the degree of infection by particular strains, physiologic races can be recognized. Well over 200 such races have been recognized in *P. graminis* f. sp. *tritici*. These are artificial entities in a sense that their recognition is quite arbitrary since the number depends upon the number of differential hosts used and the classes of pathenogenicity reactions recognized.

Necrotrophs

In marked contrast to these obligate biotrophs, most facultative necrotrophs, in addition to living parasitically, do have a free-living phase and can be grown on simple culture media. Many, the less specialized parasites, are not only able to grow on the host after they have killed it but they can also grow successfully and indefinitely as competitive saprotrophs on other dead substrates. This is not true of all necrotrophs as with increasing specialization towards parasitism there is a tendency towards a progressive loss of competitive saprotrophicability.

Secretion of Extracellular Enzymes

Necrotrophs usually kill host cells rapidly and before they penetrate them by the secretion of extracellular enzymes or both. It is these which are largely responsible for the disease symptoms seen. Their effects vary from localized necrosis to massive tissue destruction and disintegration. Many cause diseases in which the main symptom is a rapidly spreading soft, watery rot of parenchymatous tissues. This is well seen in such diseases as the soft rot of fruit, such as apples, caused by the conidial *Penicillium expansum*, or marrows, caused by the Zygomycete *Rhizopus stolonifer* and seedling diseases such as pre-emergence killing and 'damping off', caused by the Oomycete *Pythium debaryanum*. On infection the hyphae again penetrate between the cells. No haustoria are formed but the hyphal tips secrete copious amounts of a wide variety of pectolytic enzymes which attack the anhydrogalacturonic acid units making up the pectic polymers. This brings about a maceration of the parenchymatous tissues as the cells separate along the middle lamella because of the breakdown of the pectic substances which cement them together. The tissues becomc very soft, watery and lack coherence.

In pectic substances a proportion of the anhydrogalacturonic units are esterified with methanol (Fig. 4.5). Apart from pectin methyl esterase which acts upon these ester bonds and thus increases the solubility of the polymer by producing

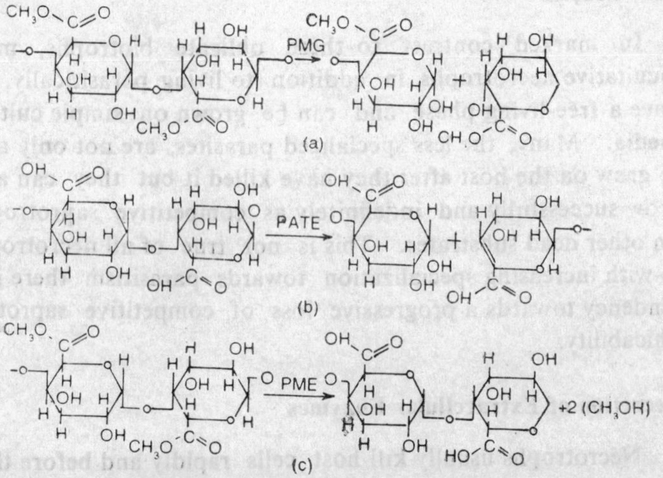

Fig. 4.5. Pectic substances and pectolytic enzymes. (*a*) Action of polymethyl galacturonase on pectin (a polymer of methylated galacturonic acid) (*b*) Action of pectic acid *trans*-eliminase on pectic acid (a polymer of galacturonic acid) (*c*) Action of pectin methylesterase on pectin (After Goodman Kiraly and Zaitlin 1967).

more free carboxyl groups, most other pectolytic enzymes are chain splitting ones which act upon the α-1,4 glycosidic linkages between adjacent residues. Some, such as polygalacturonases, rupture the bonds by hydrolysis. Others, such as pectin *trans*-eliminases, act by transelimination of a proton from C5 of one residue to give an unsaturated bond between C4 and C5 of the reaction product (Fig. 4.5). In both types of chain splitting enzymes, there are those which attack chains terminally (exo-) and those which attack chains at random (endo-), for example exo-polygalacturonase and endo-polygalacturonase. A further division can be made into those which preferentially attack esterified parts of the chain and those which preferentially attack non-esterified parts. Only pectolytic enzymes play a key role in the early and critical stages of soft-rotting. Enzymes which act upon other components of the cell wall such as cellulose and hemicelluloses are not produced initially. The pectolytic enzymes have another very important effect in that they bring about a marked increase in the permeability of the protoplasts. The cells leak so much that they lose turgor and they often die before they are separated. The enzymes diffuse out from the hyphal tips so that the cells are killed in advance of the hyphae. The fungi utilize both the residues from the hydrolysis of the pectic substances and the solutes which pour out of the affected cells.

Production of Toxins

In other diseases caused by necrotrophs, the cells are killed but the tissues, although discoloured, remain dry rather than watery and there is no cell separation. Moreover the lesions, unlike rots, are slow in spreading and discrete. In these cases the fungi produce low molecular weight and heat-stable toxins which kill protoplasts at very low concentrations. These can be demonstrated in some leaf spot diseases such as the early blight of tomatoes and potatoes caused by the conidial *Alternaria solani*. This fungus causes small black necrotic lesions on the leaves between the veins. Infected leaves become chlorotic and senesce prematurely. When the lower leaves are heavily infected, the upper leaves soon appear chlorotic. The toxin moves about the plant. A crystalline dibasic acid with the formula $C_{21}H_{30}O_8$ has been obtained from culture filtrates of *A. solani* and has been called alternaric acid (Fig. 4.6). When healthy cuttings from potato shoots are placed with their stems in solutions containing as little as

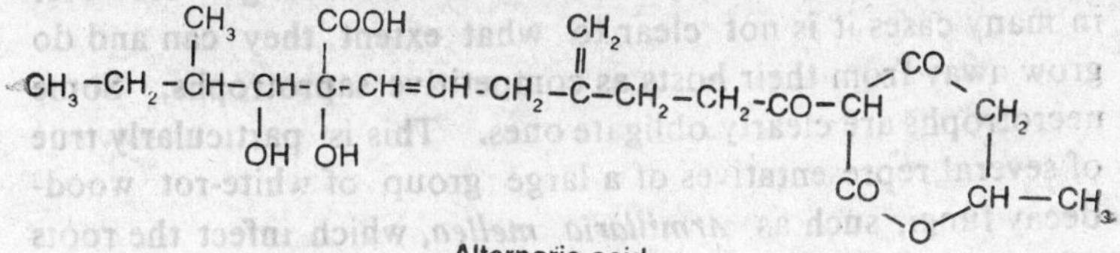

Alternaric acid

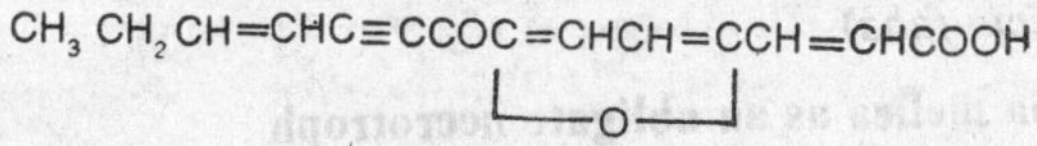

Wyerone acid

Fig. 4.6. **The phytotoxin, alternaric acid, and the phytoalexin, wyerone acid.**

20 p.p.m of the acid, typical lesions occur in the leaves after 2—3 days. It is characteristic of toxins that they are active in very low concentrations and that they are mobile within the plant. Alternaric acid is non-specific in that it will affect a variety of plants that are not parasitized by the fungus. Other toxins are host specific showing the same specificity as the parasite itself. The conidial *Helminathosporium victoriae* causes a devastating seedling blight of oats in the USA. It is particularly virulent on one oat cultivar called Victoria where it causes a wilt, necrosis at the base of the stem and reddish necrotic lesions on the leaves. It produces a toxin, a peptide linked to a tricyclic amine, named victorin. Resistant oat cultivars are not affected by it but susceptible ones, such as Victoria, are very sensitive to it. Not only does the toxin show the same specificity as the fungus itself but only toxin producing strains of the fungus cause disease symptoms. Resistance to a necrotrophic parasite is very often based upon insensitivity to a specific toxin or the ability to degrade it as soon as it is produced.

In toxin-producing necrotrophs the host tissues are again first killed and tnen exploited by the hyphae. Many toxins are also known to affect membrane permeability as do pectolytic enzymes. The rapid killing of host cells by the production of enzymes or toxins or both is a major characteristic of necrotrophs. There is a complete lack of any form of balanced relationship between host and parasite and further, unlike biotrophs, most necrotrophs have a wide host range. However in many cases it is not clear to what extent they can and do grow away from their hosts as competitive saprotrophs. Some necrotrophs are clearly obligate ones. This is particularly true of several representatives of a large group of white-rot wood-decay fungi, such as *Armillaria mellea*, which infect the roots of trees and shrubs and fungi, such as *Botrytis fabae*, which causes the chocolate spot disease of leaves and pods of broad beans (*Vicia faba*).

Armillaria mellea as an obligate necrotroph

A. mellea has been recorded as a parasite of a very wide range of dicotyledonous and coniferous trees, It is probably the most widespread and damaging root parasite of trees. It also attacks shrubs and occasionally monocotyledons, such as sugarcane, and herbaceous plants, such as potatoes, where it was formerly present on tree roots. It may rapidly kill its host by invading the cambium of the root system and spreading back to the base of the stem. The mycelium forms thick, creamy-white fan-like sheets under the bark and girdles the cambium. The tree then dies suddenly. After killing the tree, it colonizes the wood causing a white-rot with an intricate pattern of black lines, zone lines, running through it. Eventually the tree is wind-blown or falls and the fungus colonizes the whole trunk. Alternatively it may cause a butt rot without attacking the living tissues of the tree. A central conical zone of decay extends 1–2 metres upwards from the base of the stem. However the result is the same. The tree is wind-blown and the fungus colonizes the whole trunk. It persists for several years, even 10 or more, as an active saprotroph degrading the wood of the trunk. It has no free-living saprotrophic phase but is restricted to the dead tissues which it has killed. As these are usually very durable substrates it remains in the vegetative state. Eventually it produces basidiocarps and its air-borne basidiospores may infect recent wounds on other hosts at a distance. Alternatively healthy roots may come in contact with already colonized roots in the soil but once an infection site has been established the major means of local spread is by the initiation and growth of rhizomorphs from the mycelial fans. These grow out through the soil away from the dead host to infect another, perhaps metres away, obtaining the necessary nutrients directly from the dead host rather than from dead organic matter in the soil. In this case the massive tissue destruction characteristic of necrotrophy is not obvious immediately but occurs over a long period of time and is brought about by the production of cellulolytic and ligninolytic enzymes rather than pectolytic ones.

Botrytis fabae as an obligate necrotroph

Other obligate necrotrophs, such as *Botrytis fabae*, initially cause very localized tissue damage such as spots on leaves but the physiological state of the host and environmental conditions markedly influence the extent to which the parasite spreads further. When a wind-dispersed conidium of *B. fabae* lands on a broad bean leaf, it germinates, if moisture is present, by forming an appressorium from which a thin peg-like penetration hypha bores through the cuticle and cell wall. Once inside it swells and forms a short hypha often less than 100 μm long (Fig. 4.7). Within 24 h a discrete limited lesion-the chocolate spot—develops. This is without any further growth of the fungus. Some 20-50 host cells are killed around the fungus and, although there is little evidence of tissue breakdown, there is intense tissue browning. This is called the 'non-aggressive' phase of the disease. It appears that once inside the host the fungus begins to produce pectolytic enzymes. The fungus would rely upon these to provide nutrients for further growth from the hydrolysis of pectic substances and leakages of solutes from the cells. A host phenolase is activated by the action of the pectolytic enzymes and it oxidizes phenols which also leak from dying cells. The oxidized products are responsible for the brown coloration and they also inhibit the fungal pectolytic enzymes so that growth of the fungus is prevented. Other substances toxic to the fungus may also be produced. Many plants have been shown to produce a specific chemical on infection which prevents further growth of the fungus. These are called phytoalexins—'plant warding off' compounds. When *B. fabae* infects broad beans a phytoalexin called wyerone acid (Fig. 4.6) is produced and it appears that in the 'non-aggressive' phase sufficient is produced to prevent further growth.

The 'aggressive' phase only develops either on older senescing leaves or under any environmental conditions which are unfavourable to the host such as poor phosphate supply, extensive waterlogging of the soil or, in particular, very high humidities which greatly favour the fungus. The fungus then behaves as a typical nectotroph. The main symptom is a rapid and progressive rot of the leaves followed by that of the parenchymatous parts of the stems. In this phase oxidation of the phenolics cannot keep pace with the production of

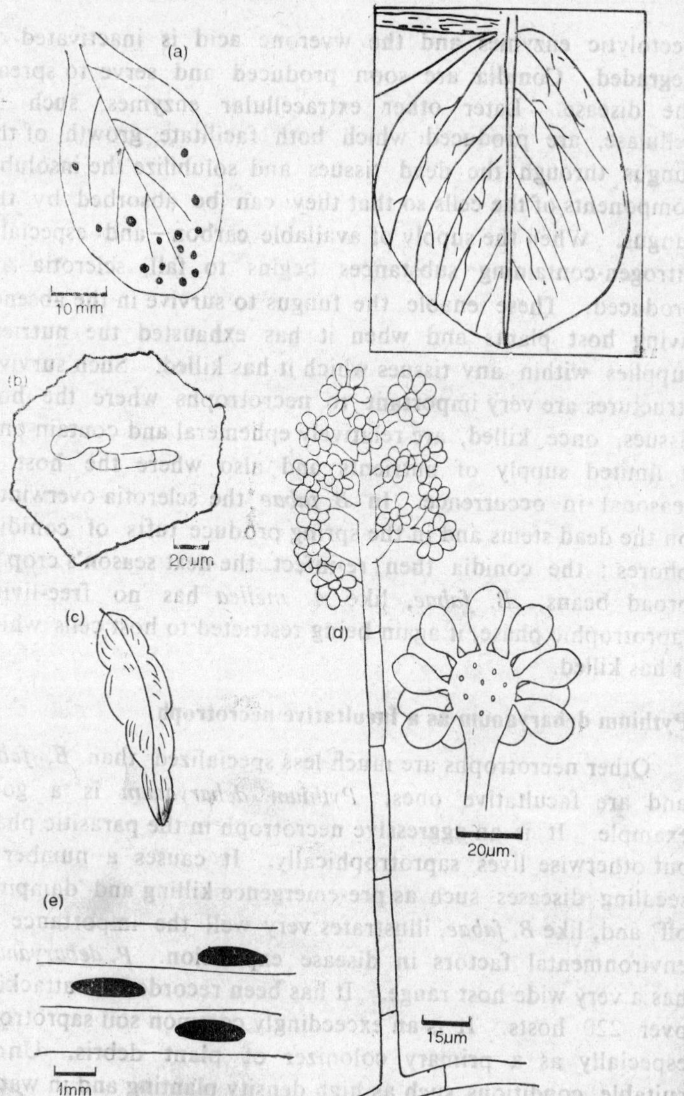

Fig. 4.7. (*Left*) cycle of *Botrytis fabae*, the cause of chocolate spot of broad beans. (*a*) Chocolate spots on a leaflet of broad bean, the non-aggressive phase. (*b*) Conidium penetration an epidermal cell. (*c*) Shrivelled and blackened leaflet, the aggressive phase. (*d*) Conidiophere and blastocondia. (*e*) Sclerotia on an old stem of broad bean.

pectolytic enzymes and the wyerone acid is inactivated or degraded. Conidia are soon produced and serve to spread the disease. Later other extracellular enzymes, such as cellulase, are produced which both facilitate growth of the fungus through the dead tissues and solubilize the insoluble components of the cells so that they can be absorbed by the fungus. When the supply of available carbon—and especially nitrogen-containing substances begins to fall, sclerotia are produced. These enable the fungus to survive in the absence living host plants and when it has exhausted the nutrient supplies within any tissues which it has killed. Such survival structures are very important to necrotrophs where the host tissues, once killed, are relatively ephemeral and contain only a limited supply of nutrients and also where the host is seasonal in occurrence. In *B. fabae* the sclerotia overwinter on the dead stems and in the spring produce tufts of conidiophores ; the conidia then re-infect the next season's crop of broad beans. *B. fabae*, like *A. mellea* has no free-living saprotrophic phase, it again being restricted to host cells which it has killed.

Pythium debaryanum as a facultative necrotroph

Other necrotrophs are much less specialized than *B. fabae* and are facultative ones. *Pythium debaryanum* is a good example. It is an aggresslve necrotroph in the parasitic phase but otherwise lives saprotrophically. It causes a number of seedling diseases such as pre-emergence killing and 'damping-off' and, like *B. fabae*, illustrates very well the importance of environmental factors in disease expression. *P. debaryanum* has a very wide host range. It has been recorded as attacking over 220 hosts. It is an exceedingly common soil saprotroph especially as a primary colonizer of plant debris. Under suitable conditions such as high density planting and in water-logged or poorly drained soils, it becomes an aggressive parasite of seedlings. In pre-emergence killing, the seedlings fail to emerge and bare patches arise in seed beds. The seeds germinate but the radicles are killed. 'Damping off' is a post-emergence condition, the term being used as it is usually associated with over-watering. Seedling emerge and after a few days there hypocotyls or plumules lose turgor, become water-soaked and collapse. One or more hyphae penetrate at or around soil level and grow between the parenchymatous cortical cells producing copious amounts of pectolytic enzymes which again separate the cells and cause them to leak and die.

P. debaryanum, as an Oomycete, produces both zoospores and oospores but also intercalary chlamydospores (Fig. 4.8).

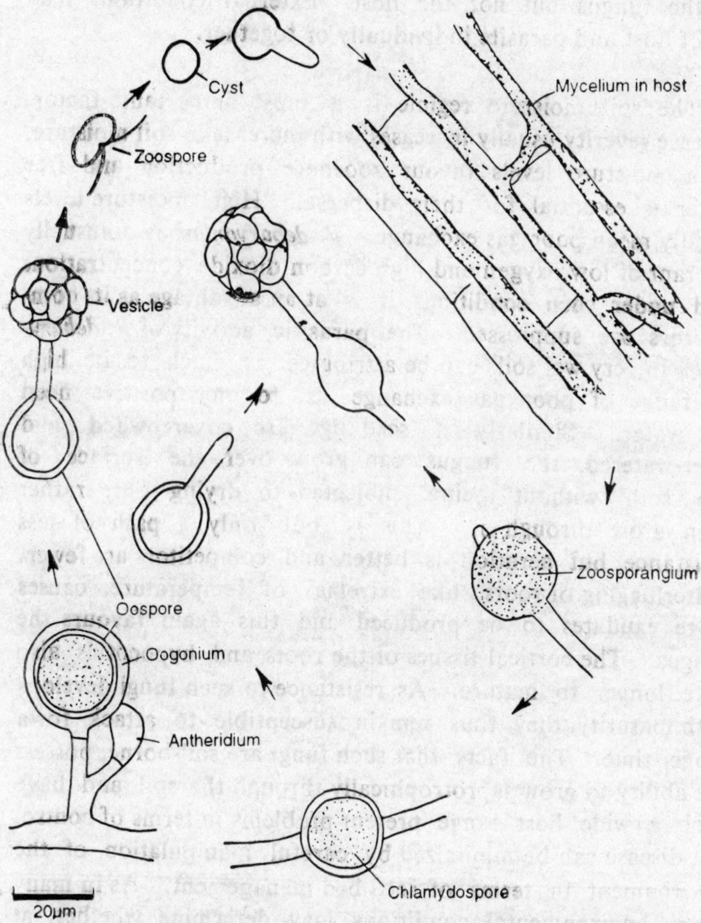

Fig. 4.8. Life cycle of *Pythium debaryanum*, the cause of seedling diseases such as pre-emergence killing and 'damping-off'.

Unlike *A. mellea*, its mycelium actually grows through the soil. Exudates, especially sugars and amino acids, from roots and hypocotyls are important in infection. They affect the fungus in a number of ways. They may stimulate resting oospores to germinate or they may induce directional growth of hyphae or directional movement of zoospores or their uptake and utilization may increase the capacity to infect. 'Damping-off' is most severe when environmental factors favour the growth of the fungus but not the host. External conditions may affect host and parasite individually or together.

The soil moisture regime is a most important factor. Disease severity usually increases with increase in soil moisture. High moisture levels favour zoospore production and free water is essential for their dispersal. High moisture levels usually mean poor gas exchange. *P. debaryanum* is unusually tolerant of low oxygen and high carbon dioxide concentrations and under such conditions it is at an advantage as its competitors are suppressed. The parasitic activity of *P. debaryanum* in very wet soils can be attributed as much to its high tolerance of poor gas exchange as to any positive need for water. Similarly if seedlings are covercowded and over-watered, the fungus can grow over the surface of the soil, without being subjected to drying out, rather than grow through it. This is not only a path of less resistance but aeration is better and competitors are fewer. Waterlogging of roots, like extremes of temperature, causes more exudates to be produced and this again favours the fungus. The cortical tissues of the roots and hypocotyls also take longer to mature. As resistance to such fungi develops with maturity, they thus remain susceptible to attack for a longer time. The facts that such fungi are soil-borne, possess the ability to grow saprotrophically through the soil and have such a wide host range present problems in terms of control but disease can be minimized by careful manipulation of the environment in terms of seed bed management. As in many cases, environmental conditions may determine whether, at any one particular moment in time, a fungus is pathogenic or not.

Nutritional relationships of Venturia inaequalis

Biotrophy and necrotrophy have been considered as two quite distinct modes of nutrition. It is better to regard them as two extremes with a whole spectrum of nutritional relationships between. A number of parasitic fungi exhibit a biotrophic phase followed by a necrotrophic one as infection of the host develops *Venturia inaequalis*, which causes apple scab, is an example. Apple scab is one of the most serious diseases which apple growers have to contend with. In the spring, young leaves are infected by ascospores released from ascocarps which have developed over the winter in dead fallen leaves. The germ tube arising from a germinating ascospore produces an appressorium from which a hypha penetrates the cuticle. Hyphal branches from this then spread between the cuticle and the outer walls of the epidermal cells (Fig. 4.9). They fuse laterally to form flat, wide strands which develop in the form of a plate under the cuticle. There may be no obvious signs of tissue damage for up to 2-3 weeks. Towards the end of this time the cuticle is raised and ruptured and conidia begin to be produced from short-conidiophores arising from the sub-cuticular mycelium. These are rain-splashed to adjacent leaves and serve to spread the disease. There is no cellular penetration but up to 40% of the cuticle may be degraded in heavily infected leaves. Haustoria are not formed. Although the fungus is outside the host cells, it induces the host to translocate metabolites to infection sites. These are the major source of nutrients for the fungus. The export of photosynthate from infected leaves is also reduced. Thus the sub-cuticular phase is biotrophic. Eventually the epidermal cells under a lesion become more and more vacuolated and discoloured by the deposition of brown pigments. Eventually they collapse and die. When they are dead, conidial production ceases. At this stage the fungus is necrotrophic but its growth is still restricted.

On senescence prior to leaf-fall, the fungus grows out into the leaf mesophyll and, once the leaves have died and fallen, the mycelium may permeate the whole leaf. This saprotrophic spread however is limited to the one infected leaf. It does not spread to others. The extent of its spread through the leaf is determined by the degree of competition which it encounters from the common primary-saprotrophs, which would also be

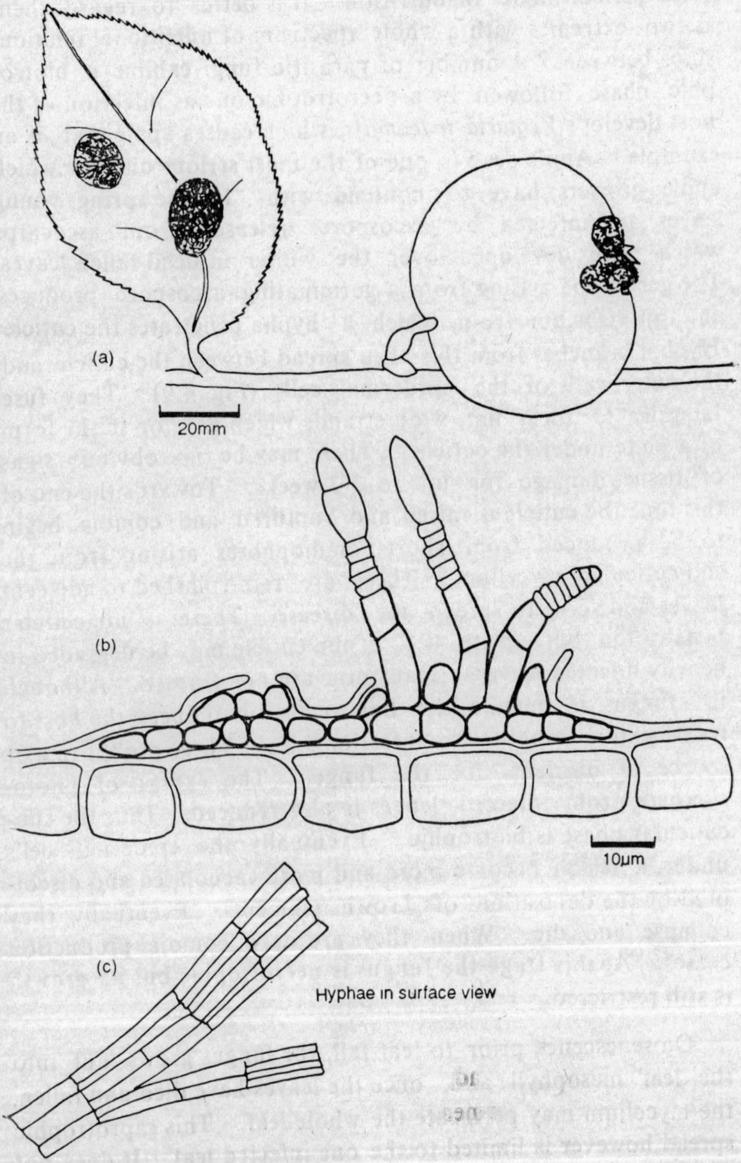

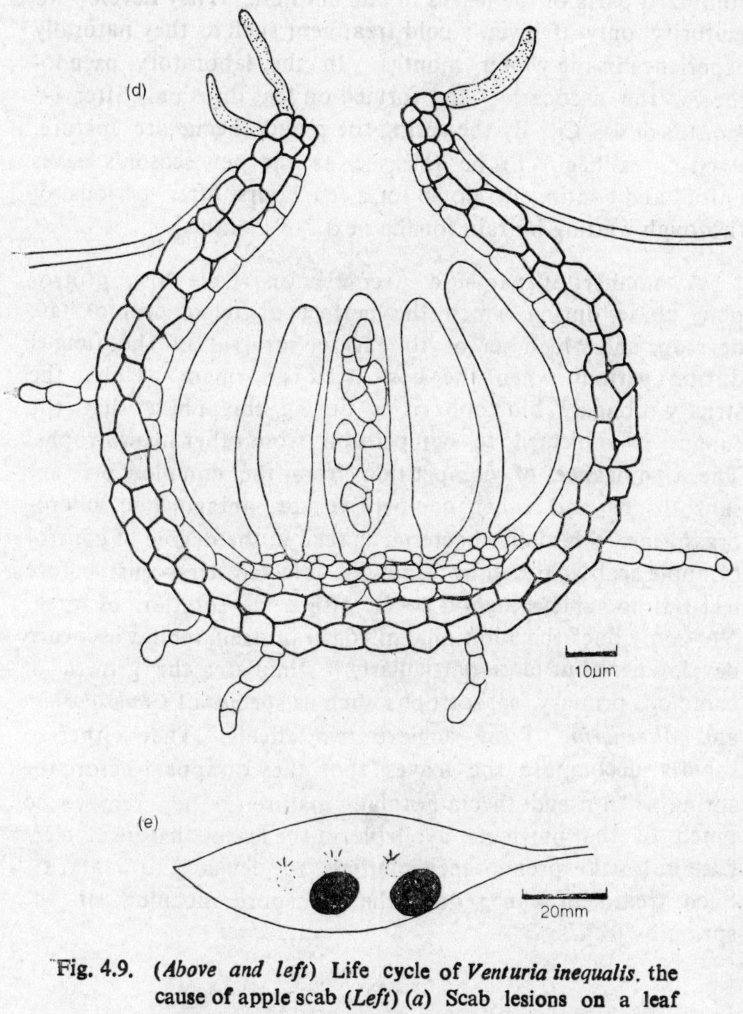

Fig. 4.9. (*Above and left*) Life cycle of *Venturia inequalis*. the cause of apple scab (*Left*) (*a*) Scab lesions on a leaf and on an apple. (*b*) Conidiophores arising from hyphae growing between the cuticle and the epidermal cells. (*c*) Surface view of part of the strands of laterally fused hyphae beneath the cuticle. (*Above*) (*d*) Vertical section through a pseudothecium. (*e*) Ascocarps on an overwintering leaf.

colonizing the leaves at that time, and leaf-inhabiting bacteria. Ascocarp initials develop in the scarb lesions and in other colonized parts of the leaves in late autumn. They develop to maturity only if given a cold treatment such as they naturally experience in the winter months. In the laboratory pseudothecia, the ascocarps, are formed on leaf discs only after 4-5 months at 4-8°C. By the spring the pseudothecia are mature. Ascospores begin to be released as the new season's leaves unfold and continue to do so for a few hours after periods of thorough wetting by rain for the next 2-3 months.

A number of parasitic Ascomycotina have this saprotrophic phase during which the perfect or teleomorphic state develops and which serves to ensure survival of the fungus during periods when the host itself is dormant. Unlike the strictly obligate biotroph, it is during this phase that the fungus is subjected to competition from other heterotrophs. The significance of competition from the common primary saprotrophs and other competitive or antagonistic micro-organisms, especially bacteria, is seen in the degree of control of apple scab achieved by spraying infected trees just before leaf-fall, to achieve a good cover, with a 5% solution of urea. Not only does the additional nitrogen in itself inhibit ascocarp development but more particularly it stimulates the growth of common primary saprotrophs such as species of *Cladosporium* and *Alternaria*. These achieve two effects. They either so rapidly decompose the leaves that they disappear before the spring when pseudothecia normally mature or they remove so much of the nutrients available in the leaves that insufficient remain for the pseudothecial initials to develop to maturity. Such treatment can reduce the ascospore inoculum in the spring by 97%.

5

Fungial Infection in Group Insects

FUNGOUS INFECTIONS

Mycoses

As far as the diseases of insects are concerned, it seems only natural that the first type of infection definitely identified as to its microbial nature was that caused by fungi. This relationship was suspected about the beginning of the nineteenth century. The conspicuous presence of a mycelium and fruiting bodies on the cadavers of dead silk worms enabled Bassi de Lodi, in 1835, to recognize the similarity between the infecting fungus and the well-known bread molds and other saprophytic fungi. As further observations were made it was discovered that some of the fungi infecting insects were obligate parasites, others were semiparasites, and still others were common saprophytic species which, under certain conditions, were able to cause a frank infection in susceptible insects. A confusing factor was the frequent development of a saprophytic fungus in the body of an insect killed by an agent other than the fungus. Such secondary invaders were often mistaken for the primary cause of the condition, and as a result the invasive properties of some of the fungi described in early reports are in doubt and these organisms probably are not pathogenic agents in the true sense of the term.

In an earlier chapter we referred to certain fungus-insect relationships that could not properly be considered in the category of disease (*e.g.*, the *Septobasidium* fungi and the

Laboulbeniaceae), and others that are of a definitely mutualistic association (*e.g.*, the ambrosia fungi of termites and wood-boring beetles). These relationships are therefore outside the limits of the usual concept as to what constitutes a fungous disease.

Nature of the Fungi Concerned. Fungi constitute a part of that subdivision of the plant kingdom known as Thallophytes, those organisms which grow in irregular plant masses (thalli) not differentiated into roots, stems, and leaves as in higher plants. The term "fungi" is a general one and may be meant to include several types of organisms that are sometimes conveniently differentiated by the terms "molds," "yeasts," "actinomycetes," "bacteria," etc. Commonly, however, the term "fungi" refers to that group of nonchlorophyll-containing organisms which usually possess the filamentous vegetative structure known as "mycelium." Or, stating it another way, the word "fungi" usually refers specifically to the molds and yeasts, which comprise the Eumycetes.

Representatives of the fungi known to cause infection and disease in insects ("*fungi Entomogeni*") are included in each of the four classes : Phycomycetes, Ascomycetes, Basidiomycetes, and Deuteromycetes (Fungi Imperfecti). Of these, the class Basidiomycetes contains the fewest species characteristically pathogenic for insects. Because, under certain conditions, some groups of yeasts produce asci and ascospores, these micro-organisms are usually included with the Ascomycetes.

For the sake of convenience we shall treat the mycoses of insects more or less according to the class to which the causative fungus belongs. There is some overlapping between the fungi belonging to the class Ascomycetes and those placed in the class Deuteromycetes. The latter group are sametimes called "Fungi Imperfecti" because they are fungi with incomplete or incompletely known life cycles. In most cases they are believed to be imperfect (asexual) stages of Ascomycetes. The details of these relationships will be elaborated upon in a later section.

In passing, mention should be made of the fact that certain artifacts, foreign bodies, or pseudofungi occasionally occur on insects and are sometimes confusing to one examining the specimens for fungi. Certain seeds and pollinia, for example, frequently are found attached to the integuments of insects in a manner closely resembling certain fungi such as Laboulbeniaceae and *Cordyceps*. Usually these foreign nonparasitic bodies may be detached mechanically from the insect rather easily, although sometimes they cling as tenaciously as might be expected of a fungus. Pollinia may be attached by adhesive discs, which will ordinarily become detached from the insect if the specimen is moistened with a drop of alcohol. Some orchid pollinia, with their stalks of varying lengths and pollen masses at the apex, are similar in appearance to some of the stalked fungi but become detached readily upon the application of alcohol. The pollen of some plants covers insects (such as wasps) visiting their flowers with a white powder that is sometimes mistaken for a fungus, but this can usually be detected upon microscopic examination.

PHYCOMYCETE INFECTIONS

Phycomycetes ("algal fungi") constitute a large and diverse group living in various habitats; some are aquatic, some amphibious, some terrestrial. They are characterized by coenocytic (nonseptate and multinucleate) mycelium and endogenous asexual spores. Septations commonly are formed where the reproductive structures are delimited; they may be formed in old hyphae of certain genera and in young thalli of certain other genera. From the standpoint of insect infections, at least four orders of the class Phycomycetes contain entomogenous members: Entomophthorales, Mucorales, Blastocladiales, and Chytridiales. As the name would imply, the most important of these is the order Entomophthorales, which is generally considered to consist of a single family, Entomophthoraceae (Empusaceae of some authors), which in turn may be divided into five and possibly six genera. Of these, the genera *Empusa*, *Entomophthora*, and *Massospora* are composed primarily of entomogenous species.

Infections caused by species of empusa and entomophthora

The literature records a relatively large number of infections of insects caused by species of *Empusa* and *Entomophthora*. Approximately 50 species of these fungi are known to occur in the United States. Some of these have been studied in considerable detail whereas others have received only texonomic descriptions. In the main, their life histories are similar, and considerable general information may be gained by a close consideration of a few of the better known species.

Thh proper use of the generic names *Empusa* and *Entomophthora* is frequently a difficult procedure, and unless the student has an understanding of the historical aspect of these names he is likely to become somewhat confused. No final official ruling h is been made on this particular nomenclatorial problem, and for the present one can only follow the usage generally employed by the leading authorities in the field. The validity of the name *Empusa*, erected by Cohn in 1855 for a fungous parasite (*Empusa muscae* Cohn) of the housefly, has been challenged by some writers because of its having been preoccupied for a genus of orchids. It was for this reason that Fresenius (1856) proposed the name *Entomophthora* to take the place of *Empusa* for the fungus. This was followed by the indiscriminate use of both names until Brefeld (1877) and Nowakowski (1884) separated them as two distinct genera, thereby recognizing the validity of the name *Empusa*. Other mycologists subsequently employed it without question. In his treatise on the Entomophthoreae of the United States, Thaxter (1888) concluded that the name had sufficient weight of authority to make it acceptable. He further pointed out that the orchidaceous genus *Empusa* is now placed as a synonym and sence seems unlikely to cause confusion.

The generic name *Entomophthora* was considered by Thaxter to be of subgeneric value in the genus *Empusa*. Since then, for reasons to be pointed out later, there has been a tendency to revert to the proposals of Brefeld and Nowakowski and to give *Entomophthora* full generic rank along with *Empusa*. The genus *Lamia*, erected by Nowakowski and considered by him to be intermediate between the two, has not found general acceptance. Unfortunately, a clear and final distinction between some of the species of *Empusa* and those of *Entomophthora* has not been made.

Nature of Empusa and Entomophthora Infections. Among the various species of *Empusa* and *Entomophthora* there is some variation as to the details of the manner in which they cause infection in their respective hosts, but in general this process is much the same throughout the two genera.

Infection of the insect by the fungi *via* the alimentary tract probably occurs only rarely, if at all. The usual route of infection is by the penetration of the integument, especially the thinner intersegmental areas of the body wall and the appendages. Soon after the fungous spore comes in contact with the integument it begins to germinate, sending out a conidial hypha that penetrates into the body cavity of the host where it develops rapidly at the expense of the softer tissues. Instead of producing a profusely branched mycelium, the hyphae usually become segmented and break down into their component cells, which are called "hyphal bodies." These short thick fragments of varied size and shape, containing a highly concentrated fatty protoplasm, multiply by budding or by fission until the insect's body cavity is filled with them. In some cases this fragmentation of the hyphae does not occur until later in the infection, after the mycelium is fairly well developed. The hyphal bodies may be very irregular in size and shape, or they may possess great regularity in this regard. The latter is frequently the case with hyphal bodies produced about the time of the host's death and just preceding spore formation.

After the hyphal bodies have been produced, if conditions of temperature and moisture are not favorable, instead of completing its development, the fungus may form chlamydospores. These spores are formed by each hyphal body which develops about itself a single wall of variable thickness depending upon the duration of this resting stage. By means of the chlamydospores the vitality of the fungus may be maintained through long periods of dormancy until the proper conditions for further growth are presented. The period from first infection to the formation of hyphal bodies or chlamydospores varies according to the host from 2 to 12 days. In general, the smaller the insect the shorter the period.

In the presence of sufficient moisture and adequately high temperature, the hyphal bodies and chlamydospores "germinate" with great rapidity. The hyphae thus produced may grow directly to the outer air and then produce a single conidium or set of conidia, according to the type of conidiophores. Sometimes, particularly when conditions of growth have been very favorable, a single primary hypha may branch indefinitely, each ultimate branch becoming a conidiophore. As a rule, the number of germinating hyphae that develop from a single hyphal body does not exceed one or two, although instances occur in which the number is considerable. In the latter case the hyphae may branch and anastomose, forming a coherent mass known as a "stroma".

As a result of the germination of the hyphal bodies either sexual or asexual resting spores (zygospores or azygospores), or conidiophores bearing conidia are produced. In the case of the latter, as explained by Thaxter, the hyphae, arising directly or indirectly from the hyphal bodies, grow rapidly outward and burst through the less resistant portions of the host's integument in spongy masses. In most instances the colour of these or bright green or dull olive. In some cases the masses project barely beyond the body wall of the host and are confined to the points of emergence. These points are usually the thin intersegmental membranes through which the fungus projects in cushionlike rings, usually formed by simple conidiophores producing few or no branches outside the host's body. Each of these conidiophores gives rise to a single conidium. In other cases the external growth may be more extended, and the masses may coalesce so as to cover the entire body with a continuous layer of conidiophores which may form a mass several times as large as the insect itself. The external parts of these conidiophores are considerably branched, and the ultimate divisions of each conidiophore are arranged in a corymbose or digitate fashion. It is this occurrence of simple and compound conidiophores in different species which has led to the establishment of the two genera *Empusa* and *Entomophthora*. The difficulty in maintaining this generic separation is that the distinction is not absolute, and intermediate forms occur (*e.g.*, in *E. culicis* Bra. and *E. apiculatus* Thax.). Compound conidiophores are sometimes found in species usually having the simple type, and simple conidiophores are commonly found in species having the compound type. In either case the growth of the conidiophores under optimum conditions of temperature and moisture takes place very rapidly and may give rise to the characteristic white masses in a few hours. Soon after the masses of conidiophores appear, the production of conidia begins.

Thaxter explains the formation, discharge, and germination of the conidia as follows: The terminal portion of the conidiophore is termed the "basidium" and is usually swollen to a greater or less extent. From the apex of this basidium, the conidium commences its formation by the process of budding. The bud, or mother cell, increases until it reaches the normal size and shape of the conidium ; then it becomes separated from the basidium by a cross partition. Within the mother cell thus formed is developed a single conidial spore. When the conidium is fully developed, the contents of the spore, as well as that of the basidium, begin to expand through the absorption of water. At first the contents of the basidium exert the greater of the two forces thus produced, and the columella is forced outward into the conidium toward which its convexity is thus turned. When the basidia are large and strong, this process may continue until the discharge of the conidium. Usually the contents of the conidium, being more dense than that of the basidium, finally exert a greater pressure and force the columella back into the basidium. As a result of the rupture, the conidium is discharged violently into the air, often to a considerable distance. The columella commonly remains unbroken by this discharge, although it may be greatly stretched and hang down from the basidium as a tonguelike projection. In some instances, however, it may be broken or remain connected to the discharged conidium.

The conidia of the different species of the tribe Entomophthoreae vary in size and form ; in fact, marked variation is sometimes apparent in the conidia of the same species. In size they range from about 10 to 75 microns in length. With regard to shape, the conidia vary from being spherical to being slender and tapering. They are usually hyaline, rarely slightly coloured, and contain fine granular protoplasm or, more commonly, coarsely granular protoplasm with large fat globules. These fat globules may be so regular in size and shape that the conidia resemble asci filled with spores. The walls of the conidium are smooth, without spines or similar modifications, and possess an adhesive quality that serves to attach them to any object with which they come in contact. The basal portion of the spore is more or less papillate.

When discharged, the conidium, if it comes in contact with a suitable host, adheres to it and sends out a germination hypha, which enters its body in the number already described. If the conidium lands in water, it gives rise to one or more hyphae that branch and elongate, growing constantly more attended, their protoplasmic contents becoming separated by successive cross partitions from the empty hyphae left behind. This growth may continue until the protoplasm is spent. The separation by cross partitions is common in the general growth of the fungus.

The process of germination is extremely variable but rarely lasts longer than a week and is usually much shorter than this. Germination ordinarily takes place soon after the conidium is discharged. If the discharged conidium falls neither on a proper host nor upon a wet surface, it proceeds to form what are called "secondary conidia," a process that provides for further dissemination if the primary spore has fallen upon a substance unsuited for its proper development. The most common method of formation, according to Thaxter, consists in the production of a hypha of variable length which, growing vertically upward, becomes swollen at its extremity into a basidium and produces a conidium usually similar to that from which it originated. This secondary conidium is discharged in the usual fashion. It, in turn, may produce tertiary conidia, and the process continues until its vitality is exhausted or until it has come in contact with a suitable host. With some species, under unfavourable moisture conditions, the appearance of the secondary conidia may differ considerably from that of the primary conidium. They are usually almond-shaped, have thick walls, and are not discharged.

Occasionally, upon the examination of a conidiophore mass, a simple hypha is seen exceeding the conidiophores in size and projecting beyond them, often to a considerable distance. Some of these hyphae, called "cystidia," are so large that they may be readily seen with the naked eye ; others are not much different from the ordinary conidiophores. The function of the cystidia is not known unless, as Thaxter (1888) suggests, they are rhizoids or hyphae of attachment but functionless because of their position. The functioning rhizoids are hyphae that attach themselves to the substratum upon which the host rests and serve to hold the fungus firmly in position. The rhizoids may be simple or branched, and their termination may be modified into an expanded suckerlike structure of attachment. Rhizoids appear to be confined to certain species and usually accompany the digitate type of conidiophores.

Formation of Resting Spores. Within the body of the insect infected with an Entomophthoreae, a process or phenomenon frequently occurs by which special spores are formed that are highly resistant to conditions that would ordinarily destroy the conidia. These "resting spores," as they are called, may be formed by an asexual process, in which case they are known as "azygospores," or they may be formed by sexual union and they are known as "zygospores." The resting spores are usually spherical, of large size, contain highly refractive fatty contents, and are surrounded by triple walls. The outer wall of the spore is thin and represents the wall of the mother cell, the second is thicker, and the innermost wall is usually as thick as or thicker than the other two combined.

Azygospores may be formed in a variety of ways. The simplest process is that by which the contents of a hyphal body become directly converted into a resting spore. Or azygaspores may be formed from hyphae of germination of arising from chalmydospores or hyphal bodies, or by direct lateral budding from them. Sometimes azygospores may be produced interstitially—between fungous cells—and this frequently results in spores having very irregular shapes.

Zygospores are also formed in a variety of ways, although sometimes the sexual nature of the spore is not well marked, in which case this may represent a transitional form from the truly sexual to the entirely asexual processes. The method of true zygospore formation is, as might be expected, by conjugation of two different hyphal outgrowths ; these meet, the intervening walls are absorbed, and the contents of the two mingle. In many cases a bud then appears on one or both of the gametes, increases rapidly and becomes the zygospore. Sometimes the spore develops as a terminal swelling from the end of one of the conjugating hyphae. Other methods of formation have been observed.

Hosts and Habitats of Entomogenous Entomophthoraceae. Any attempt to summarize the insect hosts of a particular group of microorganisms must take into account the fact that the host distribution is to some extent dependent upon the relative amount of study made of the groups concerned. In the case of the insect hosts of Entomophthoraceae, the Diptera appear to be the order of insects the members of which are found to be the most frequently infected, and from which the greatest number of species of this group of fungi have been isolated. The Hemiptera are the next greatest sufferers, followed by the Lepidoptera and Coleoptera. Certain species of Orthoptera. Hymenoptera, and Neuroptera are also known to be susceptible to these fungi. The adult stage of the host insect is affected more frequently than either the larvae or pupal stage. In insects with incomplete metamorphosis the nymph is almost as susceptible as the adult.

Although to a degree there is some specificity of hosts for each species of the tribe Entomophthoreae, this is by no means certain or uniform. Some species of Entomophthoreae infect a wide range of hosts, including insects in different orders; others have been found only on a single insect species or on a closely related group of insects. Not infrequently two different species of fungi are found on one species of host or even on a single host.

Entomogenous Entomophthoraceae are found in a variety of habitats, but usually they develop best in areas where there is a constant and rather abundant supply of moisture. The edges of ponds and brooks in shaded places are a particularly suitable environment for many of them. Some species do well in drier situations as long as periods of occasional moisture occur, enabling conidia to be produced. Damp foggy weather frequently makes the fungi more conspicuous, since the moisture causes them to become distended.

Just before dying from the infection, many hosts seek elevated positions, crawling upward on blades of grass or other food plants. This, incidentally, makes it possible for the conidia to be discharged over a considerable area. Other favorite positions assumed by the infected insects are on the undersides of leaves in shady areas or about houses.

Cultivation of Entomogenous Entomophthoraceae. Only a few species of entomogenous Entomophthoraceae have been cultivated apart from their insect hosts. The fact that success in this regard has been obtained with a few species indicates that probably most of the remaining species can be cultivated if the right combination of substratum and growing conditions can be found.

The first to obtain the complete life cycle of one of these fungi in artificial culture apparently was Speare (1912), who succeeded in thus cultivating *Entomophthora pseudococci* Speare, a parasite of mealybugs in Hawaii. As media he used potato, potato agar, oat agar, and radish, the latter two being found the most suitable. In 1929, Sawyer reported the cultivation of *Entomophthora sphaerosperma* Fres, and an unidentified species of *Empusa*, the most satisfactory media in this instance being potato, swordfish, and pork.

Sawyer's studies revealed several interesting features connected with the cultivation of the two species of entomophthoraceous fungi with which he worked. For one thing, liquid nutrient media favored luxuriant mycelial growth, whereas solid media favored the production of hyphal bodies and reproductive phases. Carbohydrates and fats did not appear to be essential to the growth of these fungi, but the substratum had to contain proteins that are quickly liquefied by the proteolytic enzymes secreted by the fungi. The hydrogen-ion concentration of the media was also important, with pH values slightly on the acid side of neutrality being the optimum for growth and development of the organism. Atmospheric humidity did not materially influence growth or reproduction, although too much moisture in the substratum inhibited latter. Conidia did not germinate below a relative humidity of about 70 per cent at 21°C. This temperature (21°C) appeared to be the most favorable for growth and reproduction ; the maximum temperature was 34°C. The exact minimum was not determined, but Sawyer found that the conidia of both species with which he worked could be frozen for several days and still germinate upon return to room temperature. Vegetative growth and the production and germination of conidia took place in total darkness as well as in light. Best growth was obtained when large quantities of inoculum were transferred to the media used. Cultures held at room temperature were best transferred once every 10 days.

The artificial propagation of entomogenous Entomophthoraceae may also be accomplished by infecting fresh hosts. Thaxter (1888) developed a method of doing this by using a tightly covered jelly tumbler in which the upper portion is separated from the lower by a round piece of wire netting. By placing the hosts to be infected in the lower of the two chambers and fastening a specimen carrying the fungus in the upper one. Thaxter found that the living hosts below acquired the discharged spores through the netting and thus became infected.

EXAMPLES OF *EMPUSA* AND *ENTOMOPHTHORA* INFECTIONS

It is not practical for us to attempt here a discussion of all the species of entomogenous Entomophthoraceae and the infections they cause. The essential relationships involved may be exemplified, however, by a brief consideration of a few of the better known species.

Since the systematics of this group of fungi is still unsettled, we shall not be too concerned about the eventual validity of the generic names to be used in certain cases. Inasmuch as most modern mycologists use the name *Empusa* to include species of Entomophthoreae having simple or unbranched conidiophores and multinucleate conidia, and the name *Entomophthora* for species with digitate or branched conidiophores and uninucleate conidia, we shall do the same. Nowakowski's genus *Lamia* is not now generally recognized, and we shall not use it here.

Empusa muscae Cohn. The first description of this fungus, commonly found infecting the housefly, *Musca domestica* Linn., was that given by DeGeer in 1782. About a century ago, Cohn (1855) gave it the name *Empusa muscae*, and it stands today as the type species of the genus. A considerable number of Diptera, in addition to the housefly, are subject to attack by this fungus, the most common of which are reported to be species of the genera *Lucilia* and *Calliphora*. Syrphidae are also susceptible. Distribution is apparently world-wide; most of the significant observations on the fungus have been made in the United States, Europe and South America.

Infected fles are usually found indoors ; they attach themselves to the walls and ceiling of houses and other buildings in the lifelike position. Close inspection of flies killed with the fungus usually reveals on the wall or window pane a distinct halo of discharged spores encircling the insect.

The life cycle of *Empusa muscae* in most respects is essentially of the general type already described in the preceding pages. The species may be distinguished by its relatively large pointed bell-shaped spores each of which usually contains a single large oil globule. The pellicle of protoplasm (from the basidium) that surrounds the spores when discharged on such a surface as that of a glass slide gives them a characteristic appearance (Fig. 5.1).

Empusa gryllt (Fres.). Fresenius (1856, 1858) originally described this fungus from a ground beetle under the name *Entomophthora Gryllt* : but since the conidiophores are of the simple type, the generic name *Empusa* is probably more applicable, and it has been recognized as such by most subsequent taxonomists. The conidia of this fungus are ovoid to pear-shaped, the conidiophores may coalesce externally when growing luxuriantly, costidia are wanting, and the resting spores are spherical and colorless. Its hosts include numerous species of Lepidoptera and Orthoptera, especially short-horned grasshoppers. Its prevalence on grasshoppers or locusts accounts

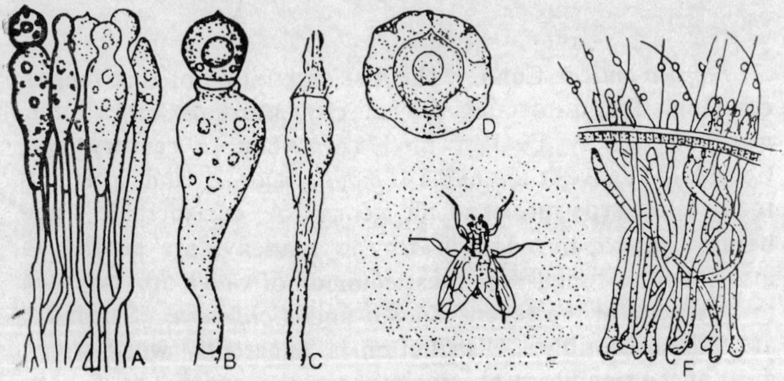

Fig. 5.1. *Empusa muscae* Cohn, the fungus commonly found infecting the housefly and other flies. *A.* A group of conidiophores showing conidia in several stages of development. *B.* Basidium bearing conidium before discharge. *C.* Basidium after discharge of conidium. *D.* Conidial spore discharged upon a glass slide and surrounded by a mass of protoplasm from the basidium. *E.* A fly infected with *E. muscae*, the ejected spores forming an aureole about the dead insect. *F.* Mycelium and conidiophores penetrating cuticle of insect (diagrammatic). (*A-D redrawn from Thaxter, 1888 : E and F redrawn from Paillot, 1933, after Brefeld*).

for the designation "common locust fungus," often applied to it.

One of the first reports in American literature having reference to the destruction of locusts by a fungous infection is that of Bruner (1883) in which the author states that numerous instances of "internal fungoid growths," had come to his attention during the preceding 12 years. In 1896, in Sourh Africa, grasshoppers had been found dying from a fungous disease. Some of the diseased specimens were sent to the United States, and a fungus was isolated. This unidentified fungus was distributed to planters, who in many cases reported considerable destruction of grasshoppers by the use of the microorganism. The possibilities attending the use of fungi in the control of grasshoppers appeared promising enough to initiate further studies, which were promoted under the direction of L. O. Howard of the Bureau of Entomology of the U.S.D.A. In 1902 Howard summarized the various reports on the effectiveness of fungi as a control agent and concluded that the results obtained did "not justify very sanguine hopes." He also pointed out that some of the cultures being distributed by various agencies and laboratories were not *Empusa grylli*. Some of them were saprophytic contaminants, and others were species of *Mucor* that did seem to have some insect-killing properties. A record was also made of the susceptibility of grasshoppers to the so-called "chinchbug fungus," *Beauveria globulifera* (Speg). In the years following Howard's report, the artificial use of *Empusa grylli*, as well as other fungi, as an agent of insect control, lost favor. Its destructiveness in natural outbreaks was recognized, but its efiectiveness was too dependent upon optimum conditions of temperature and moisture to be a practical means of artificial control.

The symptomatology of an *Empusa grylli* infection in grasshoppers is fairly characteristic. Skaife (1925), in South Africa, has described the disease somewhat as follows : The dying insects climb as high as they can on the grass stems and on the twigs of bushes with their heads pointing upward. Just before death, which usually occurs about 5 or 6 days after infection takes place, they loosen their hold with their claws and embrace the stems with their legs. After death, which occurs while they are in these elevated positions, their legs stiffen and the dead bodies remain hanging in this position for several days until they are finally blown away by the wind or washed down by the rain. Skaife observed that the majority of the insects infected with the fungus die in the late afternoon, usually between 3 and 7 P.M. An hour or so after death a fine furry or velvety growth appears growing from the intersegmental membranes, from the joints of the legs, around the neck, and at the base of the antennae. This growth usually has a white, buff, or greenish color, and it consists of innumerable club-shaped conidiophores that project from the insect's integument. At the end of each conidiophore a conidium is produced. These conidia are discharged generally in the evening when the live grasshoppers are clustered together for the night. This, of course, facilities the transmission of the fungus from the decrease insect to a healthy one. According to Skaife, there are several different strains of *Empusa grylli* varying in virulence, and apparently individual grasshoppers vary in their susceptibility to any particular strain. About 1 per cent of the dead grasshoppers fail to produce the external growth of conidiophores. When these insects are opened they are usually found to be filled with the thick-walled, resistant resting spores.

In South Africa, Skaife found that the disease develops only in those localities which had a rainfall of over 14.5 inches during the 6-month period. The disease did not appear in four areas that had received over 4 inches of rain in only 1 month. A rainfall of 14 to 15 inches over a period of 3 months, however, appears to be sufficient to start the disease in South Africa. In the United States it is generally agreed that warm humid or wet weather is necessary to enable the infection to develop. During dry weather there is practically no extension of the disease. It is unfortunate that grasshoppers do their greatest damage in the dry seasons, when conditions least favorable for the development of the fungus prevail.

Entomophthora sphaerosperma Fres. One of the best studied entomophthorous fungi is *Entomophthora sphaerosperma*, a fungus first recorded as a parasite of the cabbage-butterfly of Europe, *Pieris brassicae* (Linn.), by Fresenius (1856, 1958), who briefly described it and gave it its present name. The fungus has several times been known by other names (*Tarichium sphaerosperma* (Cohn, *Empusa radicans* Brefeld, *Entomophthora radicans* Brefeld, *Entomophthora phytonomi* Arthur, and *Empusa sphaerosperma* (Thaxter), but the name and combination originally used by Fresenius have retained their validity. It was first reported in the United States by Arthur (1886)—under the name *Entomophthora phytonomy*—as a parasite of the clover leaf weevil, *Hypera punctata* (Febr.). Since then it has been found on a wide variety of hosts in many parts of the world. In fact, its host range includes nearly every order of insects, the most common being Diptera, Hemiptera, and Lepidoptera. In numerous instances insects are subject to widespread outbreaks of infection by this fungus, which thus is frequently of considerable economic importance in the natural control of certain insect pests.

The morphology and development of *Entomophthora sphaerosperma* has been described by several authors, but one of the most complete accounts is that by Sawyer (1931). This worker succeeded in growing the fungus on swordfish and potato. From such cultures, which grew best at temperatures of from 18 to 21°C., Sawyer was able to make a careful comparison of the various stages in the life cycle of the fungus. The conidia are narrowly elliptical, with a rounded apex and a tapering base, and have an average size of 7 by 22 microns. A single membrane encloses the spore, and the apex is frequently covered by a detachable, transparent gelatinous cap. The protoplasm is finely granular and nonvacuolate in the newly formed spore although vacuoles begin to form as soon as the spore has a single centrally located nucleus. The mature condiophores are digitately branched at the distal end. A conidium develops at the end of each branch, a thin wall separating it from the conidiophore, which has also formed a cross wall at this point. After a given interval the spore is violently discharged. If the spore lands in a suitable environment germination begins with a slight outward bulging of the conidium at any point on its surface. As growth proceeds, a tube is formed at the tip of which may arise a secondary conidium ; or, if it is able to penetrate a susceptible host, the germ tubes may continue to develop, forming it mycelium and hyphal bodies. Rhizoids may also eventually be formed. These structures usually arise from the hyphae in the thoracic region of the insect and attach the host firmly to the substratum. The rhizoids constitute the only structure of the fungus that is not developed on artificial culture media as used by Sawyer. Resting spores are formed asexually (azygospores), their formation being subject to such factors as temperature and the nature of the artificial media upon which the fungus is being grown.

Although *Entomophthora sphaerosperma* is distributed widely and on many different host species, detailed studies of epizootics caused by this fungus have been made in relatively few instances. One such study was accomplished in South Africa by Ullyett and Schonken (1940) in the case of such an epizootic among larvae of the diamondback moth, *Plutella maculipennis* Curtis, which are first miners and later surface feeders on cruciferous plants. These workers observed the sporadic occurrence of the disease in fields in Transvaal where, at times, it was an important factor in the temporary control of the insect.

During the early stages of the infection the *Plutella* larva shows no significant external symptoms. As the disease becomes more advanced, however, the insect becomes restless and changes in color from green to a yellowish green. The yellow color becomes intensified, and the larva becomes increasingly sluggish in its activity as the disease progresses. With the approach of death, the body becomes somewhat distended ; and immediately after death and before the aerial hyphae appear the larva is extended on the leaf and is turgid and brittle. Upon being lifted, the body breaks easily, but there is little or no liberation of fluid. This last feature is a convenient one for making field diagnoses. Early larval instars appear to be more resistant to infection than do the later ones, fully grown larvae being very susceptible. The development of the fungus in the body of the larva is essentially the same as has been described as characteristic of the group. Germination of the conidia requires prior contact of the latter with free moisture. Thus it is that epizootics of the disease break out only after heavy rains. The germination hypha penetrates the integument of the host and enters its body cavity where growth rapidly follows, giving rise to a branching mycelium. Infection of the *Plutella* larva almost invariably takes place at the anal extremity, the hyphal threads running forward longitudinally in the body cavity. This forward growth of the mycelium apparently is a normal attribute of the fungus, since regardless of where experimental infection takes place the initial growth is toward the cephalic end of the insect taking place against the flow of the blood stream. The body cavity gradually becomes filled with the mycelium. Just before the death of the host the hyphae break up into short lengths, and immediately after death the mycelium is a compact mass of short stout lengths of hyphae (the hyphal bodies) which completely fill the body of the insect. The reproductive phase of the fungus begins with the death of the insect. Rhizoids anchor the dead larva to the leaf or other substratum, and hyphal branches emerge and ramify over the surface becoming interwoven at right angles to one another until the body is covered with a grayish-white felt of fructifying mycelium. Branched conidiophores arise from the thallus thus formed and bear the characteristic elongate conidia. When the mature conidia are discharged they come to lie on the leaf surface immediately surrounding the dead insect. It is from this source that healthy larvae pick up the spores as they crawl over them. At the completion of the aerial fruiting of the fungus, the center of the thallus turns brown in color. Within the body of the dead host azygospores are formed the hyphal bodies. These spores remain within the mummified fied host until it becomes broken up or until favorable weather conditions induce germination. Under dry conditions the viability of the azygospores is retained for long periods of time.

The effect of epizootics caused by *Entomophthora sphaerosperma* on the *Plutella maculipennis* population was examined by Ullyett and Schonken 1940) from both the theoretical and the practical standpoint. In brief, however, these workers concluded that the fungus, although it produced a decided immediate reduction in the host population during the favorable weather conditions under which it acted, was ultimately responsible for an increase in the average density of the host and, through this, for the occurrence of economic damage to the crop. The principal reason for this situation appeared to be the concurrent destructive effect on the insect parasites and predators that normally held the population density low enough to avoid serious crop damage. Thus the intervention of the disease in an existing equilibrium system resulted in the destruction of permanent mortality factors (insect parasites and predators) and their replacement by a temporary mortality factor (the fungus). When the activity of the fungus ended, the host population was able to recover more rapidly than its normal parasites and thus to attain a higher density level than before. This is an excellent example of the fact that occasionally a destructive disease may actually be a detriment rather than an aid in the long-range control of an insect pest.

An instance in which the artificial distribution of *Entomophthora sphaerosperma* affected the localized control of a destructive insect is provided by the work of Dustan (1924), who used it against the European apple sucker, *Psyllia mali* Schmidb, in Nova Scotia. This host was first found infected in considerable numbers in 1920 and again in 1921, and in the humid years of 1922 and 1923 artificial distribution of the fungus practically exterminated the insect in certain orchards. The fungus was spread by pinning leaves bearing diseased insects to leaves in the orchard being treated and by the liberation into the orchards of infected adults still capable of flying about. About a week after such "plantings," the fungus began to appear on the apple suckers in the orchards being treated. It spread very rapidly and was generally distributed throughout the orchards within 2 or 3 weeks if weather conditions were favorable. Dustan was able to initiate outbreaks of the diseases earlier than would naturally be the case by building up active epizooties in caged material early in the season and then introducing this material into the orchards.

Entomophthora aulica (Reich.). The so-called "brown-tail fungus," *Entomophthora aulicae*, attacks the caterpillars of the brown-tail month, *Nygmia phaeorrhoea* (Donov.), as well as more than a dozen other insect hosts. Thaxter (1888) considered this fungus as a strain of *Empusa grylli* (Fres.), but it has since been recognized as a distinct species. It has long been known in Europe and was first recorded in the United States by Thaxter in 1888. Both natural and artifical epizootics of the fungus on the brown-tail caterpillar have been reported.

In the spring and early summer the caterpillars leave their nests to feed on young leaves, and in the autumn they again leave before the webs of the new brood are closed for the winter. During both these periods, Speare and Colley (1912) found that the fungus could be effectively used against the insect, depending on the presence of favorable weather conditions, which appear to consist of warm nights and damp atmosphere. The natural epizootics also occur during these periods and under essentially the same conditions.

After developing methods for the artificial propagation of *Entomophthora aulicae* on brown-tail caterpillars in cages, Speare and Colley came to the following conclusions with regard to their attempts to use the fungus against this insect in the field. They were able to effect the destruction of caterpillars in great numbers and over considerable areas, although the general effectiveness of this method was variable since the factors of temperature and humidity could not be controlled. Nevertheless the introduced disease could usually be depended on to kill from 60 to 100 per cent of the caterpillars in the areas where diseased individuals were distributed. Best results were obtained in localities where the disease was not known to occur naturally. The fungus usually lives over winter from the autumn infection and reappears early in the spring, reducing the caterpillar population. For this reason the autumn infection is doubly effective. Artificial distribution appears to be most easily accomplished in sprout woodlands and in pastures where ordinary methods of control (*e.g.*, spraying and cutting) are not employed and where the caterpillar nests are more readily accessible.

Speare and Colley's (1912) report also contains a description of experiments with an unidentified *Entomophthora* on larvae of the gypsy moth, *Porthetria dispar* (Linn.). They concluded that this entomophthorods disease was not a promising one for artificial use.

Entomophthora fumosa Speare. It has long been recognized that in Florida the citrus mealybug, *Pseudococcus citri* Risso, is a pest of only secondary importance. In certain parts of California, on the other hand, it is considered one of the most serious enemies of citrus plants. The reason for this interesting difference was pointed out by Speare, who in 1922 published a revealing account on the natural control of the citrus mealybugs in Florida. According to Speare, the principal reason that *Pseudococcus citri* Risso is not an economically important pest in Florida is the fact that it is held in check by an entomogenous fungus, *Entomophthora fumosa*. The climatic conditions in Florida during the growing season are optimum for the development of the fungus, whereas this is not the case in California. In Florida the warm temperatures and the rainy season coincide ; in California they do not—the rainy season occurs during the cooler winter months. Thus it is that in Florida the fungus has an opportunity to flourish and to bring about the natural destruction of its mealybug host.

Although Speare originally gave the fungus the generic name *Entomophthora*, it would probably be more appropriate for it to bear the name *Empusa*, since the conidiophores are of the simple type. This designation depends, of course, upon the importance one is inclined to give the type of conidiophore in the generic separation of the Entomophthoraceae. The conidia of the fungus are more or less fusiform in shape, approximately 9 by 18 microns in size, and distinctly smoke-colored. Resting spores are formed and are usually spherical, black, and provided with a hyaline protuberance or appendage.

The symptoms of the disease caused by *E. fumosa* have been described by Speare (1922). One of the earliest signs of infection may be elicited by the use of a penknife or other bluntly pointed instrument which, when pressed against the body of a healthy mealybug, creates a depression that quickly returns to its original position when the pressure is withdrawn, much as a rubber ball would do. In mealybugs in the early stages of the disease, however, the body wall ruptures very easily when pressed with a penknife, allowing a droplet of milky fluid to exude. This occurs in the infected insects, even though to all outward appearance they may seem entirely healthy and active (at times their movements may be somewhat sluggish). If similar treatment is applied to a recently dead mealybug, which is also lifelike in appearance, the depressed area remains sunken. After the insect has been dead from 12 to 18 hours, its body is more or less solid to the touch, and if considerable pressure is applied to the penknife, the body will cut like a piece of cheese. No significant change in the external appearance of the mealybug takes place until about 24 hours after the first symptoms have been observed. At this time, and later, the infected insects may present one of two types of appearance. Most of the dead insects will appear to be enveloped in a dark slate-gray woolly covering, which consists of the fungal conidiophores and the developing conidia. A few of the insects, on the other hand, may appear jet black in color, sometimes almost glistening, with the body surface of the insect smooth instead of woolly. The jet-black color is due to the large numbers of spherical black resting spores formed within the body of the insect. The color of the black spores is transmitted through the thin translucent body wall still intact. In both of the two types of infection the mealybugs are attached rather lightly to the host plant by the insertion of their proboscises.

In orchards of grapefruit, Speare observed an extremely rapid spread of the fungous infection among the mealybugs during the months of June, July, and early August. In one grove, the percentage of mortality jumped from 18 to 64, an increase of 40 per cent, in but a week's time (June 22 to June 29). The role of the fungus in the natural control of mealybugs is also emphasized, in Speare's opinion, by the fact that following the application of fungicides, which destroy the entomogenous fungus as well as that of the citrus scab, the numbers of mealybugs rose sharply in the groves so treated.

Incidentally, certain Fungi Imperfecti have from time to time been reported as being parasitic on mealybugs. Species of *Aspergillas Cephalosporium* have been mentioned in this connection. Boyee and Fawcett (1947), for example, relate that during the propagation of mealybugs in California insectaries, a fungus closely related to *Aspergillus parasiticus* Speare proved to be a potentially serious parasite under conditions of high humidity and moderate to high temperatures. Efficient ventilation of the insectaries, careful irrigation of the host plants, and a lowering of the temperature to 20°C., or below, are suggested control measures. A fungus that has been named *Endosclerotium pseudococcia* Har. and McK., occasionally kills large numbers of mealybugs, including the Comstock mealybug, *Pseudococcus comstocki* Kuw., in the apple orchards of Virginia and other eastern states. When condition for growth are unfavorable, the fungus produces a highly resistant compound sclerotium which may remain viable several months.

INFECTION CAUSED BY MASSOSPORA

Of the remaining genus of Entomophthoraceae, *Massospora*, the only well-studied entomogenous species is *Massospora cicadina* Pk., named and described by Peck in 1879, although apparently seen by Leidy as early as 1850. It is a parasite of the seventeen-years locust, or periodical cicada, *Magicicada seplendecim* (Linn.), and had been observed particularly in the Eastern and Middle Western parts of the United States.

The occurrence of the fungus was subsequently recorded in the reports of a number of entomologists, but it remained for Speare (1921) to give the first adequate description of the disease itself, as well as that of the microscopic characters of the fungus. He also showed the relationship of *Massospora cicadina* to other entomogenous Entomophthorales. Goldstein's (1929) cytological study of the fungus furnished additional information on this very interesting parasite.

In some outbreaks at least, the fungus appears to be confined largely, although not exclusively, to male insects, and in the resting spore condition it usually parasitizes spent individuals, females as well as males having been found in this condition. It confines its vegetative growth to the softer tissues in the posterior segments of the cicada's body. Most of these tissues are completely destroyed, and as a result of the destruction of the flexible intersegmental membranes, the posterior abdominal segments, beginning with the last segment, slough off. The insect, however, remains alive for a considerable period, continuing to fly and crawl about. This unusual phenomenon is so striking that it is readily noticed by all observers.

One of the outstanding characteristics of *Massospora cicadina* is that, inlike species of *Empusa* and *Entomophthora*, it produces conidia within the body of its host rather than on its surface, where they are violently discharged from the conidiophores. Within the abdomen of the cicada, the conidia cohere with one another forming clumps, clusters, or a mass that is exposed when the insect's abdominal segments drop off. The movements of the insect then aid in scattering the conidia of the fungus. The conidia are oval in form, about 12 by 15 microns in size, and have a distinct but not prominent papilla. Unlike those of most other Entomophthorales, the conidial walls are regularly verrucose. Resting spores, found in certain individuals, are slightly brownish spherical bodies with reticulations that give them an appearance similar to the design on golf balls. Attempts to cultivate the fungus on artificial media have not been successful.

Just how or where the fungus survives the 16¾ years of the host's immature and subterranean existence is not known. It is possible that infection takes place while the insect is underground, or the fungus may exist on species of biennial cicadas when its regular host is absent. Its importance in the natural control of the periodical cicada is probably not great, if Speare's observation that the infection is confined largely to spent males is correct. Goldstein (1929), however, found the fungus in both males and females, and most of her specimens containing resting spores were females whose bodies still contained many eggs. Such a situation would enhance the importance of the fungus from an economic viewpoint.

INFECTIONS CAUSED BY CHYTRIDIACEOUS AND BLASTOCLADIACEOUS FUNGI

Among the more primitive of the phycomycetous fungi are those belonging to the orders Chytridiales and Blastocladiales. The "chytrids," as the former are commonly called, occur mainly in fresh water on a variety of substrata. A number of species live in the submerged and empty exuviae, or cast off integuments, of the immature stages of certain insects.

Truly entomogenous chytridiaceous fungi, however, have been reported by a number of mycologists. Wize (1904) found such a fungus in Coleoptera (*Cleonus, Anisoplia*) larvae and pupae collected in the Ukraine, and Sparrow (1939) reported on similar observations made on dipterous pupae collected in the United States by Thaxter. In both instances the fungus appears to have been the same, *Myiophagus ucrainicus* (Wize) [*Myrophagus*]. The body contents of the diseased insects are almost completely disintegrated and replaced by an orange to reddish mass of fungous material. What is apparently this same fungus has also been found in dipterous pupae in England and in Bermuda, Canada, and the United States in scale insects. Its occurrence in scale insects (purple scale, chaff scale, oyster-shell scale, red scale, long scale, and soft brown scale) is particularly interesting and the chytridiosis produced probably deserves study from the standpoint of biological control. Purple, scales have been infected by spraying them with aqueous suspensions of the fungus, and mealybugs have been found experimentally susceptible.

The entomogenous Blastocladiales are better known than are the apparently few chytrids that parasitize insects. Nevertheless these are confined largely to one group, the family Coelomomycetaceae, which parasitizes mainly mosquito larvae.

COELOMOMYCES INFECTIONS

The first recognized infection caused by a member of this group of fungi was that discovered by Keilin in 1921, occurring in larvae of the mosquito *Aedes albopictus* (Skuse) (=*Stegomyia scutellaris* Walker) collected in the Federated Malay States. To this fungus, which he believed to be related to the Chytridineae, Keilin gave the name *Coelomomyces stegomyiae*. The next year Bogoyavlensky (1922) described, under the name of *Zografia notonectae*, a fungus parasitic in the body cavity of *Notonecta* (Hemiptera) collected from ponds in Moscow. Keilin (1927), however, showed this organism to be of a nature similar to that of the fungus he studied, thus giving it the name *Coelomomyces notonectae*. Additional species have been isolated from mosquitoes (*Anopheles*, *Culex*, *Psorophora*, and *Uranolaenia*) in other parts of the world, including 2 species from India 1 species from Africa and 11 species and 2 varieties from the United States.

The Infection. The manner in which mosquito larvae become infected with *Coelomomyces* has not been determined with certainty ; *i.e.*, it is not clear whether invasion takes place via the digestive tract following the ingestion of the zoospores or the sporangia, or directly through the insect's integument. In any case the principal development of the fungus occurs within the body cavity of its host. Depending upon how far the infection has progressed, only certain regious of the body, such as the gills or the posterior segments, may appear involved, or the entire hemocoele may appear filled with the parasite. Frequently the mycelia and spores of the fungus may occur even within the head of the larva.

The hemocoele of an infected larva in the fourth instar may appear as a solid mass of sporangia, which gives the larva an opaque aspect. The color of such larvae may be dull or yellowish-white, bright yellow, orange, or even a dull reddish-brown.

In most instances the fungus completes its development during the larval stage of the mosquito. Sometimes, however, when infection occurs late in larval life the fungus completes its development in or persists through the pupal and adult stages. In surviving adult females the infection is usually confined to the ovaries. In other cases it occurs throughout the body of the adult, causing its death. Infected pupae may be unable to emerge as adults ; especially is this so once the sporangia have developed. It is probable that the majority of infections begin during the larva's first instar, since fully developed sporangia are found in the third and fourth larval instars. Since this period of larval development is usually one of only 6 to 10 days, it indicates the very rapid development of the fungus. Actually, under normal conditions in nature, it takes only 2 or 3 days for the thalli to develop into sporangia. According to Iyengar (1935), infection generally begins in the thoracie region of the larva and spreads posteriorly into the abdominal segments, travelling along the adipose tissue on which the fungus lives. "As a result of the attack by the fungus, the fat body loses its characteristic appearance and shrinks, the nuclei of the fat cells disintegrate, and finally the fat body completely disappears. In place of the fat body, a thin membrane filled with many dark-brown pigment granules remains. Only rarely does an infected larva of the fourth instar have any fat tissue. Some fat tissue may remain if the infection has been of a low intensity. One reason for an infected larva's rarely completing its metamorphosis into an adult mosquito is the suppression of the imaginal buds by the infection. In a fourth-instar infected larva, the wing and leg rudiments are underdeveloped as compared with their appearance in a healthy larva of the same instar.

No accurate survey has yet been made as to the over-all extent to which mosquito larvae are infected with *Coelomomyces* fungi. In some areas the incidence of infection appears to be very high, while in other areas infected specimens can be found only on rare occasions. Muspratt (1946*a*) estimates the mortality, in pools that were under his observation in Rhodesia from 1941 to 1945, to be as high as 95 per cent of the larvae that hatched from eggs during the rainy season. Of those larvae which reached the fourth instar, at least 9 out of 10 were infected and subsequently died. Offhand, at any rate, such data would make it appear that these fungi may possibly have some use in the biological control of mosquitoes, at least in the tropics. On the other hand, some observers have found the ratio of infected larvae to healthy ones to be so small in the areas they studied that the fungus did not appear to have much importance as a natural control agent. The biological control possibilities are under consideration by several groups of investigators as this is belong written, and the outcome of these as well as other studies must be awaited before it will be safe to draw any definite conclusions in this regard.

The Fungi. As has already been mentioned, the *Coelomomyces* fungi are Phycomycetes belonging to the family Coelomomycetaceae of the order Blastocladiales. It was at first believed that, like those of most species of Blastocladiales, the resting bodies of *Coelomomyces* required a period of desiccation before being able to germinate. Muspratt goes further to suggest that in order to bring about the germination of the resting sporangia, it may be necessary to use rain water and allow it to evaporate in the sun to about one-third of its volume before infection can be expected, the germination perhaps being regulated by a slight increase in the concentration of the soil mineral salts in solution. In 1974, however, Couch and Dodge found germination to occur in the case of three species without the previous drying of the resting bodies. These authors nevertheless feel that the fungi belong to the order Blastocladials, particularly since in structure and method of swimming the zoospores are typical of this order.

Germination of the sporangia was first witnessed by De Meillon and Muspratt (1943). These workers observed the process to be somewhat as follows. When about to germinate, a sporangium first loses its oil droplets and the interior becomes granular. A slight bulge then appears on one side. As the bulge enlarges, the outer hard shell ruptures and two thin internal membranes appear. The content of the sporangium flows out and is confined within the innermost of the thin membranes, both of which become increasingly more visible and more widely separated. The contained zoospores become less densely packed and accordingly begin to move about with increasing speed until the interior of the sporangium, and the extruded membranes, are a seething mass of zoospores. Within a few minutes thereafter the zoospores find their way out to the exterior until the sporangium and rent membranes are empty.

This whole process may take a day or two ; but once the zoospores flow into the extruded membranes, complete liberation takes place in a few minutes. The zoospores have single flagella about four times the length of the zoospores, which measure approximately 4 microns.

Further observations on the germination of the resting sporangia were reported in 1947 by Couch and Dodge. They also observed germination to begin with a swelling of the contents which causes the outer wall to split along a performed line. The contents continue to swell, bulging out through the fissure to form a dome-shaped mass surrounded by the inner spore wall. The zoospores are formed 24 to 36 hours later, and the entire exposed part of the inner wall quickly swells and dissolved to set the zoospores free. The two thin membranes described by De Meillon and Muspratt are interpreted by Couch and Dodge as a single membrane that gelatinizes. Otherwise the observations of these two pairs of workers are essentially similar.

Within the body cavity of the insect, the fungus develops a coenocytic, aseptate mycelium without rhizoids or cells walls, and surrounded only by a plasma membrane. This last characterisic enables the fungus to absorb its food directly over its entire surface. The hyphae are irregularly or rarely dichotomously branched and consist of single threads of more or less uniform diameter except at certain points. The hyphae give rise to irregularly shaped hyphal segments that are severed from the parent hyphae by division or by the thinning down of the basal connection until it is broken away by the movements of the insect's body. The sporangia are formed as are the hyphal segments. They break away from the mycelium before the formation of a definite wall and complete their development within the hemocoele of the insect. The thick-walled sporangia fill the body cavity and are usually set free upon the disintegration of the insect's body. The sporangium is surrounded by an exceedingly thin, hyaline, smooth, outer membrane derived from the old plasma membrane. The wall is two-layered, the outer layer usually being the thicker. It may be smooth, pitted, banded, striated, ridged, or otherwise sculptured or ornamented. Dehiscing occurs by a performed longitudinal slit. All these characteristics have been stated descriptively by Couch (1945), who revised the genus *Coelomomyces* Keilin and erected the family Coelomomycetaceae.

ASCOMYCETE AND DEUTEROMYCETE (FUNGI IMPERFECTI) INFECTIONS

It is more for convenience than for any other reason that we place the ascomycete and deuteromycete infection under one general heading. Most authorities recognize the majority of species now relegated to the class Deuteromycetes (Fungi Imperfecti) to be, in reality, species of Ascomycetes whose relation with an ascogenous stage has not been established. In other words, since most of the Deuteromycetes appear to be asexual (imperfect) stages of recognized or unrecognized species of Ascomycetes, it is expedient for use to consider the two groups together. Especially is this a convenient procedure in dealing with such entomogenous fungi as those which parasitize scale insects and whiteflies and which, in many cases, are known by different names in both their sexual and asexual (perfect and imperfect) stages.

As has been expressed by Wolf and Wolf (1947), the Ascomycetes "possess a multiplicity and complexity of architectural design and a seemingly infinite variety of pattern of activity that are baffling to all who attempt to catalogue or to orient them." It is impossible, therefore, to present this large group according to any one system of classification and nomenclature, and at the same time to satisfy all systematic mycolozists. For the present, the best that the insect pathologist can do is to follow the lead of those who have given the greatest attention to these entomogenous fungi until some authority makes it clear that the wrong name is being used or that the fungus is in an incorrect taxonomic location. We shall attempt to follow such a procedure in the present discussion and, inashmuch as is possible, spare the reader details of the annoying controversies or the equally trying indifference that prevail among specialists in some groups of the fungi concerned. This applies to both the Ascomycetes and the Deuteromyetes (Fungi Imperfecti). In the latter group, of course, the groupings are admittedly artificial and without real phylogenetic significance.

The Ascomycetes are characterized by the formation, somewhere in the life history, of a definite number of spores (ascospores), usually eight, contained within a unicellular sac or membrane (ascus). The ascospores are formed as the result of sexual fusion of nuclei. In addition, many Ascomycetes also produce conidia, or exogenous asexual spores, borne on stalks of mycelium known as "conidiophores." As a rule, conidia are produced in large numbers when conditions are favorable for the rapid multiplication of the fungus. Many species-form ascospores in only small numbers and frequently under particular and uncommon circumstances. The hyphae of Ascomycetes are generally septate, and cells thus formed are usually uninucleate.

Most authorities consider the Ascomycetes as consisting of two subclasses, Hemiascomycetes and Euascomycetes. In the first of these which is the more primitive, the asci occur singly or in groups, but ascocarps are lacking. In the second, which includes most of the Ascomycetes, the asci are aggregated and ascocarps are present. Ascocarps are special large fruiting organs in which the asci are formed. The ascocarp may be arranged as either a widely opened, cup-or saucer-shaped structure (the apothecium), or as a hollow sphere (the perithecium) usually having a small apical opening, or ostiolum, through which the ascospores escape at maturity. Variations of these two types occur (*e.g.*, cleistothecium and hysterothecium) but such are rare among entomogenous fungi.

INFECTIONS CAUSED BY YEASTS

Of the subclass Hemiascomycetes, the order Endomycetales (Saccharomycetsies), commonly called "yeasts," is known to contain a few species pathogenic for insects. As early as 1879 the Harvard entomologist Hagen advocated the use of beer mash, or what he termed a "yeast fungus," for the control of grasshoppers, potato beetles, phylloxerids, and other noxious insects. However, no actual attampt to control an insect in the field through the use of a yeast has been recorded. Several saprophytic yeasts have been isolated from insects, and a few have been described as infecting small groups of insects.

The first adequately described yeast infection of insects appears to have been that discovered in England by Keilin (1920) in larvae of the biting midge, *Dosyhelea obscura* Winnertz, a dipteran insect living usually in the sap that fills infected wounds of elm or horse-chestnut trees. The infecting yeast, *Monosporella unicuspidata* Keil., invades the body cavity, which becomes so filled with the parasite as to give the insect's body, especially its posterior segments, a milky appearance. Even though a great number of parasites is present, the larva is able to move until it finally dies and decomposes rapidly, setting free the resistant forms of the yeast. The fat body is apparently the only organ that is completely destroyed.

The yeast itself occurs in several forms. Young organisms appear as small oval cells, from 4 to 10 microns long, budding at one end. Later, the parasites become elongated until they reach a size of about 30 microns in length by 2.5 microns in width. These elongated forms constitute the asci, in each of which develops a long needle-shaped unicellular spore with one end sharply pointed. It is assumed that this truncated end assists the spore in penetrating through the gut wall of the insect during early stages of invasion, in a manner analogous to that observed earlier by Metchnikoff in the case of a similar organism infecting a crustacean (*Daphnia*).

A yeastlike organism, which they named *Mycoderma clayi*, was isolated in 1928 by Metalnikov, Ellinger, and Chorine from European corn-borer larvae shipped to them from Canada. The organism appeared as a large gram-positive rod, measuring 9 to 16 microns in length by 1.5 to 3 microns in width. Propagation occurs by budding. The yeast is readily cultivable and to a slight degree ferments glucose, levulose, sucrose, and glycerin. When injected into the body cavity of the larva, the cells multiply rapidly, causing a septicemia that kills the insect in from 2 to 5 days. Both *in vivo* and *in vitro* the organism has a tendency to form mycelium. Spores have not been observed.

Another *Mycoderma, M. cerevisiae* (Desm.) was cited by Burnside (1930) along with *Saccharomyces cerevisiae* Hansen and *S. ellipsoideus* Hansen as causing a dysenteric condition among caged bees that were heavily inoculated by feeding. Sometimes a condition resembling intoxication developed, from which the bees rapidly recovered. When these yeasts were inoculated directly into the body cavities of bees, death resulted in from 50 to 100 per cent of the cases. With *Saccharomyces apiculatus* Hansan death occurred in about 50 per cent of the individuals similarly inoculated. Microscopic examinations of the diseased bees showed that the yeast cells multiplied rapidly in the blood and at the time of death were present in such large numbers that the blood took on a milky appearance. The organisms appear to develop most abundantly on the muscle fibers of the thorax. The presence of a chalk-white coating on the surface of the muscles after they have dried was one of the most distinctive macroscopic symptoms of these yeast infections. Lardinois (1926) observed *S. apiculatus* constantly associated with lesions in the tissues of bees afflicted with constipation and expressed the belief that this yeast is the sole cause of "May disease."

In an attempt to find a microorganism that might be used in the control of mealybugs in Russia, Evlakhova (1939) reared *Pseudococcus citri.* Risso in the laboratory under conditions of high humidity to favor the development of disease organisms. Among those isolated was a yeastlike fungus which he named *Blastodendrion pseudococci.* Under experimental conditions of 25°C and 60 to 70 per cent humidity, a mortality of mealybugs occurred ranging up to 50 per cent ; with cultures of increased virulence a mortality of as high as 100 per cent was reached. Most of the mealybugs died in 24 hours after commencing to feed on potato stems smeared with the cultures. Spraying the mealybugs with a suspension of the organism was less effective. In causing the infection, the yeast apparently entered the insect's body through the digestive tract, penetrating into the fat body and the muscular system.

An interesting pathology involving a yeast has been described by Frobisher (1926) in *Drosophila melanogaster* Meig. being reared in the laboratory. The yeast concerned was a red torula, which was being used as food by the drosophila flies. Occasionally there appeared in the flasks used for cultivation a blue-green *Penicillium* which was capable of infecting and killing the flies. Besides the actual invasion of the insect tissues by the fungus, another mechanism seemed to be responsible for the death of the insects. The fungus appeared to grow among the torula cells in the gut and by matting them together formed a more or less solid mass that could block both elimination and engorgement. By invasion of the intestinal wall, the fungus hyphae were believed to bind this mass firmly in the intestine.

HYPOCREALES INFECTIONS

The order Hypocreales belongs to the subclass Euascomycetes and contains several genera noted for their entomogenous species. These genera fall into rather heterogeneous groups, which are nevertheless convenient for purposes of discussion. It is for this reason that we shall consider as a group the fungi parasitic on scale insects and those parasitic on whiteflies, even though some of the fungi coming under these categories are not Hypocreales. Some Hypocreales, such as *Cordyceps*, are clearly defined as taxonomic groups and may be discussed as such.

CORDYCEPS INFECTIONS

The genus *Cordyceps* contains about 200 known species, nearly all of which parasitize insects. Almost 40 recognized species occur in the United States. These have been studied principally from a taxonomic standpoint. Some species originally considered as *Cordyceps* are placed by some authors in the genus *Ophiocordyceps* which differs from the former in having clavate asci and fused overlapping ascospores. Cordyceps are cosmopolitan in distribution and occur on representatives of several orders of insects, principally Hemiptera, Diptera, Lepidoptera, Hymenoptera, and Coleoptera. The larval stage appears to be the most frequent host, but different species may occur on every stage of insect development. Information concerning the identity of the various hosts is frequently meager because of the destructive or obliterating effect the fungus has on the insect and because of the tendency early mycologists had of reporting the fungus simply on a "caterpillar," "pupa," "moth," and the like.

Associated with the genus *Cordyceps* are numerous interesting historical records. Early writers have presented rather extensive accounts of the group as it was known to them at that time. Insects parasitized by these fungi were frequently known as "vegetable wasps" and "plant worms" (or "awetos" in New Zealand). Among the most celebrated of the "vegetable wasps" were those (*Vespa* and *Polybia*) parasitized by *Cordyceps sphecocephala* (Klotzsch). An account exists in which Torrubia, a Franciscan friar, tells of finding, in 1749, some dead wasps in a field near Havana and "from the belly of every wasp a plant germinated, which grows about five spans high." As recounted by Gray, Torrubia gave a representation of two wasps lying on the ground with a tree growing out of the base of each abdomen, while three other wasps are flying around these trees, each flying insect having a similar tree affixed to it. These "trees" in all probability represent species of *Cordyceps.* On this basis it may be considered that the earliest known record of an entomogenous fungus was that made by Christian Paulinus in the beginning of the eighteenth century when he wrote that "certain trees in the island Sombrero in the East Indies have large worms attached to them under ground, in the place of roots." For additional curios of this type of historical information the reader is referred to Gray's (1858) privately printed "Notices of Insects That Are Known to Form the Bases of Fungoid Parasites."

Morphological Characteristics. The feature that characterizes the genus *Cordyceps* is the fact that the stroma arises from a sclerotium formed within the body of the insect on which the fungus is parasitic. As described by Massee (1895), who monographed the genus, this stroma usually consists of an erect stemlike sterile portion composed of a fascicle of irregularly parallel hyphae, white internally, the external or cortical hyphae being usually tinged with colour and in many species giving off numerous short lateral branches that form the minutely velvety or downy exterior of the stem.

The "head" or "club," which is the fertile portion of the stem, is usually terminal in location and may be brightly coloured. The perithecia originate side by side deep in the stroma, their shape ovate or flask-shaped, their mouths reaching the surface of the stroma. They may remain completely immersed or at maturity be quite superficial, with the entire perithecium being exposed. The latter condition causes the surface of the head to be rough, whereas when the perithecia are immersed it is smooth.

The asci contain eight spores and are very long and slender. As the spores mature, the contents of the head become swollen, and the wall of the ascus is ruptured at the apex. The slender spores, arranged in a parallel fascicle, are almost as long as the ascus, are hyaline and multiseptate or continuous. After escaping from the ascus, the multiseptate spores usually break up into their component cells, which eventually germinate. Paraphyses are absent.

Some species of *Cordyceps* are believed to have their conidial stages in such genera as *Isaria* and *Botrytis*. Such supposed relationships have not, however, been made too clear. For example, *Isaria farinosa* (Dicks) (or *Spicaria farinosa* (Fron.)) was for years considered to be the conidial stage of *Cordyceps militaris* (Lk.). This belief and its supposed proof were discounted by Petch (1936), who contends that the conidial stage of this *Cordyceps* is a *Cephalosporium*. *Isaria farinosa* is a fasciculate *Spicaria* and may occur as an *Isaria* or as a simple *Spicaria*. It has no relation to *Cordyceps militaris* which attacks larvae and pupae of Lepidoptera while *Isaria farinosa* is a general entomophyte occurring on Lepidoptera, Hymenoptera, Coleoptera, Diptera, Aphididae, and Arachnida. As Petch pointed out earlier (1934), the genus *Isaria* is an unsatisfactory one from several standpoints. If it is retained, it is a compound *Spicaria*. Petch suggests that it would be simpler to discard the genus *Isaria* and retain the name as purely a descriptive term. A relationship between the imperfect genus *Hirsutella*, once considered a basidiomycete, and certain species of *Cordyceps* has been suggested by Speare (1920*b*), and supported by Petch (1923), who reduced his genus *Trichosterigma* to synonymy with *Hirsutella*. Since then several species of *Hirsutella* have been found to be the conidial stages of species of *Cordyceps*. Such may eventually be found to be the case with other species of *Hirsutella*, but many identities as to their ascigerous stages still remain uncertain (*e.g.*, *H. subulata* Petch on the larva of the codling moth. *Carpocapsa pomonella* Linn., and *H. saussurei* (Cke.), a parasite on hornets in the tropics, and others). Other genera of Hypocreales may, however, be related to *Hirsutella*. Species of *Calonectria*, for example, parasitize leafhoppers and are the ascigerous stage of certain species of *Hirsutella*.

The first to describe the growth of any species of *Cordyceps* from spore to spore in the laboratory appears to have been Shanor (1936), who accomplished this with *Cordyceps militaris*. Living lepidopterous pupae were inoculated with hyphae and placed in moist sterile sphagnum moss. Normal mature fruiting bodies were consistently produced. No success in producing perithecia has been obtained using artifical culture media.

Pathogenesis. To illustrate the life history of most *Cordyceps* and the manner in which they infect their hosts, we might use, as did De Bary (1887), the well-known *C. militaris* as an example.

After being ejected from an ascus of the orange-coloured club-shaped stroma, the slender filiform, or rod-shaped spore, divides transversely into a large number of secondary spores. When one of these lands on the slightly moist skin of the host caterpillar, it swells slightly, becomes rounded in shape and puts out a germ tube. Sometimes the spores become partly united again by means of short connecting tubes before they germinate.

The germ tube proceeds at once to penetrate the integument, within which it enlarges into a somewhat thicker fungous hypha which ramifies its way into the deeper layers of the skin. It finally breaks through into the body cavity of the insect, insinuating itself between the muscular and fat tissue. Here the hypha breaks down into cylindrical bodies ("cylinder-gonidia") similar to the hyphal bodies already described in the case of the entomophthoraceous fungi. They continue to pass into the blood of the insect, where they elongate to twice or several times their original size, dividing repeatedly by transverse walls and by terminal and lateral sprouts. These cells disperse through the blood of the host until the body cavity gradually becomes filly with them as the quantity of blood is diminished. At this point the larva loses its normal turgidity, becomes soft and relaxed, and dies.

As soon as the insect dies, the sprout cells begin to develop rapidly at the expense of the dead body tissues, forming branched hyphae, which fill the entire body cavity except the alimentary canal and expand the body to its former size and turgidity. The various tissues of the body are more or less absorbed by the fungus, and in 1 or 2 days a body is formed that retains the approximate size and shape of the living insect but consists essentially of a mass of fungous hyphae with some remains of the insect. This fungus body has all the properties of a sclerotium and may be considered as such. If this sclerotium lies in a moist situation, it can give rise directly to fresh stromata ; if dried, it may lie dormant for several months.

Thə development of most *Cordyceps,* in their simplest form follows much the same course as the one just described by *C. militaris.* Nevertheless variations occur between species and, in fact, within the same species. The life histories of species of *Ophiocordyceps* in most respects are similar to those of *Cordyceps* species.

FUNGI PARASITIC ON SCALE INSECTS

Some of the most notable work on the fungous diseases of insects has concerned those fungi parasitic on scale insects. Not all these fungi belong to the order Hypocreales, although many of the most important ones do. Some are known only in their imperfect or conidial stages. Since those which do belong to Hypocreales have been favored with a great amount of detailed study, it is considered convenient to use this order as a center about which all the fungi parasitic on scale insects may be considered.

According to Petch (1921), the earliest record of a fungus parasitic on a scale insect was made in 1848 by Desmazieres, who collected his specimens from willow and ash at Caen, France. For many years following this, mycologists (including such systematists as Berkeley, the Tulasnes, Saccardo, and Petch) concerned themselves almost exclusively with the taxonomic aspects of these fungi. Eventually it was realized that at least some of these fungi may constitute an important natural check on certain species of scale insects. Beginning about 1912 considerable attention was focused on them in the citrus-growing areas of Florida, and attempts were made to use the fungi in the biological control of the insects. For a time it was believed that the increase of scale insects following Bordeaux spraying was the result of killing off the entomogenous fungi. That this was probably not entirely the case is indicated by the observations of a number of man, including Holloway and Young (1943), who attributed the increase to the effect of the spray residues. Apparently a complex of several fluctuating factors is involved. There does appear to be a need, however, to redetermine the true pathogenicity of at least certain of the entomogenous fungi found on scale insects. That some of these fungi are actually saprophytes or secondary parasites is a distinct possibility. In fact, Fisher (1947) in her studies of certain of the entomogenous fungi (*Microcera*, *Podonectria*, and others) of scale insects in Florida makes the statement that "No evidence has been found that any of the so-called 'friendly fungi' actually parasitize any of the scale insects."

The fungi found on scale insects belong to a number of genera ; and since many of them are important enough to deserve more than mere mention, it seems advisable to consider them according to their generic groupings, in much the same manner as that employed by Fawcett (1948). Some of the species discussed have at one time gone under other names : some have several synonyms which, for reasons of space, cannot be listed here.

Sphaerostilbe Infections. Three important species of *Sphaerostible* are found associated with scale insects : *S. aurantiicola S. flammea* Tul. and *S. coccidophthora* (Zimm). The first two species are found on insect in North and South America, the Orient, and in other parts of the world *S. coccidophthora* occurs in the Orient on coffee-tea-, and bamboo-scale insects. They are commonly called the "red-headed scale-fungi."

S. aurantiicola, which Snyder and Hansen (1945) have designated as *Nectria episphaeria* f. *coccophila* (Desm.) (= *S. coccophila* Tul. ; *N. coccophila* (Tul.)), occurs on scale insects which live on citrus and other host plants. In North America, principally in Florida and the West Indies, the best known hosts on citrus include California red scale, Florida red scale, Spanish red scale, Putnam scale, ivy scale, snow scale, green scale, purple scale, thread scale, Glover scale, chaff scale, black scale, rufous scale, and San Jose scale. It has also been reported on the citrus mealybug. It appears to be one of the most prevalent fungi on scale insects, especially in Florida, where it is reported to have more "effect" on the purple than on the Florida red scale—the opposite of the pink fungus, *Nectria diploa* Berk and Curt., which appears to be more "effective" against the red scale. The perithecia are small, globose, and orange red to blood red in color. The imperfect pustules are of the same color but are clavate or flattened pulvinate in form. The fungus has been obtained in culture from both conidia and ascospores. Working with the obscure scale, *Chrysomphalus obscurus* (Comst.), Luttrell (1944) observed that the fungus completely destroys the body of the infected scale and forms a plectenchymatous stroma between the shield of the scale and the bark of the plant host. The bark is not penetrated. On the other hand, Fisher (1948) reports that all her attempts to artificially inoculate red and purple scales with the fungus have failed.

S. flammea has been reported on the snow scale on citrus and on scale insects of many plant hosts besides citrus. The heads of the imperfect form of this fungus are red in color; those of the perfect form are bright orange-red. The latter are crowded together on a stroma which often hides the scale insect.

The imperfect, or conidial, stages of *Sphaerostible* are generally placed under the generic designation *Microcera,* which according to some authorities, is synonymous with *Fusarium.* Thus Snyder and Hansen (1945) have provided the combination *Fusarium episphaeria* f. *coccophila* for the imperfect stage of *Nectria episphaeria* f. *coccophila* (*S. aurantiicola*). Petch, however, still recognizes and uses *Microcera.*

Nectria Infections. In the United States two species of *Nectria* are of importance : *N. diploa* Berk. and Curt., and *N. vilis* (Syd.). Concerning the latter, only the imperfect form *Tubercularia coccicola* Stev. has been reported from North America, where it occurs on the snow scale, purple scale, and probably others. *N. barbata* Petch occurs on scale insects on citrus in Ceylon.

Nectria diploa, commonly known as the "pink fungus" in Florida, where it was discovered in 1912, forms pustules somewhat similar to those of *Sphaerostilbe.* The stroma is light rose or pink in color. A pink or reddish border surrounds the scale on which it has grown. It occurs on numerous scale insects in various parts of the world. It has its imperfect stage in *Microcerca* (or Petch's *Pseudomicrocera*). The Florida red scale has been reported to be particularly susceptible to its action, and Watson (1913) claimed success in spreading it artificially to good effect. The fungus also occurs on purple scale but with less frequency than on the red scale.

Podonectria, Lisea, and *Tarrubiella* Infections. In 1921 Petch proposed the generic name *Podonectria* (= *Berkelella*) to include three species of fungi commonly called the "white-headed scale-fungi" : *P. coccicola* (E. & E.), *P. aurantii* (Hoeh.), *P. echinata* Petch. All are found on scale insects.

P. coccicola was one of the first entomogenous fungi observed in Florida where it occurs on Glover scale, purple scale, and chaff scale. It has also been found in South America, the Orient, and elsewhere. Watson and Berger (1937) are of the opinion that it was this fungus which saved the citrus industry of Florida in the 1830's when Glover scale (or the "long scale") was introduced into that state. The trees were at first killed back each year by the scale until presumably the fungus came to the rescue and saved them. Others feel that the fungus is in reality only a secondary parasite or a saprophyta which grows on the armors of the scales. Fisher (1947) observed it growing in such a location on purple scale that had been killed by the endoparasitic chytrid, *Myiophagus* sp. It usually appears as small whitish heads growing out from the scales. The imperfect form is *Tetracrium coccicolum* Hoeh.

P. aurantii is similar in appearance to *P. coccicola* and is found in Brazil on citrus and in Formosa on the black parlatoria scale. *P. echinata* occurs in Ceylon on *Lepidosaphes* sp.

Lisea parlatoria Zimm, has been observed on the black parlatoria scale on citrus leaves in Java. The perithecial heads are dark violet to black in color.

Torrubiella lecanli Johnst. occurs on the hemispherical scale in Cuba and Puerto Rico. Its perithecia are of a vivid yellow color. Incidentally, at least three species of *Torrubiella* have been reported from North America. One species, *T. gibellulae* Petch, a parasite of arachmids, is known to be the perfect stage of *Gibellula aranearum* (Schw.). A considerable number of species originally placed in the genus *Torrubiella* are in reality members of other genera or synonyms of other species.

Hypocrella and *Aschersonia* Infections. Members of the genus *Hypocrella*, included in the genus *Hypocrea* by early mycologists, have their imperfect stages in *Aschersonia*. The genus *Aschersonia*, as such, will be treated more fully in our section dealing with the fungi parasitic on whiteflies.

As has been pointed out by Petch (1921), even though species of *Hypocrella* so closely resemble the corresponding species of *Aschersonia* that it is not possible to decide which a given stroma is without sectioning it, yet it was apparently not until 1896 that any relation between the two was suggested. Since then it has definitely been ascertained that *Aschersonia* is the conidial form of *Hypocrella*. Nevertheless, the *Aschersonia* stage is found much more frequently than the *Hypocrella* stage. An *Aschersonia* stroma usually does not subsequently become perithecial although there are exceptions to this. In some gathering all the stromata will be *Hypocrella* ; in others all *Aschersonia*. Just what conditions govern the production of either stage is not known. Occasionally both stages may be found in the same stroma. It is by such findings that it is possible to correlate species of *Aschersonia* with their *Hypocrella* stages.

In 1921 Petch wrote that 70 species of *Hypocrella* and 60 species of *Aschersonia* have been described. Of these, however, only a total number of 54 names covering 42 species are valid. In the group found on scale insects there were then known to be 20 species of *Hypocrella* ; the corresponding *Aschersonia* was known in 11 cases. In 1947 Petch informed the writer that because certain material was not available to him during World War I, these figures should actually have been slightly different. At any rate, in the scale-insect group of fungi, Petch states that there are now known to be 22 species of *Hypocrella*, the corresponding *Aschersonia* being known in 12 cases, and 3 unattached species of *Aschersonia*. Charles (1941*b*) lists 8 species of *Hypocrella* known in the United States, at least 5 of which have known *Aschersonia* stages.

Whereas species of *Sphaerostilbe* and *Nectira* occur on the armored scales (Diaspididae), species of *Aschersonia*, on the other hand, infect soft scales (Coccidae) and whiteflies (Aleyrodidae). In the latter case, there is also a difference between the two families in that those *Aschersonia* parasitic on whiteflies have paraphyses (elongated sterile cells) in the pycnidium, while those parasitic on the coccids have no paraphyses. It might further the mentioned that considerable care should be exercised when examining scale-infested trees to decide just which insect a *Hypocrella* is infecting. If an armored scale and a soft scale occur together on the same plant, the fungus may destory all the soft scales, leaving only the armored scale. Cursory examination may then lead the observer to think that the armored scales he sees were the hosts of the *Hypocrella*.

As far as scale insects are concerned, the better known *Hypocrella* and their *Aschersonia* counterparts include *Hypocrella epiphylla* (Mass.) (*Aschersonia cubensis* Berk. and Curt.), *H. turbinata* Berk. (*A. turbinata* Berk.), and *H. javanica* (P. & S.) (*A. coffeae* P. Henn.).

Myriangium Infections. Although not a Hypocreales, but of the order Myriangiales, the genus *Myriangium* merits consideration at this point because it is one of the important parasites of scale insects throughout the tropics. Mordern descriptions of the genus have been presented by Petch (1924, 1946) and by Miller (1938, 1940). In the words of the latter author, the plant body of *Myriangium* consists of a black pseudoparenchymatous storma, which later gives rise to a peculiar type of apothecium with an apically delimited fertile region, consisting of asci at different levels embedded in coalesced fungous tissue. Apparently there is no conidial stage. The stroma is superficial on the bark of the tree, and under each storma are several dead scales, penetrated and covered by mycelium. There is a definite relationship between the death of the scale and that of the limb and the *Myriangium*. Live stromata are not found on dead branches.

The commonly known species of *Myriangium* include *M. duriaei* Mont. and Berk. *M. floridanum* Hoch. and *M. montagnei* Berk. Several authors have cited these species to be

effective natural control agents, especially against citrus scales.

Other Fungous Parasites of Scale Insects. In addition to the species of Hypocreales and Myriangiales mentioned in the preceding paragraphs, a number of fungi parasitic on scale insects exist in other groups of fungi. Most of them remain among the Fungi Imperfecti.

Cephalosporium lecanii Zimm. occurs on several species of scale insects in the Americas and in certain islands of the South Pacific. In Florida it has been found on a number of scales on citrus and from a natural-control standpoint has been reported to be an effective parasite of the pyriform scale. In Brazil it is considered to be important in the control of the green scale on coffee plants, where it affects all the internal parts of the insect, including the eggs. The fungus grows around and over the scale insects. It is at first white but finally becomes plae yellow or lemon yellow in color. It usually has a powdery or mealy appearance, which is due to the numerous minute heads or spore clusters that develop on the conidiophores breaking out through the scale. Eventually it may appear smooth and waxy because of the fusion of the conidial heads. The fungus grows well on artificial media. Some authorities consider this fungus to belong in the genus *Verticillium* along with another species, *V. cinnamomeum* Petch, which is found on scale insects and whiteflies in Florida. Other species of *Cephalosporium* have been reported on scales as well as other insects in countries other than the United States.

Similar to *Virticillum* is the genus *Cladobotyrum*, which contains at least one species, *C. heterocladum* (Penz.) parasitic on scale insects. It was described originally as *Verticillium* on the brown scale in Italy, and this or a closely related species has been so designed in Florida.

Several species of *Fusariam* have been reported on scale insects in various parts of the world. *Beauveria globulijera* (Speg.), the so-called "chinch-bug fungus," is sometimes found on the brown scale. *Spicaria javanica* Bally occurs on several scales in Ceylon as well as on the cottony cushion scale there and in Florida, where its control value has been noted when conditions were optimum. *Aspergillus depauperatus* Petch (considered to be a strain of the *A. restrictus* Smith series not far from *A. gracilis* Bainier) has been identified on specimens of California red scale on citrus in Palestine, and on other species in Enland and Ceylon. In certain parts of the world representatives of the genera *Pegiotrichum*, *Acrostalagmus*, and *Rhinotrichum* are found parasitzing scale insects. Species of *Cladosporium* are also found in this relationship but probably only as saprophytes or weak parasites.

In moist seasons along the California coast, a white powdery fungus appears on the black scale, sometimes killing considerable numbers of them. This fungus has been known as "*Isaria*" although no true *Isaria* stage has been observed. Quayle and Tylor (1915) were able to kill a fair percentage of scales with this fungus in moist chambers in the laboratory, but attempts to initiate outbreak in the field were unsuccessful. Incidentally, as early as 1898 attemps were made to control the black scale in California by means of fungi, with favourable results being repected (Woodbridge, 1906).

Although on the whole the relationship between scale insects and species of *Septobasidium* (Basidiomycetes) is one of mutualistic symbiosis, enough of any colony of scale insects are parasitized to warrant the mention of the genus along with other fungi parasitic on this group of insects.

FUNGI PRASITIC ON WHITEFLIES

A considerable number of species of fungi are parasitic on the immature stages on several species of whiteflies (Aleyrodidae). Particular attention has been given to these combinations of parasitism in Florida, where the two most prominent species of whiteflies on citrus are the citrus whitefly, *Dialeurodes citri* (R. & H.), and the cloudy-winged whitefly, *Dialeurodes citrifolii* (Morg.). At least five species of fungi occur commonly on the nymphs of these two species. These are of enough importance to merit separate discussions.

Aschersonia aleyrodis Webber. The so-called "red aschersonia," *Aschersonia aleyrodis,* occurs on a number of species of whiteflies other than its principal hosts in this country. The fungus has its perfect stage in *Hypocrella libera* Syd. The imperfect stage forms a raised, flattened, pulvinate, red to pinkish-buff, stromatic pustule. It was 1st described in 1897 by Webber, who observed that, although all immature stages of the insect are readily attacked, those attacked while young were most abundant. In the area of Panasofkee, Florida, he noticed that the whitefly population decreased markedly with the spread of the fungus. In previous years the whitefles and the accompanying sooty mold has been so abunbant as to require the washing of all fruits. A few years after the fungus had extended its effectiveness no fruits had to be washed.

Soon after a nymph becomes infected with the fungus, it becomes swollen and may secrete more honeydew than usual. With the aid of a lens the intricately woven hyphae may be seen within the body of the insect. As the fungus develops internally, the interior organs appear to contract away from the margin. Soon after this the insect dies and the hyphae break through the body wall and form a dense marginal fringe around the edge of the insect. Subsequently the red pustule with its spores develops. As Fawcet (1944, 1948) has pointed out, what is often referred to by entomologists as "natural" or "unexplained mortality" may in reality be caused by the fungus in the early stages of its growth within the body of the insect. If conditions are such that the nymphs should become completly dried soon after death, the development of the fungus would be stopped before the latter became visible and in this way passes for unexplained mortality."

The fungus was first cultivated by Fawcett (1907, 1908), using sweet-potato strips. Berger (1919, 1921) used this method in modified form for the production of large quantities of the fungus by inoculating sterilized sterilized sweet-potato slices in pint bottles. Matured cultures were obtained in this way in from 30 to 40 days and could be held in storage for extended periods of time. These bottles were then sent out to the growers who, after shaking the culture in water and filtering the suspension through cheesecloth, distributed the fungus artificially by spraying. One pint of culture was recommened for each acre of the orchard. The season in which the most successful results could be expected was during the rainy moist-weather period of June and July—Florida's rainy season.

The grower did not necessarily have to depend upon these cultures for a supply of the fungus. This could usually be secured by obtaining infested leaves from another orchard. Such leaves, bearing the fungus-infected insects, could be placed in water and the resulting spore suspension used as a spray. Before the spray method wall was understood, leaves bearing pustules of the fungus were pinned to trees infested with whiteflies. This method, through reasonably effective, was so slow that the spray method soon replaced it.

Aschersonia goldiana Sacs. and Ellis. The yellow aschersonia, *A. goldiana*, for a number of years was identified in Florida as *A. flavocitrina*, P. Henn. It has been reported on species of whitefles in Florida, the West Indies, Panama, and on an unidentified insect in Venezuela. In Florida it is parasite especially on the cloudy-winged whitefly, *Dialeurodes citrifolli* (Morg.), although it is occasionally found on other species.

A. goldiana resembles the red aschersonia in habits and general appearance, except that its pustules are yellowish in colour. It can be grown on sweet-potato slices in much the same way as that used for the red aschersonia. It does not appear to be so effective a control agent as was first expected and is useful only in combating the cloudy-winged whitefly, for which it is probably no more effective than the red aschersonia. For interesting field reports on this and other entomogenous fungi, the reader is advised to consult the Annual Reports, as well as other publications of the Florida Agricultural Experiment Station during the early 1900's.

Aegerita webberi Faw. This fungus, commonly called "Webber's brown fungus," occurs in the United States, the West Indies, Ceylone, India, and New Zealand. In Florida its principal hosts are the *Dialeurodes* species mentioned earlier and *Aleurothrixus howardi* (Quaint). In Cuba the citrus blackfly, *Aleurocanthus woglumi* Ashby, is attacked. The fungus was first discovered in its sterile form in 1896 by Webber ; its spore stage was found by Fawcett (1910*b*), who named it *Aegerita webberi.* It is a deuteromycete, the perfect form not having been found.

In its sterile form the mature fungus consists of a chocolate-brown stroma covering the entire insect, from the margin of which extend colorless thick-walled hyphae. Later in the development of fungus, usually during the summer or fall, the marginal hyphae grow out long and colorless, extending not only over the undersurface of the leaf but also around the edges and upon its upper surface. The hyhae may sometimes extend down the petiole and along the stem to the next leaf. Every larva in the area covered becomes infected. When the hyphae become very abundant, the fungus may almost entirely cover the leaf surface with silky grayish-brown strands. The sporodochia, which consist of aggregations of conidialike, inflated, spherical cells, and borne on the upper surface of the leaf, on short lateral hyphae. These aggregations usually remain in union and function as a spore. Their light weight and radiating appendages aid considerably in their dissemination by the wind.

The fact that *A. webberi* may spread by means of superficial hyphae spreading over the surface of the leaves, as well as by sporelike aggregations of cells carried by air currents or by insects, reported makes it one of the most efficient parasites of whitefly nymphs, provided that conditions for growth are favourable. Its ability to spread rapidly was one of the characteristics noticed by Webber when he first found this fungus in part of a 5-acre orchard in Manatee, Florida. At that time millions of live nymphs could be found in the trees of this orchard where the fungus had not spread. There was no trace of the fungus in adjoining orchards. None months later the fungus had spread so rapidly that it was difficult to find a living specimen of the whitefly in this 5-acre ordhard.

Trees in the adjoining orchards remained heavily infested with the insect. Within an additional year's time, the fungus had spread over a radius of approximately 2 miles, killing all immature stages of the insect as it spread. Although the accuracy of these observations has been questioned. Webber himself did not see fit to alter his opinion. That the sporodochia themselves might spread the fungus was shown experimentally by Fawcett (1910*b*) when he inoculated nymphs by drawing over them a camel's-hair brush moistened with water containing the aggregated sporelike cells. The insects showed signs of infection in 9 days, and in 16 days the stromata brust through the edges of the nymphs.

Since no satisfactory method of growing *A. webberi* in pure culture in large quantities has been developed, its distribution is usually effected by one of the three following methods (1) pinning onto leaves containing whitefly larvae other leaves bearing infected insects and mature pustules ; (2) washing off or triturating the pustules in water and spraying this suspension on the infested trees ; (3) planting young fungus-bearing trees in such a way as to have their leaves intermingle with the whitefly-infested trees. In certain areas and under optimum conditions satisfactory control of whitefly larvae was obtained using these methods. Natural outbreaks, especially in certain low-lying hammock groves, are sometimes equally effective, making artificial remedial measures unnecessary.

Fusarium aleyrodis Petch. This fungus has already been referred to in connection with scale insects. It is best known, however, for its parasitization of whiteflies, *Dialeurodes citri* (R. & H.), and *D. citrifolii* (Morg.). Because it present a delicate fringe of white hyphae growing outward from the edges of the larvae, *Fusarium aleyrodis* is commonly known as the "white-fringe fungus." It is probably the best known of about six identified and a considerable number of unidentified entomogeneous species of *Fusarium* in North America.

The hyphae at first bear one, two, or three-celled conidia, oval to fusiform in shape. Later pinkish spore masses are formed on the edge of the larvae. The fungus is readily cultivable in pure culture.

The role of this fungus in the control of whiteflies in Florida does not appear to be so definite as is the case with some of the others. Under certain condition, Watson (1913) found that 98 per cent of the larvae showing a "natural mortality" were infected with the fungus even though it was not apparent to the unaided eye. Adult whiteflies were also observed to harbor the fungus, which Watson succeed in isolating from them. The eggs of the insects apparently also become parasitized and destroyed by the fungus.

Other Fungi Parasitic on Whiteflies. Similar in general appearance to the pustules of Webber's brown fungus are those of the cinnamon fungus, *Verticillium cinnamomeum* Petch, first seen in Florida in 1905. It has been found on both whiteflies and scale insects. Several species of *Aschersonia* and *Hypocrella* have been reported on aleyrodids in various parts of the world, including the West Indies and the Orient. According to Petch (1947), in the whitefly group of fungi, there are 13 species of *Hypocrella* of which the Aschersonia stages are known, and 14 unattached species of *Aschersonia.* Another Hypocreales, *Stereocrea aurantiaca* Petch, occurs on whiteflies in Florida and is the perfect stage of an *Aschersonia.* In Ceylon, *Rhinotrichum album* Petch has been observed on a species of aleyrodid.

An unidentified species of *Sporotrichum* has been isolated from the adult winged stage as well as from the nymphs of whiteflies in Florida, and it was thought to be responsible for the death of large number of them.

OTHER ASCOMYCETE INFECTIONS

In the past some authors have considered many of the entomogenous fungi that we have discussed under Hypocreales as belonging to the order Sphaeriales. Modern treatments however, clearly distinguish between the two orders. Both are extremely large groups, but as now constitued, most of the entomogenous species fall in the order Hypocreales. Only a few are included in the Sphaeriales. The entomogenous members of some general, *e.g.*, *Sphaeria*, have been placed in synonymy with certain Hypocreales, *e.g.*, *Ophicordyceps*.

As for as North America is concerned, the remaining orders of Ascomycetes contain relatively few species of fungi definitely pathogenic for insects. Species such as *Scorias spongiosa* (Schw.) in the order Dothideales have been found associated with insects but are probably saprophytic in nature, since the initial substratum frequently consists of insect secretions. A species of *Eurotium* (order Eurotiales) has been listed by Charles (1941*b*) on a chalcid. Some primitive Eurctiales grow saprophytically on the pupal cases of certain Lepidoptera. Since this order includes many species of *Aspergillus* and *Penicillium*, ordinarily known in their imperfect or conidial stages it is entirely possible that some of these entomogenous Fungi Imperfecti will eventually be grouped here. The genera *Cenangium* and *Scleroderris* of the order Helotiales have at times been used to include entomogenous members, most of which have now been placed in synonymy with other Ascomycetes.

Although members of the large and important order Laboulbeniales may be considered as cutaneous parasites, they are generally thought of as commensals.

THE MUSCARDINE DISEASES

Since the term "muscardine" has been indiscriminately used in entomological literature to mean almost any type of fungous infection, it appears desirable to consider the origin and proper use of this word.

As far as its modern usage is concerned, the term "muscardine" apparently originated in the Italian language with the word "*moscardino*," meaning a musk comfit, grape, pear, and the like, or any of the various plants with musk-scented foliage or flowers. ("Musk," incidentally, refers to the odorous substance from the abdomen of the male musk deer, used as a basis for perfumes.) The French have the words "*muscadin,*" meaning a musk lozenge, and "*muscardin,*" which, in addition to referring to the dormouse (*Muscardinus*), also means a comfit or bonbon. Because the bodies of the insects infected with the fungi we are about to describe are transformed into white mummified specimens resembling in appearance comfits or bonbons, the natives of France referred to them as "*muscardin.*" French scientists added a final "e" to the word and used it in referring to the fungus concerned. It has since been taken over as a bona fide English word, and English dictionaries and encyclopedias furnish us with at least three meanings or uses of the word : (1) as a noun meaning the fungus ; (2) as a noun meaning the disease caused by the fungus ; (3) as an adjective, "muscardined."

The word "muscardine" was first used as it applied to the well-known disease of the silkworm, and also referred specifically to the fungus *Beauveria bassiana* (Bals.). Soon thereafter it was also used in reference to the fungus we now know as *Metarrhizium anisopliae* (Metch.), and because of the green colour of the spores it was called "green muscardine." In addition to these two infections, it might be permissible to use the term generally in connection with those mycoses in which the fruiting bodies arise on the exterior of the insect, producing a thick covering about the animal. The fungi concerned are known primarily in their imperfect stages (Fungi Imperfecti).

MUSCARDINE OF THE SILKWORM

Muscardine of the silkworm, *Bombyx mori* (Linn.), occurs throughout the world wherever this insect is reared. The disease has been particularly important in France and Italy, where tremendous losses have been experienced. Even after the nature of the muscardine and its control were understood, losses were considerable. For example, in northern Italy, around 1925, approximately 11 million pounds of cocoons were being lost every year from muscardine alone. Outbreaks still occur, but for the most part they are sporadic. In Italy, incidentally, the disease was first called "*mal del segno*" (the disease having a sign) ; later and at precent it is known as "*calcino*" (calcium ; white powder).

Early History. It is not strange that muscardine was the first acknowledged disease of the silkworm, since it is so easily recognized. The white mummies into which the diseased larvae are transformed are very distinctive and noticeable, and early sericulturists were able to ascertain their presence readily.

Until 1835 it was generally believed that the disease was not of a contagious nature but that it was caused by a variety of agencies, notably those pertaining to meteorological conditions and rearing techniques. For example, Boissier des Sauvages (1763) believed that it was due to a particular state of the atmosphere which precedes storms and which is called "*touffe*" (*i.e.*, "wisps of heat"). Nysten (1808) attributed the disease to defective incubation. Dandolo (1825) declared that it resulted from abnormal physiological conditions. The white efflorescence that develops on cadavers was, according to him, "the original mineral." Other causes were postulated by other writers, but most of them were variations of these amicrobic influences.

The credit for first showing that the disease is a contagious one and that it is parasitic in origin is generally given to Bassi de Lodi (1835-1839), who showed that the affliction was caused by a fungus that multiplied in and on the body of the silkworm. This discovery naturally created quite a sensation among sericulturists and microbiologists alike. Other investigators were drawn to this interesting case of parasitism. The fungus itself was studied and described by Balsamo, who gave it the name *Botrytis bassiana*, the specific name honoring Bassi. Audoin (1837 *a,b*) elaborated on Bassi's discovery and demonstrated that this "cryptogam" would reproduce the disease when inoculated artificially into the body cavity or the fatty tissue of a healthy silkworm. This was true regardless of the fungus's development, however, certain humidity requirements were necessary. Audoin also observed that the malady was not peculiar to the silkworm but that many other insect species were also susceptible. In 1839 Johanys was able to cultivate the fungus on nonliving organic media.

The contagiousness of muscardine was not, however, accepted by all biologists and sericulturists without question. It took some years before the relation between the fungus and its host was thoroughly understood. Such a recognized authority as Guerin-Meneville (1848), for example, doubted the contagiousness of the infection and believed that the manifestations of the fungus were merely symptoms or signs of careless handling. Vittadini (1853), however, did a satisfactory job of refuting Guerin-Meneville's arguments and of showing that the fungus was, in fact, the true cause of muscardine.

Among the more modern workers (*i.e.*, since 1900) on this silkworm disease or on the fungus itself are included such names as Quajat, Verson, Conte and Levrat, Beauverie, Arnaud, Paillot, and Masera. These and others have contributed to our knowledge of muscardine until now it is a well-understood infection and one that can be coped with intelligently.

The Causative Fungus. As we have already mentioned, Balsamo placed the fungus responsible for muscardine in the genus *Botrytis* of which *Botrytis cinerea* Pers., the cause of a vine disease, is the type species. In 1911 Beauverie (1914) studied the fungus of muscardine, comparing it with yet another closely related species (*B. effusa*) found on silkworms, and he showed that the two species possessed common properties and that it was necessary to create a new group to include them. Shortly thereafter (1912) Vuillemin, who in 1910 was preoccupied with modifying the classification of Hyphomycetes of the Fungi Imperfecti, created the genus *Beauveria* of which the species *bassiana* became the type. Although Clements and Shear (1931) have since placed the genus *Beauveria* in synonymy with *Phymatotrichum*, most authorities retain the former name. Therefore, according to present usage, the name of the fungus responsible for muscardine of the silkworm is *Beauvria bassiana* (Balsamo) Vuillemin. Following the original taxonomic work on *B. bassiana*, a number of workers made comparative studies of the several species in this genus, including *Beauveria effusa B. densa* (Lk.) (=*B. tenella* Del.), and *B. globulifera* (Speg.), all of which are capable of infecting the silkworm. The distinct identity of the genus is now generally recognized.

The morphological and physiological characteristics of *Beauveria bassiana* have been studied in culture as well as in nature. The fungus grows well, at an optimum temperature of approximately 28°C., on most of the artificial media used to cultivate fungi. It characteristically produces a flat, mealy, chalky, pulverulent growth, with spore formation taking place in from 3 to 7 days. When the spore is placed in water it becomes swollen in 24 to 48 hours and puts out one or more slender thin-walled germ tubes. About 32 hours later the germ tubes range from a few microns to about 80 microns in length, and the branches arising from these are very short. Conidial development begins at this stage, and later, when conidia are abundantly produced, they are borne in rather compact globose heads, either on the main hyphal branches or on short laterals that are usually at right angles to the main axis. This branching may be repeated, forming compact heads.

The Disease in the Silkworm. Infection of the insect by *Beauveria bassiana* begins soon after the animal's integument becomes contaminated with the spores of the fungus. Certain workers believe that the insect also becomes infected by way of the tracheal openings or the digestive tract. Infection by these routes, however, apparently occurs infrequently. It appears that in most instances the infection occurs by a direct penetration of the integument by the infecting germ tube. The germ tube is produced within about 2 days after the spore or conidium lands on the insect. Infection is facilitated by the presence of warm temperatures and humid atmosphere ; although if these conditions are maintained for more than 24 hours, the rate of infection is usually retarded. As the mycelial filament penetrates the chitin the latter appears to be dissolved or digested. The filament penetrates through the cuticle and is met by an increased number of blood cells which become intermingled with the fungous threads. Phagocytosis may take place to some extent, but apparently it does not afford much real protection, since the fungus usually keeps on developing even at the expense of the blood cells.

During this invasion the hypodermis is destroyed in the area immediately surrounding the infecting hypha. Hypodermal cells adjacent to this area also show pathological manifestations. For example, large vacuoles appear in the distal ends of the cells *i.e.*, the end next to the newly forming cuticle. Furthermore these cells do not stain so deeply with acid fuchsin as do normal cells.

Once within the body cavity of the silkworm, the fungus continues its development in a characteristic manner. As early as 1853 Vittadini gave a fairly accurate description of this process. He showed that in the blood of the insect the fungus multiplies in the form of short filaments or hyphal bodies. The volume of the blood diminishes, and the blood cells are destroyed in proportion to the development of the disease. The physical-chemical properties of the blood change, as is indicated by the fact that the acidity diminishes and approaches neutrality. The formation of unidentified crystals in the blood has also been reported. As the infection proceeds, the circulation is slowed, then stops, and the consistency of the body becomes pasty. General paralysis sets in, followed by the death of the insect. According to Paillot (1930), invasion of most of the caterpillar's solid tissues, such as the fat body, does not occur until after the death of the insect.

After death the insect's body takes on a reddish tinge and becomes more and more hardened. The exact nature of this red coloration has been a subject for argument. Perroncito as well as Masera believe it to be due to the presence of the red-pigmented bacterium, *Serratia marcescens* Biz. Within 24 to 48 hours after death, the body is covered with a white network, which finally takes on a mealy aspect after the formation of the conidia. At the same time there appears a white inflorescence of an essentially crystalline nature. Verson gave the composition of this material as a double oxalate of magnesium and ammonium. Its true nature and identity have not been confirmed. A common post-mortem change seen in some insects infected with *B. bassiana* is the liquefaction of the internal tissues. This liquefaction is usually not followed by the formation of spores by the fungus.

That the lethal action of *B. bassiana* may not be entirely parasitic is indicated by Dresner's (1947) observation that a steam-acetone extraction of mycelium produces a substance that, even when greatly diluted, has a marked insecticidal action against certain mosquito larvae. He also noted that the germinating fungus secretes a chemical that possesses a knockdown effect on houseflies. Within half an hour after being dusted with spores of the fungus, the flies began dropping from the walls and ceiling of the moist chamber in which the tests were run. Within 3 hours there was 100 per cent knock-down with no subsequent recovery. A very high relative humidity appeared to be necessary to bring about the phenomenon.

Individual silkworms show very little true immunity against the fungus. The fact that larvae of the fifth instar are more sensitive to infection than are those of the first instar does not in itself represent a true immunity in the latter but rather is the result of mechanical and physiological barriers. Apparently there has been a diminution in the severity of the disease since the early part of the nineteenth century. Some workers have postulated that this may be due to a decrease in the virulence of the fungus, a developed immunity in the insects, or both. There is not proof that any of these occurs, however, and it is more likely that the decrease in incidence has been the result of better handling of the silkworms and the adoption of more sanitary precautions. Strains of *Beauveria bassiana* kept on artificial media have been known to retain their virulence for extremely long periods of time. Frequent transfer does lower its virulence somewhat, and it is therefore better to keep a single culture for a long period of time than to transfer it frequently if the maintenance of a high virulence is desired.

Transmission of Fungus. *Beauveria bassiana* is transmitted from one silkworm to another principally through contact and contamination. The fact that infection occurs by the cutaneous route enhances the effectiveness of this mode of transmission. The fungus is not transmitted through the egg. Adult moths, however, may contract the disease, and, in fact, death may occur more rapidly among them than among the caterpillars. Althouh not of much consequence, transmission from one generation to the next may occur when conidia contaminate the exterior of the eggs and thus infect the larvae that hatch from them. Such contaminated eggs may be disinfected and the danger of infecti on thus removed.

Theprincipal means by which the fungus is transmitted from one generation of insects to the next is made possible by the great longevity of the conidia or spores. These remain alive from one season to the next in the rearing rooms, and it is by such preservation that the disease is carried from generation to generation in most cases. Although in moist air the duration of vitality would be considerably shorter, it has been shown that the spores remain viable in dry air for at least5 y ears.

Methods of Combating Muscardine of the Silkworm. Measures employed in the control of silkworm muscardine may be divided into two groups : those taken during the course of rearing and those taken between rearing seasons or without reward to these seasons.

When the disease strikes during the rearing season, the sick caterpillars should be removed and burned before the appearance of the conidia, which are formed about 48 hours after the death of the insect. The ailing larvae may be picked out by hand, but a more convenient method is that of using a wire mesh or perforated paper as a barrier. The healthy caterpillars have the ability and energy to crawl through the openings in this barrier to fresh food ; the diseased ones remain behind and with the litter should be collected and burned.

Numerous attempts have been made to apply chemical and gaseous disinfectants without interfering with the vitality and development of the silkworms. Formaldehyde vapors and burning sulfur have been used in Europe with varying degrees of success. They are somewhat impractical because of the care necessary to maintain a dosage which is effective and yet which is not harmful to the insects. For example, the burning of 100 grams of sulfur per cubic meter is effective, but it also is injurious to the insects. Smaller dosage, 20 to 30 grams per 100 cubic meters, repeated daily for considerable periods has been recommended. Similar difficulty has been had with the use of other compounds. Masera (1940) found "Procid," a mixture of copper oxychloride and hygroscopic substances, to be very effective against the fungus, but toxic against larvae that ate it. Similar results were obtained with sodium bisulfite.

Between rearing seasons, the equipment and surroundings should be thoroughly cleaned and disinfected. Almost any effective fungicide may be used to destroy the fungi, as long as it is capable of destroying the resistant spores. For many years fumigation with sulfur in the presence of water vapor was used, especially where expense was a factor. Formaldehyde is perhaps one of the most effective vapors, since it is penetrating as well as fungicidal. Commercial formalin is useful in disinfecting the rearing equipment, as are numerous manufactured fungicides. A commonly recommended disinfectant is copper sulfate used in a 5 per cent solution. Some writers suggest a much lower dilution (*e.g.*, 1 pound of copper sulfate in 100 gallons of water). Pulverized copper sulfate powder added to walks and pathways leading to the nurseries is another precautionary measure.

In countries like France and Italy, where the disease has been particularly severe, laws and regulations have been established to aid in its control. When the disease makes its appearance in a silkworm nursery it must be reported to the proper officials. Quarantine placards are posted, and no one associated with a diseased colony is permitted to visit a healthy one.

The measures mentioned here for the control of silkworm muscardine may, to a considerable extent, be applied to most fungous diseases of other insects. Modern insectaries, as well as individual entomologists, are occasionally troubled by the occurrence of fungous diseases among stocks of insects being reared for experimental and other purposes. The experiences sericulturists with muscardine provide information that may have general application.

BEAUVERIA BASSIANA INFECTIONS IN OTHER INSECTS

In North America, at least 30 species of insect have been reported as hosts of *Beauveria bassiana*. Numerous additional species have been cited from Europe and other parts of the world. Some of these insects represent serious agricultural pests, and their susceptibility to this fungus deserves further mention.

Infection in the Corn Borer. The European corn borer, *Pyrausta nubilalis* (Hbn.), is readily attacked and killed by *Beauveria bassiana*. This was shown with certainty when Metalnikov and Toumanoff (1928) showed that *B. bassiana*, as well as certain other fungi, was capable of causing disease and death in the corn borer when experimentally infected. Corn-borer larvae were much more sensitive to infection than were larvae of the wax moth, *Galleria mellonella* (Linn.), which was also susceptible. The first report of *B. bassiana* on the corn borer in North America was that by Lefebvre (1931), who observed the mycosis among laboratory specimens at Arlington, Massachusetts, and imported from Manchuria. He was able to distribute the disease readily both in the laboratory and in the field. In 1934 Bartlett and Lefebvre reported on rather extensive field tests using the fungus against the corn borer in eastern United States with encouraging results. A considerable reduction in the larval population was obtained after fields of infested corn and weeds were dusted with a mixture of spores and flour. Canadian workers then undertook similar field trials in Ontario, obtaining controls of 60 to 70 per cent. They emphasized the fact that the time of application of the fungus is of much greater importance than is the rate of application. A similar emphasis was made by Steyaert (1935) in the case of *Stephanoperes hampei* Ferr., a pest of coffee in the

Belgian Congo. In this instance infection with *B. bassiana* must be effected before the insect has tunneled into the berry.

The pathogenesis of the infection in the corn borer has been worked out by Lefebvre (1934). The germinating spore on the surface of the larva produces an infection hypha, which penetrates the integument of the insect at any point except the head. This penetration appears to be assisted by the production, on the part of the hypha, of substances that dissolve the chitinous layer of the body wall and by mechanical pressure exerted by the fungus. Lefebvre asserts that he found definite indications that infection may also take place by way of the alimentary tract. After infection through either point of entry, the fat body is the first structure to be attacked. The glandular structures and ganglia are next affected, while the last tissues to be attacked are the nervous system, the gonads, and the muscles. The fungus frequently penetrates the tracheae, greatly impairing the process of respiration.

The external appearance of a diseased corn borer is similar to that of a silkworm infected with the same fungus. Such a larva first becomes sluggish in movement, fails to respond to most external stimuli and in most instances turns a characteristic pink color. The larva remains soft and pliable until the mycelium has ramified and grown throughout the various parts of its body. Following this the body becomes rigid and mummified. It can now be readily broken, and the whitened chalky contents can be crumbled into a powder. No external signs of the fungus are evident as long as the mummified larva is kept in a dry atmosphere. Soon after exposure to moist air, however, the white mycelium becomes apparent over the surface of the insect. Within a day or two conidia are produced abundantly giving the insect a mealy, powdery appearance.

Certain phases in the course of infection taken by entomogenous fungi in the corn borer have also been given attention by Toumanoff (1928, 1933), who worked not only with *Beauveria bassiana* but also with *B. globulifera* (Speg.), *Aspergillus flavus* Link and *Spicaria farinosa* (Fron.). Larvae infected with either of the last two species move slowly or make but slight convulsive movements when touched. Small black spots appear on the skin ; filaments of the fungus are found under these spots in a kind of abscess under the destroyed hypodermis. The filaments ramify into the body wall and into the abdominal cavity, eventually destroying all the tissues.

Infections in Other Insects. A number of other economically important insects are known to be susceptible to *Beauveria bassiana.* For example, Jaynes and Marucci (1947) found considerable numbers of codling-moth larvae, *Carpocapsa pomonella* (Linn.), dead of a *B. bassiana* infection in New Jersey apple orchards. Laboratory tests showed the fungus to be highly pathogenic when spores of the organism were either dusted upon or inoculated into healthy hibernating or freshly spun-up summer larvae. In the field, artificial dissemination of spores on the foliage brought about a significant increase in mortality. Spore dust or spray applied to infested fruit prevented the newly hatched larvae from making a successful entrance and also killed the larvae after entry. To a certain degree, therefore, fungus is an agent of natural control on all stages of the larval population. It appears to be particularly important in cool wet seasons. Infected codling-moth larvae are characterized by the dark-brown lesions over the epidermis and a weakened condition of the insect, which becomes typically mummified after the penetraiing hyphae have consumed and displaced the body tissues. The fungus is capable of killing the larvae even when the moisture content of the air is low, but the external mycelial growth with its consequent sporulation is not produced unless considerable amounts of moisture are present. The fungi within the mummified larvae will remain viable for months and give rise to the white mycelial growth when moisture is supplied. Jaynes and Marucci assert that in a few infectivity tests larvae showing only a few lession were able to recover. Since such recovery is rarely seen in insects, an investigation into its nature appears warranted.

Although not observed in large-scale epizootics, *B. bassiana* has been found pathogenic for such insect pests as the

Japanese beetle and the Colorado potato beetle. The latter insect is host to another *Beauveria*, *B. doryphorae* Poi. and Pat., in Europe where a number of important pests, such as *Pieris brassicae* (Linn.), are hosts to *B. bassiana*. The course of the infection in *Pieris* has been described by Arnaud (1927).

WHITE MUSCARDINE OF THE CHINCH BUG

Although at least six species of fungi have been reported as parasitic on the chinch bug. *Blissus leucopterus* (Say), only two have been found to be of significance as far as actual epizootics are concerned : *Beauveria globulifera* (Speg.) and *Empusa aphidis* Hoff. The latter is also a common entomophthoraceous parasite of aphids. It is frequently called the "gray fungus" to differentiate it from *Beauveria globulifera*, the "white fungus."

The chinch bug, one of the most injurious insect pests of cereal crops in the United States, was first noticed in 1783 in North Carolina. In 1840 it was reported in Illinois, and it has been under observation ever since throughout the central Mississippi Valley states, rendering its greatest damage in the Corn Belt.

Apparently the first observer to record the presence of a white fungus on the chinch bug was Shimer, in 1865, at Mount Carroll, Illinois. Although the identity of the fungus was not determined, Shimer credited it with causing the destruction of vast numbers of the insect, especially in low, creek-bottom land during moist warm weather. His belief that it was a contagious disease did not gain ready acceptance by most entomologists, and several prominent ones scoffed at the idea. Some years later, in 1882, Forbes, in Illinois, and Popenoe, in Kansas, confirmed Shimer's observations when they reported the susceptibility of the chinch bug to fungous attack. The fungus seen by Forbes and Popenoe was the so-called "gray fungus," *Empusa aphidis* (Hoff.).

The white fungus, *Beauveria globulifera* (Speg.), was observed on chinch bugs by Forbes in 1887, in Clinton County,

Illinois. Shortly thereafter it was reported from Minnesota, Iowa, Ohio, and Kansas. Thus more than 100 years elapsed between the time the chinch bug was discovered and the time the insect was found infected with the white fungus.

The Fungus. Beauveria globulifera. (Speg.) Pic. (=*Sporotrichum globuliferum* Speg.), first described in 1880 from a wireworm adult from Argentina, has been reported on approximately 75 species of insects in the United States and on numerous additional species in other parts of the world. Its parasitism does not appear to be so specialized as is that of many entomogenous fungi, since its hosts are in numerous families and in several orders of insects. It also grows on corn and certain other plants but not so well as on insects.

On its insect host, *B. globulifera* usually appears as a loose white cottony or mealy growth, at times almost completely enveloping the insect. At short irregular intervals the conidiophores bear minute heads, sessile in attachment and creamy-white in color, composed of conidia closely packed into a nearly spherical form. Some of the characters of this fungus on artificial media have been described by Pettit (1895), Lefebvre (1931*b*), and others. They are essentially the same as those of *B. bassiana* (Bals.).

Since temperature, humidity, and possibly light are variable factors that are able to cause marked changes in the activity of the fungus a consideration of their effects is warranted. Such a study was made by Headlee and McColloch in 1913; and, because their data might in a general way apply to other muscardine fungi, it is quoted from their publication :

Temperature. Field temperatures may influence the fungus in two ways—the extremes may cripple and destroy it, and the optimum encourage and further its growth.

Careful laboratory studies at this station have shown that the exposure of spores of the fungus to 104°F. to 105°F. in a saturated atmosphere for twenty-four hours does not prevent strong growth, that an exposure for forty-eight hours destroys most of the spores, and that an exposure for seventy hours does not kill them all. Spores allowed to develop for forty-eight hours, and then exposed to this temperature and moisture, perish. Spores previous to germination, in a comparatively dry atmosphere, may be exposed for five hours to as high as 209°F. without injuring the germination.

Spores, dry or wet, may be exposed over night to low temperature, even when the mercury reaches as low as—18°F., without apparent injury. Spores freshly sown were exposed to changing temperatures from January 5 to February 12, during which time the temperature changed from freezing to thawing, or *vice versa*, twenty-one time—the daily mean was below 32°F. for nine days, and the minimum temperature 15°—without injury. Growing fungus threads subjected to repeated freezing and thawing temperatures are not destroyed.

While these facts render it quite possible that the fungus growth, encouraged by moist, cool weather in exposed and cultivated fields, might be destroyed by succeeding hot weather, for the exposed dusty soil will reach about 135°F. at the surface, it is very unlikely that the ordinary foci of the disease, sheltered as they are beneath heavy growth of weeds and grass, would be even seriously injured, for in such location the temperature rarely or never reaches 91°F. It is not likely that the fungus, even in cultivated fields, would be eradicated by dry, hot weather, because while exposed mycelium would perish, the dry spores can withstand far higher temperatures than they would experience, and in the process of cultivation much of the fungus must be thrown beneath the surface of the ground, where the temperature rarely exceeds 100°F.

In view of the fact that the maximum air temperature of the last twentyfive-year period at Manhattan has been 113°F. and the minimum—35°F., it does not seem probable that extremes of temperature such as the fungus is likely to experience in Kansas will ever seriouly reduce, not to say eradicate it.

On bee-both agar the fungus makes good growth at 70°F. and 80°F., less vigorous growth at 90°F., and no growth at all at 50°F. and 100°F. The growth comes a little more quickly at 80°F. than at 70°F. The optimum temperature for its growth and spore production on this medium is therefore between 70°F. and 80°F., probably about 75°F.

As might be expected, the temperature for best growth on chinch bugs is the temperature for best growth on this medium. Only three temperatures were given careful trial—50°F., 70°F., and 90°F. The bugs succumbed to the disease most readily in 70°F.

In general, the average mean temperature in this state [Kansas] is such that the fungus can grow well from April to October, while it is more or less completely dormant from October to April.

Moisture. Moisture of two sorts appears to influence the growth of the chinch-bug fungus—relative humidity and water.

The fungus will not grow in a relative humidity of 90 per cent or less, but will remain dormant for an indefinite period. We have kept it in dried corn-meal culture for more than eighteen months, and found it perfectly virile at the end of that time ; found it would grow readily and would destroy chinch bugs. It is hardly conceivable that the chinch-bug fungus could have too much moisture in a state of nature. Dashing aad washing rains might carry much of it away, but would serve merely to distribute it, and enough would in all probability be left to carry the disease on. The best degree of moisture for germination seems to be a film of water, although germination will take place in a relative humidity of a little less than 100 per cent.

These facts indicate that while the fungus can not grow in dry weather, it is unlikely to be eradicated by any extreme of moisture it is likely to experience in Kansas. As a matter of fact, this ageees with universal experience with this fungus. It thrives in wet weather, disappears during dry periods and springs up again on the advent of sufficient moisture.

Light. General germination of the spores in a film of water required under a temperature of 75°F., twelve days, while subjected to normal daylight coming through north windows, and when in complete darkness germination became general in one day less. In a confirmatory experiment general germination was made simultaneously in thirteen days. Ordinary daylight is thus shown not to be seriously hostile. This also is borne out by field experience.

The Infection in Chinch Bugs. The course of the infection with *Beauveria globulifera* in the chinch bug is very much like that which takes place with similar fungi in other insects. Under very humid conditions, the conidium lying on the integument of the chinch bug germinates by sending out a tube that penetrates the body wall of the insect. The bug dies of the infection in about 3 days. Hyphae fill the body cavity of the bug, finally penetrating to the outside, where the body is covered with the typical white mycelial growth. Numerous conidia are formed on the conidiophores, and when these land on other insects the cycle is repeated if conditions of temperature and humidity are favorable. As might be expected, the disease spreads most rapidly when the bugs are crowded together or are very numerous on the host plants.

According to Smith (1933), the fungus also kills chinch bugs, beetles, and other insects during the winter and in the early spring which they are still in hibernating quarters, in grass clumps, under stones, and in wheat fields. The fungus thus can overwinter in protected situations that supply the proper moisture conditions, and it is carried to the field by the spring migration of the chinch bugs.

As demonstrated by Snow (1896), healthy chinch bugs may be infected with *B. globulifera* experimentally. This may be accomplished simply by dusting the insects with the fungous spores. Chinch bugs of all ages are attacked, but the older the bug the more easily it succumbs. Bugs that have been weakened or injured from any cause appear to be more susceptible to attack than are those which are in perfect condition. Adults of the overwintering second brood are more resistant to the fungus than are the adults of the first brood.

Use as Control Agent. The efforts of early workers to use *B. globulifera* in the control of the chinch bug constitute an interesting and important phase in the history of microbial control. The failure of the fungus to fulfill the high hopes of those who advocated its use serves as a warning to be remembered whenever excessive enthusiasm is expressed over the control value of newly discovered insect pathogenes, especially when this control depends upon the artificial distribution of the microorganism.

The first attempt to bring about an outbreak of the disease by artificial dissemination was that by Lugger, in 1888, who distributed diseased bugs in several localities in Minnesota. Although the experiment appeared successful, Lugger suspected that, since the disease spread so rapidly, the spores of the fungus were already present in the test fields and that he had only reintroduced them.

In 1888 F.H. Snow began his observation and experiments in Kansas ; these extended through 1896. In 1891 the Kansas state legislature established an experiment station at the University of Kansas to propagate the fungus and to distribute it free of charge. It was placed under Snow's direction. Almost 50,000 packages of the fungus were distributed by this organization, but the true value of the program was never ascertained with certainty. The reports of observers in 1891 and 1892 were very favorable, whereas those during succeeding years were less favorable. The probability that the fungus was widely distributed naturally could not be ruled out.

Distribution programs were carried on in states other than Kansas, but in each case the work was eventually dropped. Lugger in Minnesota tried it again in 1895 but abandoned the method by 1902. It was similarly abandoned in Illinois, Nebraska, Missouri, Ohio, and Oklahoma. The artificial distribution of the fungus did not appear to affect materially the incidence of the disease in chinch-bug populations or the effectiveness of the control brought about by the infection.

An appraisal of the effectiveness of these distribution programs was made easier by the work of Billings and Glenn, who, in 1911, published the results of their comprehensive observations on the disease. The following points were among those set forth in their conclusions :

1. The chinch-bug fungus is present naturally in fields everywhere throughout the infested area in Kansas.

2. It is present in such great abundance that any artificial distribution of infection in a field would be too insignificant, by comparison, to be of practical use.

3. Its distribution naturally through a field is much more uniform than any artificial distribution can be made.

4. The amount of fungus used experimentally in both wheat and corn fields was so far in excess of any that would be used by the farmer in infecting his own fields that he could not reasonably expect to succeed.

5. The fungus above little tendency to spread from centers of artificial infection. The apparent rapid spread of the fungus is due to favourable conditions bringing it into activity simultaneously over considerable stretches of territory.

6. In fields where the natural presence of the fungus is plainly evident its effect on the bugs cannot be accelerated to any appreciable degree by the artificial introduction of spores.

7. In fields where the fungus is not in evidence spores introduced artificially have no measurable effect.

8. Apparent absence of fungus among chinch bugs in a field is evidence of unfavorable conditions rather than lack of fungus spores.

9. All the benefits of the Sporotrichum [Beauveria] disease of chinch bugs may be realized by merely letting the fungus naturally present in the soil do the work of extermination as far as it will.

10. Moisture conditions have much to do with the appearance of chinch-bug disease in a field ; artificial infection nothing.

11. Spent adult chinch bugs succumb to attack more readily than younger ones, but as the old bugs have finished depositing their eggs, their loss by fungous disease accomplishes little else than increasing the amount of the infectious material.

12. Laboratory experiments can be made to prove that artificial infection accomplishes results upon bugs confined in cramped quarters and without food, but In the field, where fresh and usually drier air prevails and food is abundant, an entirely different situation is presented.

13. Advocating artificial infection or encouraging it by sending out diseased chinch bugs does not serve the best interests of the farmer, since his attention is thus diverted from other and more efficient methods of combating the pests.

14. The reported success of former year on the part of farmers is believed to be due to the following causes (1) failure to recognize spontaneous outbreaks of the disease because of previous artificial sowing of infection, and also failure to use check, or untreated fields as a basis of comparison, thus claiming the outbreak as directly due to artificial infection ; (2) failure to distinguish the skin of molted bugs from dead bugs ; (3) mistaking the scattering of chinch bugs in cornfields for evidence of their death by fungous disease when carcasses were not present as proof.

Perhaps one of the most significant conclusions arrived at by Billings and Glenn was that in fields where the fungus was naturally present, its effect on the bugs could not be accelerated appreciably by the artificial introduction of spores. Fawcett (1944) has accounted for this fact by the suggestion that the wind-borne spores are produced in such great profusion, providing wide distribution in a short time, that the saturation point for infection is readily attained and that artificial distribution is therefore of little account. It should be remembered, however, that the conclusions of Billings and Glenn apply particularly to the chinch bug, to *B. globulifera*, and to the localities in which they worked. It would be unsafe to make sweeping generalizations regarding all insect mycoses solely on the basis of their findings.

What may be accepted as the most general currently accepted opinion with regard to the control value of the fungus against the chinch bug has been expressed by Packard and Benton (1937). These authorities state that *B. globulifera* is probably the most destructive natural enemy of the chinch bug ; that it is generally present in fields throughout the country but that its effectiveness depends upon the weather ; and that, since it has been proved that the spores of the fungus are present wherever the bugs are common, its artificial dissemination as a control measure is needless.

GREEN MUSCARDINE

Upon examining containers of soil in which had been placed some larvae of the wheat cockchafer, *Anisoplia austriaca* Hbst., Metchnikoff, in 1879 observed some of the specimens to be dead. Two days after the insects died a white fungous growth appeared on their bodies, being particularly noticeable around the spiracles. This growth soon covered the animals entirely except the head. Later it took on a green color, eventually becoming a dark blackish-green. In the blood of the insects oval fungous bodies could be seen, which, as the infection progressed, filled the body cavity as a mycelial mass. For this fungus Metchnikoff first proposed the name *Entomophthora anisopliae*. Later, at the suggestion of the botanist Cienkowsky, he called it *Isaria destructor*. Delacroix (1893) published on the fungus under the name *Oospora destructor*. A few other synonyms (*e.g.*, *Metarrhizium cicadinum* v. Hoh.) are to be found in the literature. Although one of the Fungi Imperfecti and not an entomophthoraceous fungus, the original specific name remains valid, and the organism is now as *Metarrhizium anisopliae* (Metch.) Sor., as it was designated by Sorokin (1879). It is commonly referred to as the "green-muscardine fungus," being characterized by the dark-green color of its conidia, or spores. The green color, however, does not appear to be characteristic of all species of the genus, since another species, *Metarrhizium album* Petch, parasitic on leafhoppers in Ceylon, is white in color, and *Metarrhizium brunneum* Petch, parasitic on a cicadellid in the Philippine Islands, is yellow to brown. On the other hand, a fourth species, *Metarrhizium glutinosum*

Pope, isolated from deteriorated cotton, has dusky olive green to olivaceous black conidia.

Metchnikoff's 1879 report is of particular significance, since in an addendum to his article he describes how he experimentally initiated the disease in healthy larvae. It was this experiment that may be credited with establishing the idea of intentionally causing a disease in insect pests of economic importance. Metchnikoff mixed spores of the fungus with the soil and placed the healthy larvae therein. After 10 days, 8 out of 9 larvae were dead of the fungus. In 1880 he reported a similar experiment in which larvae and adults of the sugar-beet curculio, *Cleonus punctiventris* Germ, exposed to spores of the fungus died within 12 days.

Since Metchnikoff's early observations a large number of additional insect species have been reported as hosts of the green-muscardine fungus. This includes approximately 70 species in North America, with probably an equal or greater or number from other parts of the world. Apparently the first record of the fungus in the United States is that of Pettit (1895), who found it parasitic on the wheat wireworm *Agriotes mancus* (Say) in New York State.

Characteristics of the Fungus. The green-muscardine fungus *Metarrhizium anisopliae* is usually considered to have a systematic position near *Penicillium* in the family Moniliaceae. Its perfect stage is not known, and the report that it is a *Cordyceps* is generally discounted. Pettit (1895) describes a variety *americana* and Johnson (1915) records forma *major* and forma *minor* occurring on different hosts. Most of the variations that have been noticed are concerned with differences in spora sizes and in coloration.

Metarrhizium anisopliae grows readily on artificial media and develops particularly well on potato media. Ordinarily it is easiest to start a culture by using the spores of the fungus. With proper techniques, however, cultures may be initiated with nonfruiting forms. Upon germination the spore swells somewhat and a germ tube appears at one end and sometimes at both ends of the spore. As described by Speare (1912), this germ tube elongates, and branching occurs after about 30 hours and continues for several days. Certain of the branches extend upward and become the sporophores. The latter are usually closely packed with their branches intertwined. After about a week they begin to form spores which are at first white, but gradually, as they mature, appear olive green in color. The spores become partly cut off from the grouped sporophores and cohere in prismatic masses. Spore formation occurs by the simple abstriction of elongated buds which appear at the distal ends of the branched sporophores. Other buds form below the original one, and this continues until chains of spores are formed. So closely do the chains of spores arising from adjacent sporophores cohere that in old cultures they may form a crust over the surface of the medium and may be removed in large flakes. The sporophores are short and are branched in a manner somewhat resembling the branching in *Penicillium*, although the long unbranched basal portion of the sporophores usually associated with this genus is not seen in *Metarrhizium*. The spores may vary considerably in size but are usually within a range of from 5 to 7.5 microns in length to 2.3 to 3.7 microns in width.

Although only a meager amount of information is available relative to the physiological properties of *M. anisopliae*, it may be assumed that they are in general similar to those of the white-muscardine fungi already described. High humidities and warm temperatures promote the growth and development of the fungus. In culture, a temperature between 24 and 26°C appears to be optimum, the entire range for normal development being between 10 and 30°C. The ability of the spores to germinate is destroyed at temperatures between 55 and 60°C. for 5 minutes. The range of hydrogen-ion concentrations for the normal growth of the fungus is from pH 4.7 to 10. The optimum value lies between pH 6.9 and 7.4. Although a medium containing organic matter is more favourable for the growth of *M. anisopliae* than is one having an inorganic nitrogen source, both types of medium can support its growth. No special requirements are necessary as far as the organism's carbon source is concerned. The growth and fructification of the fungus are retarded by rays of the sun. The spores may be kept in a dry condition for 3 years or more.

The Infection in Insects. From the knowledge already gained during studies on white muscardine of the silkworm, a general idea of the course of such infections in insects was at hand from the time *Metarrhizium anisopliae* was discovered. Not until 1926, however, was much attention paid to the details of the pathogenesis of *M. anisopliae* infections in insects. In that year Glaser reported on his studies of the green-muscardine disease in silkworms. This was followed 3 years later by similarly detailed reports, by other authors, of the infection as it occurs in the European corn borer.

The silkworm, *Bombyx mori* (Linn,), acquires the infection via penetration of the integument by the germinating hypha. It does not appear to be susceptible via the digestive tract. After the fungus enters the body cavity it develops in the blood and does not penetrate the tissues until after the death of the insect. A reaction on the part of the host occurs as is indicated by large masses of phagocytes that congregate around and between the fungous threads. The phagocytes are unable to offer very much protection, however, since the parasite is too large for ingestion. Four or five days after infection the larvae begin to lose their appetites, assume a yellowish color, and appear sick. In 5 or 6 days they become extremely feeble and die, after which they become rigid in form and brownish-yellow in color. The fungus not invades nearly all body tissues.

After the silkworm dies, certain of the hyphae penetrate outward, and 7 or 8 days after the beginning of the infection the hyphae break through the integument everywhere. The larva soon becomes covered with a white down consisting of hyphae destined shortly to become the sporophores. A day or two later the rather short and branched sporophores begin to form spores. These soon cover almost the entire body of the silkworm, and as they mature, the spores become olive green in color, cohering in masses which may be removed in large flakes. The silkworm usually dies in the larval stage. Sometimes, especially among those infected during the last instar, some individuals will spin a cocoon and succumb within. In such cases the white down appears, but the fungus rarely fruits.

In 1929 Wallengren and Johansson published a description of the manner in which *M. anisopliae* infects the European corn borer, *Pyrausta nubilalis* (Hbn.), and in so doing added considerably to the information supplied by Glaser in connection with his silkworm studies. Infection of the corn borer takes place without any great difficulty ; the conidia fasten very easily to the intersegmental folds and from there penetrate the integument. The intersegmental folds have a thinner chitinous layer than the other parts of the segments and therefore offer a more favorable place for infection. The germinating hyphae, however, can pierce the fairly thick integument on the dorsal side of the segments.

Upon examining under the microscope a piece of integument or skin dissected from a corn-borer larva, one can see that the areas immediately surrounding the bases of the hairs are darker and brownish and present a granular appearance produced by a number of small, round, dark pigment bodies, gathered close to each other immediately under the surface of the chitin. The rest of the skin has similar granules farther apart. Histologically, the body wall shows a lamellose structure, the lamellae lying parallel to the surface. The black pigmentary bodies lie immediately under the outer heavily sclerotized surface membrane. From each pigment granule emanate one or more threads, clustered together, which continue down into the cuticle to about a third of its thickness. The threads converge so that the entire formation resembles an inverted pyramid. In the areas around the bases of the hairs, however, the threads are not gathered in pyramids but run straight down into the cuticle ; they appear to be somewhat shorter than those in the pyramids. With this histological picture in mind, let us now consider the details of infection as conceived by Wallengren and Johansson.

With the proper conditions of temperature and humidity, the fungous spores, or conidia, fastened on the surface of the integument send out germinating hyphae. At those places where the conidia germinate the cuticle assumes a yellowish color. Under this yellow spot the pigmentary bodies become clotted and fall to pieces, and the threads disintegrate. Each hypha pierces the cuticle and begins its entrance into the body wall at the point of one of the thread pyramids, which disintegrates and becomes granular. The outer layer of the cuticle is more or less destroyed, and the area concerned assumes a more yellowish or dark-brown color. These colored increase in size as the infection progresses. It is possible for the infection to occur without causing the brown spots. Also, after the hypha has penetrated the skin, the fungus appears to be able to continue its growth without further discoloring of the integument.

The growing conidial hyphae apparently secrete some sort of chitin dissolving substance, possibly an enzyme, which opens or facilitates the way for penetration. The hyphae also exert a rather great pressure on the substratum. In sections, the cuticular lamellae can be seen to be pressed down when the hyphae are developing at right angles to it. The hyphae often grow for relatively long distances between the lamellae, evidently because growth in this direction offers the least resistance, but sooner or later they turn inward, perforate the endocuticula and hypodermis, and enter the body cavity. According to Wallengren and Johansson, the hyphae seem to have a special preference for the adipose tissue, but they also invade the muscular and central nervous systems. It is not entirely clear, however, whether these workers refer to premortal or postmortal conditions. Since their description at this point apparently refers to living insects, this observation is in conflict with Glaser's assertion that in the silkworm the fungus does not enter the internal tissues until after the death of the insect. In any case, after the corn borer dies, rapid growth of the fungus takes place, filling the interior of the body. Hyphae force their way through the body wall and appear as a white fluff on the surface of the integument.

Finally, the characteristically green conidia are formed. Conidia also formed within the body of the insect.

The symptomatology of the mycosis in the corn borer is similar to that which characterizes the infection in most larvae. The first symptoms include the appearance of yellow or brown spots on the cuticle ; these usually spread over other parts of the body. The fungus seems to have some toxic influence on the host. Marked nervous disturbances become evident in the later stages of the disease. The larvae lose their appetites, become apathetic, make no spontaneous movements, and their irritability decreases more and more. Finally the insects lose their righting reflex, and death soon follows. The body finally becomes stiff and mummified. Under conditions of an atmosphere saturated with moisture and a temperature of about 25°C., death occurs in approximately 4 days. Occasionally an infected larva, under conditions not entirely optimum, will survive for a considerably longer period of time. It is unknown for larvae to survive longer than 26 days, however, once the fungus has penetrated the bodywall. In corn-borer larvae the mycosis produced by *M. anisopliae* under experimental conditions has a death rate of 100 per cent.

Use as Control Agent. Following Metchnikoff's (1879, 1880) discovery that the wheat cockchafer, *Anisoplia austriaca* Hbst., and the sugar-beet curculio, *Cleonus punctiventris* Germ., could be artificially infected with *M. anisopliae*, other workers soon considered using the fungus as a control agent on a wide scale. In southern Russia, entomological commissions were established in Kharkov and Odessa to investigate the practicability of artificially distributing the fungus to combat these insects, particularly the cockchafer. In 1886 and 1888, Krassilstschik undertook to use the fungus for combating the sugar-beet curculio, reporting field mortalities of from 50 to 80 per cent, In his laboratory at Smela, near Kiev, he produced spores of the fungus in large enough quantities to make field distribution practical, although still on a rather limited scale.

Further attempts to use *M. anisopliae* against various insect pests were made during the early years of the twentieth century in such parts of the world as Java, Hawaii, Trinidad, and Puerto Rico. In Trinidad, for example, Rorer (1910, 1913) used the fungus against the froghopper. *Tomaspis varia*, a pest of sugarcane with apparently favorable results. The spores were usually distributed in the form of dusts (mixture of flour and spores) applied to the sugar-cane plants at the rate of 2 or 3 pounds to the acre. Specially constructed large culture cabinets were devised to produce the spore material in great quantities.

In some areas, as in Java and in Puerto Rico, the fungus, although an efficient parasite of several noxious insects, did not prove to be an entirely practical means of control. Since it was rather widely distributed in nature, and since its effectiveness depended upon the existence of favorable climatic conditions, many workers decided that attempts to distribute the fungus artificially would be useless. In some localities excellent results were obtained after intial distributions, with a declining effectiveness upon subsequent distributions. The reasons for this are not clear, except that the initial epizootics might have indicateed the absence of or the rather low number of naturally existing spores in the area concerned.

In 1913, Friederichs reported excellent results with the fungus when directed against the rhinoceros beetle, *Oryctes rhinoceros* Linn., a pest of coconut. Infected beetles were placed in trap piles of rotten coconut husks and other debris, placed here and there throughout the coconut groves. The beetles gathered in these piles for egg laying. When the larvae hatched, nearly all of them were apparently attacked and killed by the fungus. According to Simmons (1939), however, "no efficient control seems to have come from the introduction" of the fungus into Samoa, although grubs are frequently found killed by the organism, and some officials maintain that it is still a beneficial agency. Simmons points out that regular provision of infected breeding traps is required and that this necessitates constant careful supervision or the results are likely to be the reverse of what is desired.

In Europe a number of attempts were made to use the green-muscardine fungus against the European corn borer with rather encouraging but not convincing results. Hergula (1931), for example, found that, when infested corn plants were dusted experimentally with fungus-spore mixtures, the infestation was reduced to such a point that the damage to the plant by the insect was practically negligible. Data on practical large-scale field application were not gathered.

From the reports and information available, it appears that the value of *M. anisopliae* as a control agent is largely in the realm of natural control. The introduction of the fungus into areas, localities, or islands where it does not occur may be entirely worth while. Once the organism is established, however, it appears that, with certain exceptions, the artificial distribution of spores is useless. If and when favorable wheather conditions prevail, the fungus if present, exercises practically its full effectiveness without the help of man. There seems to be little reason for doubting that the sporadic and small but continuous occurrence of the disease among certain insects has its beneficial effect even though this effect may rarely be enough to bring the insects under control from an economic standpoint.

SOROSPORELLA INFECTION

Although the infection we are about to consider—one caused by *Soroporella uvella* (Krass.) Gd.—is sometimes referred to as the "red muscardine" this terminology is not applicable in the same sense in which we have been using it in the forgoing discussions. Unlike the true muscardines, the fruiting bodies of the fungus in this infection are not as a rule produced as a covering on the exterior of the insects; and, instead of becoming mummified and hardened, the insects disintegrate soon after death.

One of the most thorough studies of the fungus and the infection it causes has been performed by Speare (1920). Much of the discussion to follows will be based upon reports of this mycoentomologist.

The Discovery and Systematic Positton of the Fungus. The fungus under consideration was first observed in Russia by

Krassilstschik in 1886. He observed it in the same insect, *Cleonus punctiventris* Germ., on which he did so much work with the green-muscardine fungus, *Metarrhizium anisopliae* (Metch.). Krassilstschik named the newly discovered fungus *Tarichium uvella.* Two years later Sorokin (1888), another Russian scientist, found a fungous parasite of the cutworm *Euxoa segetum* (Schiff.), which he named *Sorosporella agrotidis.* In 1889 Giard pointed out the identity of *Tarichium uvella* and *Sorosporella agrotidis.* Since the genus *Tarichium* was at that time employed for species of *Empusa* and *Entomophthora* in which only resting spores were known, and since the resting spores of the fungus under consideration were unlike those of any known *Empusa* or *Entomophthora,* Giard proposed that the generic name *Sorosporella* be used together with the specific name *uvella.* Giard's proposal has been generally accepted, even though this worker adhered to the belief that it was an Entomophthorales. That the fungus does not belong to this order was shown by Speare (1920), who considered it a verticilliacious Hyphomycete (Moniliales ; Fungi Imperfecti). This remains the systematic position generally accorded it today.

Probable synonyms include *Acremonium cleoni* Wize (1905) and *Fusarium acremoniopsis* Vincens (1915). *Massospora staritizii* Bresadola (1892) appears almost certainly to be a *Sorosporella,* but since several of its characters are quite different from those of *S. uvella,* it is perhaps another species of this genus.

Symptoms and Pathology. Noctuid larvae parasitized by *Sorosporella uvella* exhibit no symptoms of disease during the first few days after being infected. Three or four days before death, however, they lose their appetites and become sluggish in movement. In some host species a change in outward appearance may be observed 1 or 2 days before death. Such larvae may turn a creamy-white color. Sometimes red-colored patches appear a few hours before death, either posteriorly, anteriorly, or more commonly nearly the middle of the body.

Soon after death, which usually occurs within 7 to 10 days after infection, the creamy-white color changes to pink. The pink color appears simultaneously over the entire body, becoming more and more intense until the final development of the fungus is reached and the insect is of a characteristic brick-red color. The body of the insect appears shrunken and wrinkled and somewhat flattened. A longitudinal, ventral, furrowlike depression is nearly always present in its abdomen. The cadaver is soft and pliable, and if a portion of the integument is indented, it remains sunken with little or no reaction. Unlike the case in many other fungous infections, the body is never hard and sclerotiumlike. No fluid exudes if it is pricked with a needle. If the body is torn open entirely, the internal fungus mass is seen to be coherent, of a shiny creamy or pink color, and of a gelatinous consistency.

Later the host's body becomes more shrunken and the reddish color more intense. The body wall is now rather brittle, the slightest shock serving to rupture it. The spores of the fungus become brick red in color, dry, dustlike, and less coherent than previously. Virtually nothing remains of the host's internal organs, and the bony appears as a small sac filled with dust.

The histopathology of the disease has been studied to some extent by Speare (1920*a*) in connection with his observations on the yeastlike cells or "blastocysts," which represent the vegetative stage of *S. uvella*. These blastocysts multiply in the hemolymph by yeastlike germination until enormous numbers of them are floating in the circulating blood. They are found throughout the body cavity, within the heart, and wherever it is possible for the blood to penetrate. The hemolymph becomes so loaded with blastocysts that not only is the circulation impeded but its life-sustaining elements seem to become depleted.

The blastocysts apparently never actually invade or penetrate into the living tissues or organs of the host. There appears to be a substance secreted by the fungus which causes the host's tissues to break down. The softer tissues disintegrate first, usually beginning with the fat body. A colony of fungus cells will multiply rapidly near the fat cells; and, after the wall of the fat body has broken down at the surface, the colony will tend to enter and take the form of the organ. As a result, lamellalike or vermiform convoluted colonies are formed. After the far body, the muscles, nerve fibers, malphighian tubes, alimentary tract, and all the hypodermal tissues are gradually destroyed, until only fragments of the tracheae can be seen. The chitinous body wall, originally very thick, becomes very thin--apparently because of the action of a solvent of some kind.

Phagocytosis of the blastocysts is known to occur. This may consist simply of the enqulfment of the parasites within the cytoplasm of individual phagocytic leucocytes, or it may take the form of phagocytic complexes within which the blastocysts are included ; these complexes usually consist of masses of phagocytes arranged in concentric layers about a number of blastocysts. The number of blastocysts contained within a single phagocyte varies from 1 to 15. When large numbers have been phagocytosed, the leucocytes are gradually destroyed. From the information at hand it would appear that the phenomenon of phagocytosis offers little, if any, real protection against the fungus once the infection is under way in a susceptible host. When a small number of *Sorosporella* conidia are inoculated into the body cavity of a relatively nonsusceptible host, such as the silkworm, something occurs, not known to be phagocytosis, which renders the fungus cells innocuous and causes them to disappear.

Life History of Fungus. When the blood of a cutworm infected with *Sorosporella uvella* is examined, 6 or 7 days after inoculation. the yeastlike cells, or blastocysts, will be seen floating free in the hemolymph. These blastocysts are the early vegetative stages in the development of the fungus. When young, the cells are elliptical in form, are hyaline, measure 8 by 5 microns in size, and in blood smears occur singly or coherent in pairs, rarely in threes. They multiply by a yeastlike germination ; budlike outgrowths appear that grow rapidly until they reach the approximate size and form of their parents, then break away and proceed to form new cells by the same process. Multiplication continues until large numbers of coherent blastocysts are present in the form of colonies throughout the body cavity of the insect. Small colonies coalesce so that large masses are formed which ultimately develop into resting-spore masses. When these spore masses mature they eventually appear typically as brick-red "dust" within the cadaver of the host insect. It is in this condition that specimens of the diseased insects are most often collected in the field. The resting spores, or chlamydospores, mark the termination of the development of the fungus.

In order to determine the origin, nature, and function of the resting spores, Speare (*1920a*) conducted a number of experiments dealing with the organism's life history. Under the stimulus of moisture and suitable temperature, the protoplasm of the resting spores swells, producing budlike protuberances which soon assume the shape of a germ tube, branch freely, and become septate. The fully developed conidiophores are supplied with bottle-shaped branchlets or sterigmata at the tips of which are borne the conidia, which are quite unlike the resting spores from which they arose. The conidia are elliptical in form, thin-walled, vacuolate at each pole, and measure 4 to 6 by 9 to 11 microns in size. The development and structure of the conidiophores and conidia are typical of the verticillate Moniliales, of which the genus *Sorosporella* is a member.

Germination of the resting spores in unbroken larvae may be induced by placing the insects in circumstances that provide the proper moisture and temperature requirements. In a few days an external fungous growth on the unbroken integument of the host becomes visible. The germinations thus obtained are confined only to those spores which lie near the surface, immediately beneath the integument. When, however, the infected larvae are torn open and the red spore masses are freely exposed to the air and light, the germination may involve practically all the spores. The germ tubes from adjacent spores may cohere in such a way that *Isaria*-like fascicles of conidiophores are produced. It appears, therefore, that although the germ tubes are capable of breaking out through the cuticula of the insect, the process of germination is facilitated considerably when the larva becomes disintegrated in the soil.

When the conidia are placed in contact with healthy insects, either externally or internally, infection may be readily induced. In all probability the conidia send out germ tubes that penetrate into the body cavity of the insect where they produce bodies that give rise to the yeastlike blastocysts. Thus, in summarizing what is known of the life cycle of *Sorosporella uvella* it might be briefly said that the thick-walled resting spores germinate, producing conidiophores upon which are borne thinwalled conidia ; the conidia in turn probably send out germ tubes, which, after invading the host insect, give rise to the vegetative cells known as "blastocysts," which multiply rapidly, ultimately becoming resting spores.

The resting spores do not necessarily require a long period of rest before germination, as is the case with certain other fungi. Germination may take place at once if suitable conditions are provided. The spores may be viable after being held in a dried state for 14 months, and they are able to withstand the low temperatures that prevail during the winter months in northeastern United States. Thus the resting spores are able to tide the fungus over periods of unfavorable conditions. The thinwalled conidia, on the other hand, apparently have the function of spreading the organism rapidly while favorable conditions persist.

S. uvella can be cultivated on a variety of nonliving media. Although its normal method of vegetative development within infected insects is by means of the yeastlike budding cells, on media a semifilamentous growth is obtained As in nature, no perfect or ascigerous stage has sufur been observed in any of the artificial cultures.

Distribution in Nature. Since its original discovery, *S. uvella* has been reported as a parasite of a number of lepidopterous and coleopterous insects in North America and in Europe. It has been reported from cutworms in Canada, and in the United States at least 10 insect species are known to be susceptible.

In the United States it was originally collected in the field on the striped cutworm, *Euxoa tessellata* (Harr.). It has also been found killing the corn earworm, *Heliothis armigera* Hb., in Virginia. Barber and Dicke (1937) report that high humus-content soil appears to be very favorable to the development of the fungus and that, during a 5-year experimentation period, no moths of the corn earworm emerged from hibernation in this soil. In Florida considerable reductions in populations of mole crickets (*Scaptetiscus*) have been reported through the agency *S. uvella* and the green-muscardine fungus, *Metarrhizium anisopliae* (Metch.). It is probably more widespread than is generally recognized ; the relative inconspicuousness of the fungus probably accounts for its being overlooked in many cases.

In Moravia the poppy-root weevil, *Stenocarus fuliginosus* Marsh, is occasionally found parasitized by the fungus, as is the sugar-beet curculio, *Cleonus punctiventris* Germ. Some authorities maintain that *S. uvella* is a more effective enemy of *Cleonus* than is *M. anisopliae*. Along with the green-muscardine fungus, *S. uvella* kills a considerable number of the larvae of the black beet weevil, *Psalidium maxillosum* Fabr., in the Russian province of Krasnodar. Most *Sorosporella* infections, however, are found in lepidopterous insects, the noctuids, *e.g.*, *Euxoa segetum* (Schiff.), being particularly susceptible. An accurate estimate as to the probable amount of natural control effected by this fungus has not been made. Since it has been shown that infection of larvae may result by various methods of inoculation (direct contact, feeding, and spray methods), it can be assumed that transmission in nature is probably readily maintained as long as all environmental conditions continue to be satisfactory.

MYCOSES OF BEES

A number of species of Fungi Imperfecti are known to be capable of parasitizing bees, particularly the honeybee, *Apis mellifera* Linn. That this beneficial insect suffered from mycoses was not generally realized until after the bacterial diseases of brood were recognized. The first comprehensive studies on the fungous diseases of bees were accomplished in Europe. Very little was done in this connection in North America until Burnside (1930) undertook a study of the fungi associated with the honeybee and their pathogenicity for this insect. Although there have been subsequent investigations and reports on this subject, our information is still meager, and much remains to be learned of the epizootiology of these diseases One reason for this lack of attention is probably the fact that fungous diseases of bees appear to be less serious and less destructive than the common bacterial diseases of these insects. Furthermore the mycoses rarely become epizootic in intensity, although such outbreaks may occur among wild bees.

Fungi Concerned. In Europe two species of fungi, *Aspergillus flavus* Link and *Pericystis apis* Maassen, are recognized as the etiological agents of diseases of brood and adult bees. Since *A flavus* attacks worker brood and adult bees, it is generally considered of greater economic importance than *P. apis*, which attacks only drone brood. When affecting brood the *A. flavus* infection is commonly known as "stone brood" (*Steinbrut*), and the *P. apis* infection as "chalk brood" (*Kalkbrut*). Other fungi that have been reported from Europe as capable of infecting and killing bees and their brood include *Trichoderma lignorum* Tode, *Mucor mucedo* Linn., and the yeast *Saccharomyces apiculatus* Hansen. Although not ordinarily pathogens of live bees, *Penicillium glaucum* Link and *Pericystis alvei* Betts are two of the most common fungi in European hives.

In the United States, *Aspergillus flavus* attacks brood of the honeybee more frequently than other fungi. It has also been recorded on five or six other species of insects in this country. *Pericystis apis* is not known to occur in North America. In addition to *A. flavus* other forms of *Aspergillus* known to attack bees include *A. fumigatus* Fres., *A. nidulans* (Eid.), *A. niger* Tieg., *A. glaucus* Link, and *A. ochraceus* Wilh. Burnside (1935) found *Mucor hiemalis* Weh. pathogenic for young bees when the latter are exposed to a temperature of about 20°C. By inoculation a 100 per cent mortality may be obtained with this fungus, of which the chlamydospores from infected bees appeared to be the most virulent form. A number of yeasts (*Saccharomyces*) are known to be pathogenic to the honeybee when introduced into its body cavity.

Nature of the Mycoses. During his experiments on the pathogenicity of various fungi associated with bees, Burnside (1930) observed several interesting aspects relating to the nature of the mycoses concerned. We shall repeat here the statement of some of his findings.

In the first place, Burnside observed that the greater number of fungous species that cause diseases of adult bees also attack the brood. He found this to be the case with most of the species or strains of *Aspergillus* used in his investigations. When inoculated experimentally with a pathogenic *Aspergillus*, brood is attacked and killed more quickly than are adult bees, though the loss of brood resulting from *Aspergillus* mycosis is much less than that of adult bees. In general, mycoses of bees reach their greatest significance with adult workers. Throughout the active season, an appreciable number of these are killed by pathogenic fungi, principally by the yellow-green spored aspergilli. All races of bees common in the United States are susceptible.

The frequency with which bees are attacked by fungi in nature, according to Burnside, appears to depend chiefly upon the virulence of the pathogenic species and upon their dispersion. Conditions that favor abundant growth of pathogenic fungi in nature are conducive to the spread of fungous diseases. The fact that brood is rarely attacked can probably be explained by the small probability that the larval food will contain a sufficient number of viable spores to cause infection. The pathogenicity of a fungus apparently is determined by the ability of its spores and mycelium to resist the action of the intestinal fluids within the digestive tract of bees. Most writers on the subject seem to believe that infection is initiated by ingested spores, the infection progressing from the alimentary tract to the body cavity and internal tissues.

The appearance of brood dead of a fungous disease is fairly characteristic and easy to identify. The larva becomes noticeably harder soon after it dies, and the glistening white colour changes to a dull creamy white. Later the dead insect becomes shrunken and wrinkled. The anterior end dries most rapidly, often curving upward at first, later tending to straighten out again. The fungus grows through the body wall, at first just back of the head in the form of a white ring or collar, a day or two later it covers the entire insect. As the fungus continues its development, spores are formed on the external mycelium giving it a green, yellow, black, or other colour depending on the particular species of fungus concerned. As the spores and insect careass become old and dry, the colour usually fades.

When adult bees are infected with a fungus they become restless and weak. The weakness increases until death occurs. Some of the infected bees may die in the hive, but usually they fly or crawl away to considerable distances before dying. An increased firmness of the insect's body can sometimes be noticed at the time of death. It is usually more noticeable a few hours later. Under suitable conditions the external production of spores takes place, giving the insect a characteristic mealy appearance. Since nonpathogenic or saprophytic fungi may develop on the cadaver of a bee dead from another cause, care should be exercised in ascribing the death of the insect to the fungus present, unless the examination is made directly after the death of the bee.

The exact cause of death has not been determined in the case of most of the mycoses, although the destruction to the host's tissue caused simply by the growth and development of the fungus is probably sufficient in most instances. There is evidence, in certain instances at least, that in addition to this physical action there is a chemical or toxic action exerted by fungus. Undoubtedly a certain amount of digestion of the insect's tissues caused by the activity of fungous enzymes takes place. *Aspergillus flavus* has been shown to produce a transient toxic substance that is capable of causing a fatal poisoning in bees. In fact, Toumanoff (1931) believes that the pathogenic action of this fungus is due almost entirely to its toxic action.

Economic Importance. The importance of mycoses among bees is difficult to estimate accurately. Larvae may be carried out of the hive soon after they become infected, and adult bees usually die away from the hive. The mortality also varies somewhat according to the species of fungus involved. Some species attack adult bees in nature with considerable frequency; others attack the insects so rarely that under normal conditions their importance is negligible. Within the hive, the death rate may vary according to the condition and inherent strength of the colony. If wintering conditions are poor, allowing moisture to collect in the hive, and if the colony is weak, mycoses are likely to be a more serious threat than otherwise.

Iu general, it may be said that only slight losses of brood are caused by fungous diseases, and these are of negligible economic significance. Losses of adult bees from fungus infection are also of little economic importance except that, when pathogenic fungi grow within the hive on combs, frames, dead bees, and the like, late in the winter or early in spring, significant losses are likely to occur. Such losses, however, can be largely remedied by exercising a few simple precautions.

Great care should be taken to ensure the proper ventilation of the hive and to provide for the escape from the hive of metabolic water vapor. Bees dead of fungi or covered with spores should be cleaned from the hive whenever this can be done without exposing the colony to severe weather conditions. All combs and equipment covered with fungi or "mold" should be washed or dipped for a few minutes in 20 per cent formalin, or other effective fungicide.

OTHER FUNGI-IMPERFECTI INFECTIONS

The entomogenous Fungi Imperfecti are so numerous that it has been necessary to limit detailed discussion to a few of the better known examples. Furthermore it is impractical even to mention examples of all the remaining genera of imperfect fungi known to include entomogenous species.

Representatives of some of these miscellaneous genera include *Vermicularia cicadina* Ell. and Kell on cicadas (some authors regard *Vermicularia* and *Colletotrichum* to be of the same generic type), *Monilia penicillioides* Del. on scarabaeids, *Trichoderma viride* Pers. on bees and culicids, *Cylindrodendrum suffultum* Petch on tipulid pupae, *Acrostalagmus aphidum* Oud. and *Cladosporium aphidis* Thuem. on aphids, *Acremoniella verrucosa* Togn. on the clover leaf weevil, *Stemphylium botryosum* Wallr. on coccids, and *Synnematium jonesii* Speare on various Hemiptera.

Approximately six species of *Penicillium* have been reported as being parasitic on North American insects. About the same number of species of *Sporotrichum* have been similarly reported. The student should remember that the literature contains mention of a number of fungi described or cited as species of *Sporatrichum*, most of which are in reality synonymous with species of other genera such as *Beauveria* and *Acremonium* (*e.g.*, *B. globulifera* (Speg.) and *A. tenuipes* Petch). The genus *Rhinotrichum* contains at least one American species of entomogenous fungus (*R. depauperatum* Charles) parasitic on the avocado red mite. Members of the genus *Hormodendrum* are rarely pathogenic for healthy insects but are found occasionally on weakened or dead insects. In the United States, *Macrosporium* has been reported on certain Coleoptera, Lepidoptera, and Hemiptera. *Syngliocladium* has been found on wireworms.

The genus *Spicaria* contains numerous entomogenous species, one of which, *S. farinosa* (Fron) Vuill., was mentioned earlier in this chapter. It was with this species, as the type, that Petch (1934) proposed that the name *Isaria* be discarded in favor of *Spicaria*, *Isaria farinosa* (Dicks. Fron becoming *Spicaria farinosa* (Fron) Vuill. The generic name *Isaria* had been used generally by mycologists in connection with the conidial stages of *Cordyceps*. About a dozen species of *Spicaria* have been reported as parasites of insects in North America. Among these, in addition to *S. farinosa*, are

S. heliothis Charles on the corn earworm, *S. canadensis* Vuill. on larvae of the satin moth, and *S. rileyi* (Farl.) on a number of Lepidoptera and Coleoptera. The genus *Nomuraea* is generally considered as a synonym of *Spicaria.* Another genus whose species are imperfect stages of *Cordyceps* is *Hymenostilbe*, of which at least six species have been found on North American insects.

BASIDIOMYCETE INFECTIONS

The most remarkable association between insects and Basidiomycetes concerns those fungi belonging to the genus *Septobasidium.* Beneath the stromata of these fungi live certain scale insects, some of which are parasitized by the fungi. Because of the latter fact some authors consider the general relationship between the fungi and the insects to be one of parasitism. On the other hand, one of the leading students of the genus *Septobasidium*, Couch (1931), considers the relationship to be one of mutual symbiotism, in which the parasitization of some of the insects is for the good of the colony. Since the parasitism that does take place is also discussed there, we shall omit further reference to the genus in the present chapter.

The number of known basidiomycete fungi truly parasitic on insects is comparatively small. Certain species in the genera *Helicobasidium Corticium*, *Thelophora*, *Hymenochaete*, and *Daedalea* are now considered synonymous with members of the genus *Septobasidium.* Intermediate between the rusts and the genus *Septobasidium* has been erected the genus *Uredinella* of which the type species, *U. coccidiophaga* Couch, has been found on specimens of *Aspidiotus.* The imperfect genus *Hirsutella* was at one time thought to be a basidiomycete (Agaricales), but most of the species now contained in it are considered to be the imperfect stages of certain Ascomycetes, *e.g.*, certain *Cordyceps.*

6

Root Infecting Fungi

This group of fungi includes nearly all the specialized root pathogens apart from the vascular wilt fungi. In the ectotrophic pathogens, infection of the root tissues is typically preceded by the superficial extension of mycelium over the root-surface ; the interval between the advancing margin of epiphytic mycelium and the infection that follows it can vary from a millimetre or so with some pathogens of herbaceous plants to several metres with some fungi infecting roots of tree species. For *Rhizoctonia solani*, root infection by unspecialized fungal pathogens may also be preceded by superficial mycelial growth and attachment to the root surface. But because such unspecialized pathogens are usually limited by mature tissue-resistance to immature, senescent, wounded or unhealthy parts of the root system, this superficial spread of mycelium does not continue indefinitely ; sooner or later, it is halted by host resistance. The specialized pathogens, on the other hand, are not thus halted by the development of effective defence mechanisms in older parts of the root system ; provided always that soil conditions are favourable for ectotrophic infection, an essential proviso, they can continue to spread from the original foci of infection, perhaps on outlying parts of the root system, up to the root collar and stem base. The continuous, ectotrophic infection-habit of these specialized pathogens is thus no more and no less than the outward and visible sign of an infectivity that is adequate to overcome host resistance to infection throughout all parts of the root system.

There is an obvious sperficial resemblance between the infection-habit of these specialized pathogens and that of the ectotrophic mycorrhizal fungi forming symbiotic associations with many species of forest tree ; it was this resemblance that led us to propose the name 'ectotrophic' for these specialized pathogenic fungi as well. The use of this term 'ectotrophic' does indeed provide a convenient and unequivocal morphological description of the infection-habit of the pathogens, but it cannot be taken to imply any relationship with the ectotrophic mycorrhizal fungi that is more profound than a similarity in morphology of the infection-habit. It can be suggested that the ectotrophic mycorrhizal fungi have evolved from the specialized root pathogens, according to the general postulates about the evolution of parasitism. No less plausible, however, was Harley's (1948) suggestion that ectotrophic mycorrhizal fungi may have evolved from saprophytic root-surface fungi. Such a view is not appropriate for the ectotrophic infection-habit of the mycorrhizal fungi ; in this apparently harmonious hostparasite relationship, there is no evidence that the fungus provokes any significant degree of host resistance, at least so far as the limited degree of infection that actually develops is concerned. Harley (1969) has described the morphogenesis and behaviour of various types of ectotrophic mycorrhizal association in critical detail. But a single example will suffice here to make this point. In his work on the ectotropic mycorrhizal association of Scots pine, Robertson (1954) demonstrated that although infection of the short root is truly ectotrophic (*i.e.* they are enveloped by the usual sheath of fungal mycelium), that of the long roots proceeds *internally*, as an extension of the Hartig net intercellularly in the cortex (Fig. 6.1).

From Robertson's evidence, it is clear that the ectotrophic infection-habit is not essential for successful extension of infection along the young long roots of Scots pine. We shall discuss later parallel situations with ectotrophic fungal pathogens, in which a decline of host resistance to a low level

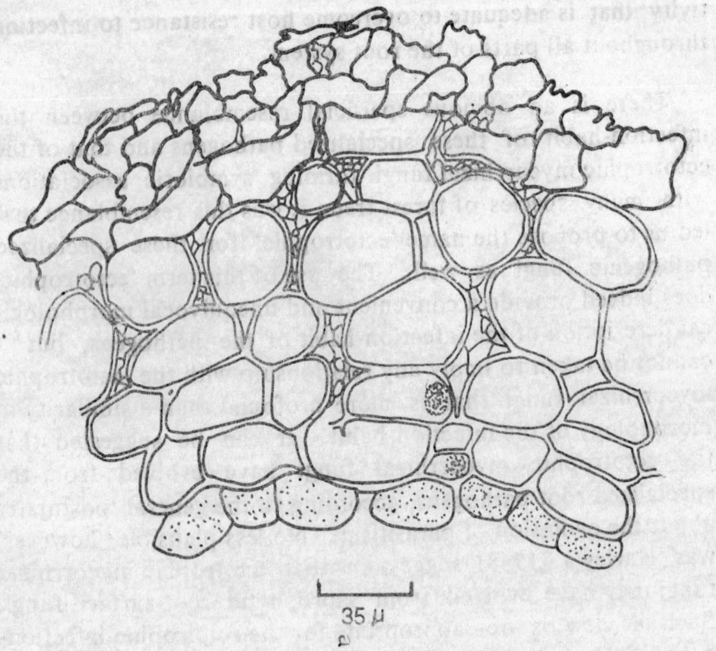

Fig. 6.1. Transverse section of cortex of mycorrhizal long root of *Pinus-syloestris* approx. 8 mm behind the growing apex. Note Hartig net of mycelium between cortical cells.

permits a purely internal extension of infection along roots of much impaired vigour. This suggests that the ectotrophic infection-habit is not imposed upon ectotrophic mycorrhizal fungi, as it seems to be on ectotrophic pathogens, by the occurrence of a strong provoked resistance to infection. It thus seems most probable that the external fungal sheath of ectotrophic mycorrhizas has developed purely as an absorbing organ for the taking-up of nutrient salts from the soil. This seems to have been a roundabout way for a plant pathologist to reach a conclusion that most mycorrhizal workers have probably always taken for granted.

INFECTIVE MECHANISM OF THE ECTOTROPHIC INFECTION-HABIT

The essential features of the ectotrophic infection-habit amongst these specialized pathogens are well shown by

Ophiobolus graminis, restricted to the Gramineae and causing the take-all disease of cultivated cereals. The fungus progresses at first epiphytically over the root system as a sparse network of dark-coloured runner hyphae, from which hyaline branches quickly penetrate and infect the underlying cortex and eventually the vascular cylinder of the roots (Fig. 6.2).

In one early work on the ectotrophic infection-habit of *O. graminis*, one could note briefly two observations, the full significance of which one was not realized at that time. First, it was observed that as the fungus travelled down a seminal

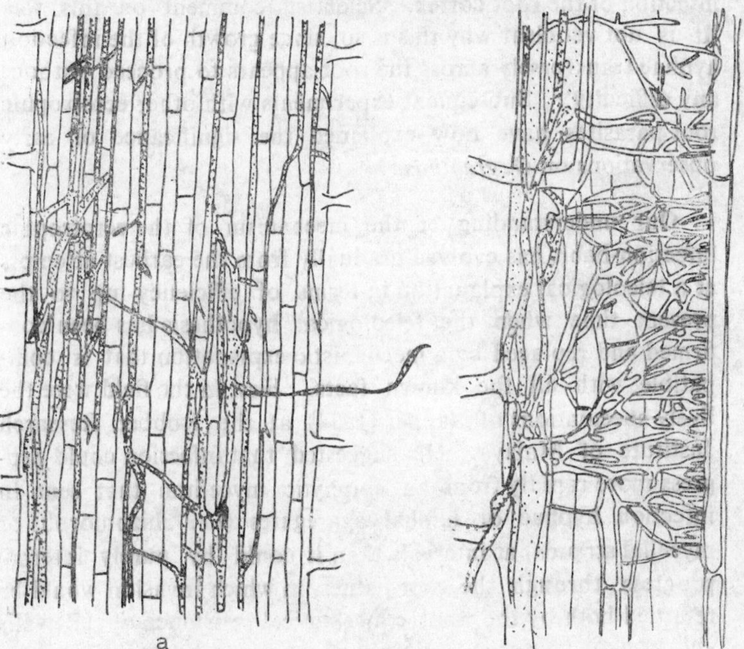

Fig. 6.2. Ectotrophic infection of wheat seedling root by *Ophiobolus graminis*: (*a*) in surface view; (*b*) in longitudinal section. (After drawing made by G. Samuel, in Garrett, 1934).

root from the point of inoculation below the germinating seed, it often abandoned the ectotrophic habit on the distal part of the root and the runner hyphae thereafter travelled intercellularly one or two cells deep within the cortex Scientist attributed this to an impairment in resistance to infection of the distal part of the root, resulting from its virtual disconnection from the rest of the plant through destructive infection of the proximal part. The significance of this observation will become more apparent later, though they did not appreciate its full implications at that time. Secondly, it was noted that if ectotrophic infection of a root was inhibited by sufficiently unfavourable conditions, such as too extreme a degree of soil acidity or too poor a degree of soil aeration, then *O. graminis* also failed to extend appreciably further by purely internal infection of the root cortex. Scientists comment on this was 'It is not yet clear why this is so, since growth of the infection hyphae transversely across the root appears to proceed without any difficulty'. Subsequent experiments with other ectotrophic root parasites have now explained the significance of early observations on *O. graminis.*

Our understanding of the mechanism of the ectotrophic infection-habit has evolved gradually from the earliest attempts at a teleological explanation in terms of efficiency up to the present time when the teleological hypothesis has been confirmed and replaced by a mechanistic explanation that is compatible with all the known facts. First in the field were the lively speculations of Napper (1934) at the Rubber Research Institute of Malaya. He suggested that infection could progress more rapidly from an epiphytic mycelium that sent in infection hyphae or hyphal aggregates (*i.e.* rhizomorphs or mycelial strands) at intervals than it could by purely internal progress through the root tissues, in which invasion would be retarded both by the passive mechanical resistance of cell walls and by active defence-responses of the living tissues. Scientents postulated that the ectotrophic infection-habit provided a mechanism whereby the parasite could initiate a series of infections in comparatively rapid succession along the length of the root, thus diluting and eventually overwhelming the active defence-response of the root to infection. This hypothesis thus clearly implied synergism between a series of successive infections in overcoming active tissue-resistance of the root.

At that time, this explanation rested largely upon evidence collected by Rishbeth (1951) upon the ectotrophic infection-habit of *Fomes annosus* upon roots of *Pinus sylvestris* (Scots

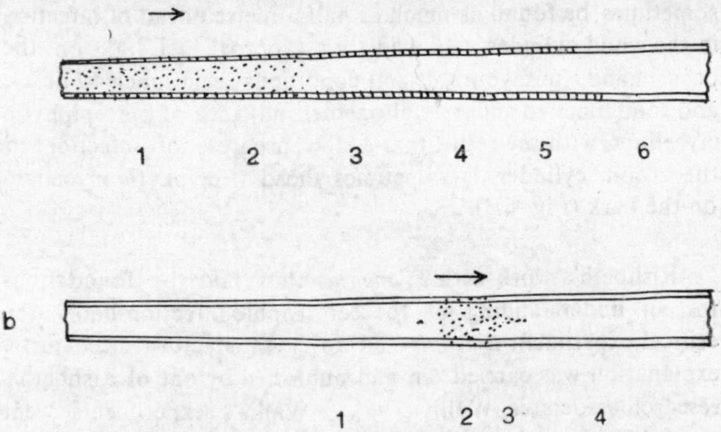

Fig. 6.3. Diagram representing two modes of infection of pine roots by *Fomes annosus.*

(a) Infected root from alkaline soil

Region	Extent (cm)	Characteristics of region
1	>15	Little resin exudation ; wood dry ; *F. annosus* hyphae abundant.
2	10	Marked resin exudation ; wood resinous ; hyphae less-abundant.
3	10	Marked resin exudation ; wood resinous ; hyphae in brak only.
4	17	Local resin exudation ; wood not resinous ; hyphae in bark only.
5	25	Bark scales whitish, peeling ; a few superficial hyphae present.
6	—	Root normal, bark scales intact.
		(b) Infected root from acid soil
1	>15	Bark split by resin ; wood resinous ; secondary fungi present only.
2	4	Marked resin exudation ; wood resinous ; *F. annosus* hyphae in bark and wood.
3	2	Slight resin exudation ; wood resinous ; hyphae in centre of wood only.
4	—	Root normal.

▭ Indicates region where *F. annosus* is present.
→ Indicates direction of invasion by *F. annosus.*

pine) and *P. nigra* var. *calabrica* (Corsican pine). In alkaline soils, this fungus grows freely on the outside of pine roots and the sheet-like mycelial aggregates amongst the bark scales can sometimes be found as much as half a metre ahead of infection in the wood cylinder. In acid soils (approx. pH 4.5), on the other hand, unfavourable soil conditions seem much to delay, and sometimes to suppress altogether, advance of the epiphytic mycelium, with the result that a slow progress of infection in the wood cylinder is sometimes ahead of epiphytic mycelium on the bark (Fig. 6.3).

Rishbeth's work with *F. annosus* thus laid the foundations for an understanding of the ectotrophic infection-habit ; the critical experiment that has led to a satisfactory mechanistic explanation was carried out and published by one of Rishbeth's research associates, Wallis (1961). Wallis's experiments went very directly to the point at issue, *i.e.* whether the ectotrophic tnfection-habit is imposed upon the fungus by the active defence-response of the living root tissues. Wallis therefore made use of the earlier finding by R. Leach (1939) in East Africa that severing a tree root from the parent tree greatly reduces its power of active resistance to infection and spread by *Armillaria mellea.* In Wallis's experiments, roots of 20-year-old Scots pine trees growing on alkaline (pH 7.0-7.5) and acid (pH 3.8-4.3) soils, respectively, lying at a depth of 15-25 cm, were exposed and brushed free of soil. At 70-90 cm from the base of the tree, a root section 8-12 cm long was cut out and replaced by a woody inoculum block permeated by mycelium of *F. annosus.* Twelve months after inoculation, both external spread of *F. annosus* over the bark and the extent of internal infection of the wood cylinder were measured for all the treatment series ; results for dominant trees in normal vigour are summarized here in Table 6.1.

Wallis's results as summarized in Table 6.1 are so clear-cut as to need little comment. Behaviour of *F. annosus* on the proximal root portions, still connected with the parent tree and thus still in possession of full vigour and a full active defence-response to infection, is typical of that earlier observed by

Rishbeth ; ectotrophic advance of the fungus on the outside of the roots in alkaline soils is much ahead of that in acid soils,

Table 6.1. Extent of external mycelial spread and of internal infection of wood cylinder by Fomes annosus on roots of Scots pine (measurements in cm).

	Acid soil		*Alkaline soil*	
	Outer bark scales	*Wood*	*Outer bark scales*	*Wood*
Proximal root portions (connected)	5	3	15	0·7
Distal root portions (disconnected)	15	48	56	75

and it is only a matter of time before internal infection of the wood cylinder catches up with it. In the distal root portions, disconnection from the parent tree has evidently reduced the active defence response to a level at which *F. annosus* has been able to spread more rapidly *within the wood cylinder* than it could do outside the root, even under the more favourable conditions for external spread, *i.e.* in the alkaline soil.

From the results of Wallis's experiments, it has thus become clear that the ectotrophic infection-habit is imposed upon these specialized root pathogens by the active defence-response of roots in possession of their full vigour. If no active defence-response is provoked, then internal infection can proceed without obvious hindrance, as I have already suggested in explanation of Robertson's (1954) observation on the internal extension of the Hartig net in the long roots of Scots pine (Fig. 6.1). The level of active defence-response of living root tissues varies to some extent in different host-parasite combinations. As we have noted earlier, it seems to be sufficient to prevent a purely internal extension of infection by *Ophiobolus graminis* within wheat roots. The same conclusion apparently holds for infection of roots of the rubber tree by *Fomes lignosus* ; thus John (1958) demonstrated that infection of rubber roots by this fungus could be completely halted merely by removing the epiphytic rhizomorphs at weekly intervals whilst leaving the original inoculum undisturbed. In contrast to these two pathogens, *Fomes annosus* can sometimes progress slowly along a pine root by internal infection of the wood cylinder, even though such internal infection are quite often brought to a halt by the active defence response of the living host tissues ; the nature of this defence-response of pine roots will shortly come up for discussion.

Following upon this recent work, it has now become possible to explain the mechanism of the ectotrophic habit on the assumption that the active defence-response of the host root is diluted and finally overcome by synergism between the effects of separate infections arising in succession from the externally travelling mycelium on the outside of the root. We can best illustrate this by an analogy with the epidemic development of the chocolate spot disease of field beans (*Vicia faba*), caused by species of *Botrytis* and chiefly by *B. fabae*. Wilson (1937) successfully elucidated the epidemiology of this disease by distinguishing between its *non-aggressive* and *aggressive* phases. The non-aggressive phase consists of scattered leaf-spots, or primary lesions, caused by single-spore infections ; if the primary lesions are scattered thinly on the leaf, their further expansion is arrested by the host defence response, and they remain static. But as Wilson demonstrated, one essential condition for development of the aggressive phase appears to be a degree of crowding of the primary lesions that exceeds a certain critical value ; above this critical value synergism between the effects of the individual infections operates in such a way as to overcome host resistance. The fungus is thus enabled to break out of the primary lesions and to initiate the aggressive phase of rapid spread through leaf and stem tissues, which continues unless and until interrupted by a spell of warm, dry weather. At the Cambridge Botany School, we have confirmed and extended Wilson's findings in a yearly demonstration experiment for students set up by Mr W.J. Bean, which has continued for nearly 20 years. In Fig. 6.4 we have provided a diagrammatic representation showing in parallel what scientists conceive to be the sequence of events following upon : (*a*) a single isolated infection ; (*b*) a multiple series of infections, either contemporaneous or shortly consecutive first for the chocolate spot disease of beans and secondly for a typical, ectotrophic root-infecting fungal pathogen.

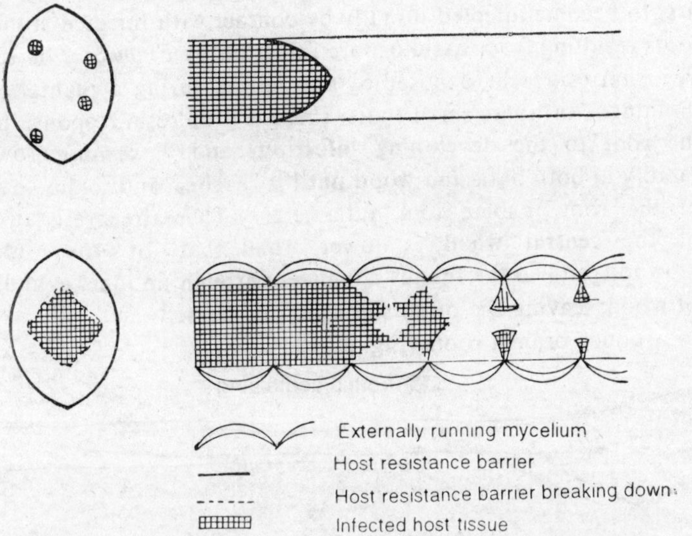

Fig. 6.4. Diagram illustrating possible mechanism whereby the ectotrophic infection habit enables a pathogenic root-infecting fungus to overcome host resistance. Compare this with a similar synergistic effect of multiple infections by a leaf-infecting fungal pathogen, which enabies its invasion of the leaf to become progressive.

Non-Ectotrophic Infections

Under this heading it is appropriate to describe the behaviour of *Fomes annosus* on some species of conifer other than those of *Pinus*, as elucidated by Rishbeth (1951). In these other conifers, such as Douglas fir (*Pseudotsuga taxifolia*), larch (*Larix decidua*) and spruce (*Picea abies*), resistance to infection of the inner wood tissues is lower than that in species of pine, and so infection follows a very different course. In these conifers, as in pines, exposure of very young trees to heavy or multiple infections can result in outright killing ; in older trees, however, infection leads to development of butt rot, which can cause much heavier losses of marketable timber than does the more dramatic tree-killing as manifested in pine plantations. It trees, some 20-30 years old, of Douglas fir, larch or spruce, the following infection sequence has been described by Rishbeth. The youngest roots are generally the first to become infected, usually by contact with infected stump roots resulting from basidiospore infection by *F. annosus* of the tree stump surfaces, exposed by tree felling during a plantation thinning. In these small roots, there is little resin response by the root to the developing infection, and *F. annosus* grows rapidly in both bark and wood until it reaches and older part of the root of some 2 cm in diameter. Thereafter, resistance of the central wood is lower than that of the outer sapwood, and so the fungus advances through an inner cylinder of wood, leaving the outer sapwood uninfected and still able to produce branch roots (Fig. 6.5).

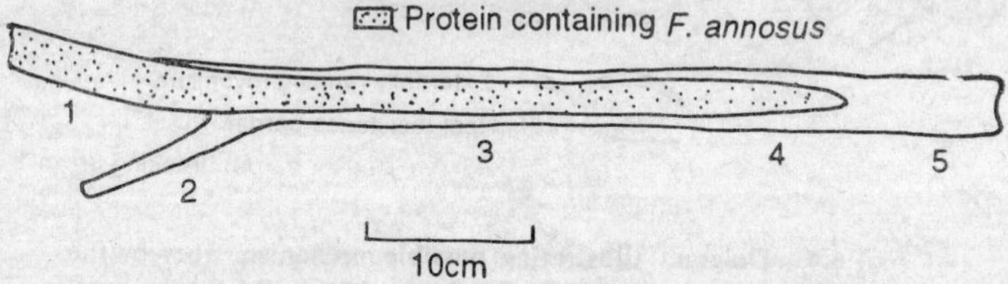

Fig. 6.5. Infection of larch root by *Fomes annosus*, of the type that leads to eventual development of butt rot. 1. Wood very rotted, yellow; bark, dead, brown; *F. annosus* hyphae abundant in both. 2. Small living root. 3. Most of the wood resinous, red-brown, hyphae abundant; outermost wood healthy; bark discoloured, bright red in parts, otherwise healthy, hyphae local on surface only. 4. As last, but smaller region of central wood with hyphae. 5. Whole root healthy. (After J. Rishbeth, 1951 *c*).

As *F. annosus* approaches the root collar, or butt of the tree as it is sometimes called, the boundary between the central cylinder of infected wood and the outer uninfected wood is often marked by a conspicuous layer of resin (Fig. 6.6).

It is thus apparent that in species of conifer susceptible to the butt-rot type of infection, internal spread of *F. annosus* is made possible by the fact that the resistance of the inner cylinder of wood is sufficiently lower than that of the outer sapwood in these species, and is also lower than that of any part of the wood cylinder in pine species. The reasons for these differences have been elucidated in an investigation of outstanding interest by Gibbs (1968), working as one of Rishbeth's research associates. By direct inoculation of *F. annosus* into the wood cylinder of the roots, Gibbs was able to estimate the resistance of the wood in Douglas fir, Sitka spruce (*Picea sitchensis*), Scots pine, and Corsican pine, respectively ; the degree of resistance was estimated as the reciprocal

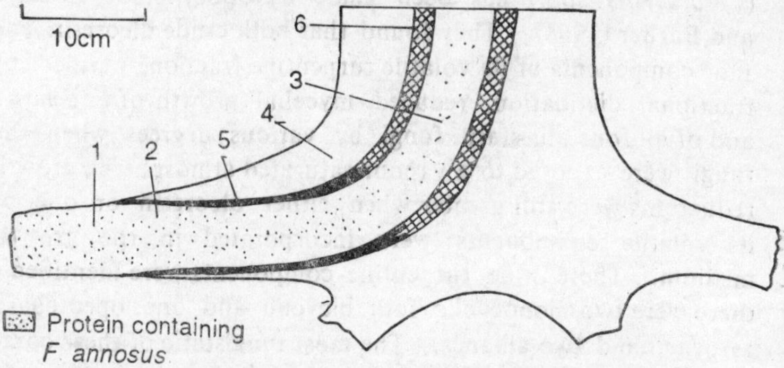

Fig. 6.6. Diagrammatic section of lower butt of Douglas fir. showing incipient development of butt rot, with *Fomes annosus* entering from a lateral root. 1. Wood extensively rotted, yellow. 2. Wood, partially rotted, yellow-brown. 3. Wood with incipient rot, red-brown. 4. Wood healthy, white. 5. Resin layer. 6. Wood resinous, deeply stained.

of the distance travelled by *F. annosus* in the wood over a given period. Dougals fir showed least resistance, and the two pine species the most ; the degree of resistance up the ascending series from Douglas fir to the pines was closely correlated with the complexity and activity of the resin canal system. This is nicely illustrated, as Gibbs has pointed out, by commercial use of these various genera of conifers for commercial production of resin ; various species of *Pinus* are widely employed for resin tapping, those of *Picea* much more rarely, and those o *Larix* and *Pseudotsuga* scarcely at all. Gibbs further noted from the results of his inoculation experiment that in spruce and Douglas fir, *F. annosus* failed to infect the outer sapwood and was restricted to the inner wood, in which a high proportion of the resin ducts become atrophied. In other experiments, with Scots and Corsican pine, Gibbs recorded daily resin yields by tapping ; resin yield, tree vigour and degree of resistance to *F. annosus* all varied in parallel under the conditions obtaining in these trials.

An analysis of the toxicity of resin (in this paper designated as 'oleoresin') to *F. annosus* and four timber bluestain fungi (*Ceratocystis* spp.) has been made by Cobb, Krstic, Zavarin and Barber (1968). They found that both crude oleoresin and nine components of its volatile turpentine fraction, purified by fractional distillation, reduced mycelial growth of *F. annosus* and of all four bluestain fungi by various degrees when the fungi were exposed to a vapour-saturated atmosphere ; growth reductions were still greater when either oleoresin or one of its volatile components were incorporated in the growth medium. These nine turpentine components were identified ; there were two monocyclic, four bicyclic and one open-chain terpene, and two alkanes. The most fungistatic of these components was the alkane *n*-heptane, which in a vapour-saturated atmosphere completely inhibited mycelial growth of both *F. annosus* and *C. pilulifera*.

Mycelial aggregation into sheets, strands and rhizomorphs

Ectotrophic pathogens that infect woody perennials, *i.e.* bushes and trees, typically show aggregation of the epiphytic mycelium into mycelial sheets or strands, or into rhizomorphs. They thus contrast with fungi infecting herbaceous species, such as *Ophiobolus graminis*, in which the epiphytic mycelium is not organized into hyphal aggregates but consists of separated individual hyphae, This distinction is not affected by the fact that *Armillaria mellea*, a typical ectotrophic parasite of tree root systems, will infect potato tubers if opportunity offers ; were *A. mellea* limited to potato plants and other herbaceous species, it is unlikely that it would have evolved so powerful an offensive armament as the fungal rhizomorph. Evidence of three kinds suggests that these various forms of hyphal aggregation have evolved in response to the need for an inoculum potential adequate to overcome host resistance to infection by the roots of woody perennials, and particularly by those of the larger arborescent species. First, mycelial aggregates as organs of infection are not found in root-infecting fungi restricted to herbaceous hosts. Secondly, many experiments have now shown that there is a critical level of inoculum potential necessary for initiation of a progressive infection ; the older, larger and more highly organized host roots demand a higher level of inculum potential for establishment of a progressive infection than do younger, smaller and simpler roots. Thirdly, there is no evidence for the supposition that the older, cork-covered parts of tree root systems can be successfully infected either by spores or by individual hyphae in unorganized mycelium, though there is much evidence for the contrary view that infection can be accomplished only by mycelial aggregates growing out from a food base generating adequate inoculum potential.

Morphogenesis and behaviour of mycelial strands and rhizomorphs

Mature mycelial strands are superficially difficult to distinguish from rhizomorphs ; both types of organ may reach a diameter varying from one to several millimetres, and a mature strand may have a surface as smooth as that of a rhizomorph. These two types of fungal organ, however, develop in very different ways and whether a particular organ is a strand or a rhizomorph can be determined only by a careful study of its morphogenesis. A mycelial strand is built up gradually, sometimes by the chance meeting of individual hyphae or minor strands ; more commonly, however, strands develop acropetally by the ensheathigg growth of branches of the parent leading hypha or hyphae. At least two types of strand morphogenesis have been described, and further variants on the general theme are likely to be discovered. Rhizomorphs, on the other hand, are morphogenetically similar to the roots of a higher plant ; they grow from an apical meristem, though this meristem is of a type quite different from that found in any higher plant.

In consequence of this difference in morphogenesis, mycelial strands and rhizomorphs also differ in the way that their inoculum potenttal develops. The diameter of a rhizomorph attains its maximum value just behind the growing apex, and so the apical region possesses, to the maximum possible degree, 'instant inoculum potential'. A mycelial strand, in contrast, makes its first contact with a host surface with the arrival of its leading hypha (or associated leading hyphae) ; thereafter, both strand diameter and inoculum potential increase acropetally up the still developing strand, as further constituent hyphae grow up towards the strand apex from the rear.

Mycelial strands

The morphogenesis of the mycelial strands produced by *Helicobasidium purpureum*, causing the violet root rot disease not only of sugar beet but also of various woody perennial species, has been analysed by Valder (1958). Strands develop from a food base lying in or on the soil when environmental conditions are suitable for mycelial growth. The food base employed by Valder in this study was a disc of nutrient agar colonized by *H. purpureum*, but under natural conditions both infected roots and sclerotia provide food bases for strand development. From such a food base, there first develops a sparse growth of robust, coloured hyphae over and through the soil ; further hyphae growing out from the food base sooner or later encounter one of these leading hyphae and then follow it in associated growth. By continued accretion of these 'following' hyphae and also of small stands, each of the original leading hyphae may develop into a main strand, the base of which comes to resemble a river delta. Owing to this mode of morphogenesis, these strands are widest at the base and taper off towards the apex. Anastomoses occur between strands to give network, especially near the food base. Branching of strands can occur, due to the branching of the original leading hypha, with the branch attracting some of the follower hyphae away from their parent hypha. Valder further observed that coherence of the strands of *H. purpureum* was brought about in three ways : (1) by an interweaving growth of the associated main hyphae ; (2) by the binding action of short branches of limited growth, sometimes hooked at the apex ; (3) by anastomoses between hyphae (Fig. 6.7).

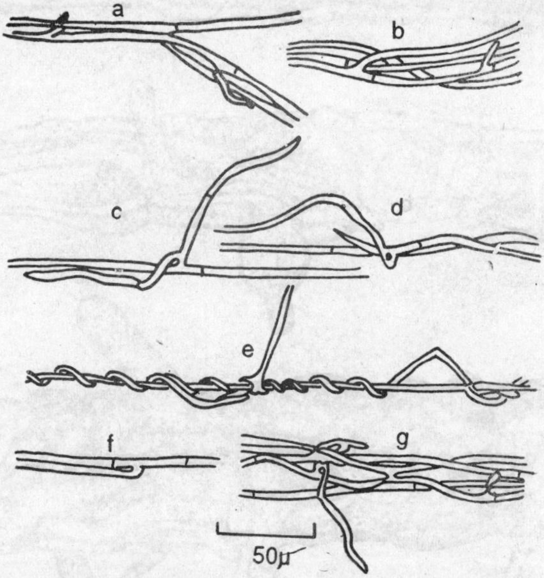

Fig. 6.7. Morphogenesis of mycelial strands in *Helicobasidium purpureum.* The general direction of hyphai growth is from left to right. Note interweaving growth of main hyphae, binding action of short side-branches of limited growth (*a,b,g*), and the locking action of hyphal anastomosis (*g*).

A second and commoner type of strand morphogenesis is exemplified by that of *Phymatotrichum omnivorum* (Shear) Duggar, which causes Texas root rot of cotton and other dicotyledonous species, including a number of woody perennials. Rogers and Watkins (1938) have described how the strands of this fungus develop around single, exceptionally wide hyphae, which become ensheathed by their own branch hyphae. By subsequent adjustment of position and septation in the ensheathing branch hyphae, a compact cortex of several layers in thickness is eventually formed around the single large hypha at the centre of each strand (Fig. 6.8).

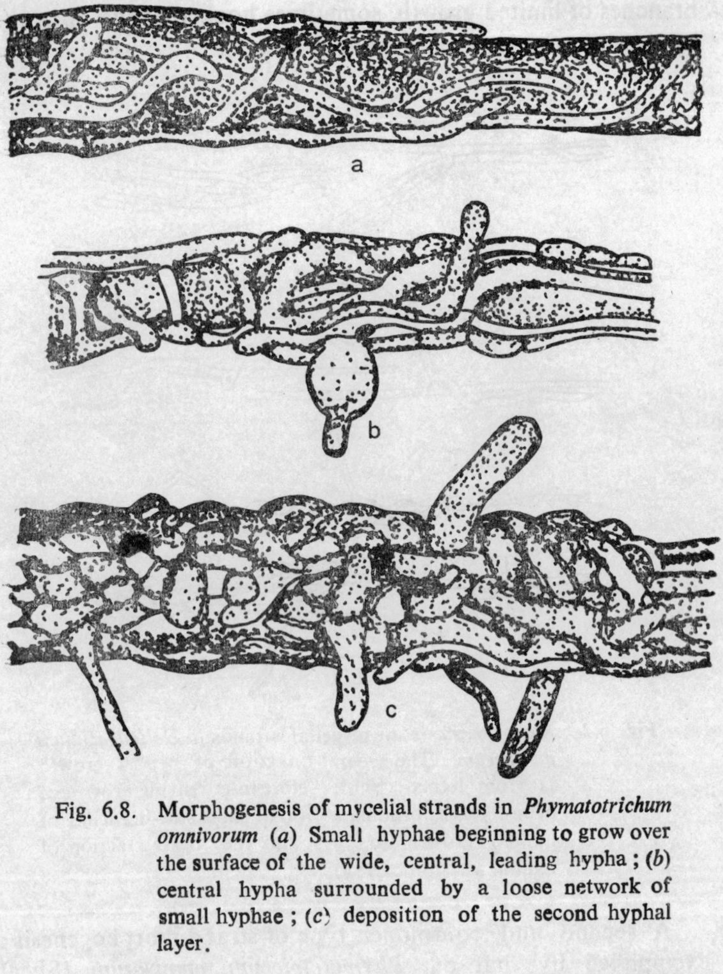

Fig. 6.8. Morphogenesis of mycelial strands in *Phymatotrichum omnivorum* (*a*) Small hyphae beginning to grow over the surface of the wide, central, leading hypha ; (*b*) central hypha surrounded by a loose network of small hyphae ; (*c*) deposition of the second hyphal layer.

A type of strand morphogenesis very similar to that in *P. omnivorum* has been described for two well known saprophytes : *Merulius lacrymans* Fr., causing dry rot of house timber and *Agaricus bisporus* (J. Lange) Pilat, the cultivated mushroom (Mathew, 1961).

Mycelial strands are typically the form in which a strand-producing fungus migrates outwards from a food base through the mineral matrix of the soil, which approximates to a non-nutrient medium, and at first epiphytically over root surfaces. When growing on a nutrient medium, such fungi either do not produce distinguishable strands, except amongst the aerial mycelium out of contact with the substrate, or they produce them only during the later stages of substrate colonization. In earlier accounts of strand morphogenesis scientistshad postulated an attraction between hyphae of the same mycelium or species as responsible for strand morphogenesis, and had suggested that this attraction might be similar to, or even identical with, that responsible for anastomoses between hyphae. Absence of strands, or delayed production, no nutrient media had then to be explained by a repulsion between hyphae in such a situation that was strong enough to override the postulated attraction. Causes of repulsion were not far to seek ; on a nutrient medium every hypha acts as a diffusion sink for nutrients and as a source of growth products, some of which are likely to become fungistatic above a certain concentration. The consequences of these two processes were thought to cause hyphae to keep as far away as possible from one another on a nutrient medium, and thus to favour a dispersed type of growth.

Very recently, Dr Sarah Day (1969), carried out an experiments with *Merulius lacrymans* to test this hypothesis ; she argued that if it were correct, then *M. lacrymans* should from strands in a liquid nutrient medium that was renewed sufficiently rapidly by a drip feed device to prevent diffusion shells, either of nutrients or of fungal growth products, forming around individual hyphae. The results of this experiment were the reverse of those expected ; mycelial strands formed in the still but not in the replacement cultures. From a consideration of these results and of many observations by earlier workers, Day was able to formulate a new hypothesis that seems to account satisfactorily for all the known facts. This hypothesis postulates that the leading hyphae, around which strands are formed, exude nutrients when they are translocating in much the same way as do young roots, and that it is this nutrient exudation that causes the chemotropic response of associated growth in the branch hyphae. An indication of nutrient exudation by hyphae of *Rhizoctonia solani* in soil was earlier obtained by Thornton (1953), who observed these hyphae to be invested by actinomycete filaments. But Day herself has obtained direct evidence of nutrient exudation by translocating hyphae of *M. lacrymans* from the results of experiments with radioactive isotopes.

Day's hypothesis thus provides an explanation for the regular occurrence of strand formation by strand-forming fungi when they are growing from a food base through natural soil or through any other non-nutrient media, such as those employed by G.M. Butler (1957) for *M. lacrymans*. The delay in stranding when a fungal colony is growing out over a nutrient medium can be explained by the masking of nutrient exudation from the leading hyphae through the initially high concentrations of nutrients in the medium ; only when this initially high nutrient level has been sufficiently lowered by fungal growth will the masking effect be removed. So this hypothesis can account for the fact that strand formation by *M. lacrymans*, and by some other fungi too, does occur on nutrient media in due course, though its intensity varies with nutrient concentration and balance, as Day has shown.

An objection that can be raised against this hypothesis is that both strand formation and hyphal anastomosis occur only between mycelia of the same species, whereas the attraction provided by hyphal exudation is unlikely to be so specific, or, indeed, to have any significant degree of specificity at all. The answer to this objection is that the hyphae of a growing fungus are already loosely associated in their colonial growth, and that hyphal exudation needs only to hold together as a strand those hyphae that either arise in juxtaposition, *i.e.* branches of a leading hypha, or else approach one another by chance, an in strand formation by *Helicobasidium purpureum*. The same argument also applies to the mechanism of anastomosis ; a nutrient stimulus from hyphal exudation will serve to prolong a hyphal association brought about by mere chance in the first place. Indeed, hyphal association represents the first stage only in the development of anastomosis ; as Flentje (1969) has described for *Rhizoctonia solani*, various causes of failure are spaced in the time sequence from initial juxtaposition of hyphal growth up to the ultimate consummation of anastomosis between compatible hyphae.

Rhizomorph of Armillaria mellea

The best known rhizomorph is that produced by *Armillaria mellea*. If we assume the mean diameter of the straight running, unbranched, constituent hyphae to be about 10μ, then we can calculate that a rhizomorph of 1 mm diameter will be made up of several thousand hyphae, even after allowing generously for interhyphal space and for the central open channel running down the mature part of the rhizomorph. Growth of the rhizomorph, like that of a root, occurs from an apical meristem, and this implies a mechanism for coordinating apical growth of the constituent hyphae. When a fungal colony is growing on a nutrient agar, the rhizomorphs produce branches and the density of branching increases with the concentration of nutrients in the medium up to an optimum value. On such a nutrient medium, every rhizomorph eventually produces a mantle of fringing mycelium, which further enhances its resemblance to a root with its mantle of root hairs. Fig. 6.9, after R. Hartig, shows the apical region of rhizomorph of *A. mellea*.

Much has been learnt about the behaviour of rhizomorphs of *A. mellea* through simple experiments on the nutrient agar plate. When a 4 mm agar disc taken from the growing margin of a colony of *A. mellea* is laid at the centre of a fresh plate of nutrient agar, an outwards growth of unorganized mycelium can be observed within 24 h at 25°C ; after 7 days, a circle of rhizomorph initials can be seen around the margin of the original inoculum disc. In an experiments, the young rhizomorphs, which started growth later than the colony of unorganized mycelium, grew some five times as quickly as the mycelial colony, and hence rapidly outgrew it. From the older region of the rhizomorphs fringing mycelium grew out, so that a mature rhizomorphic colony on an agar plate was deeply lobed. Scientists observations were confirmed and amplified by Snider (1959), working with a number of single-basidiospore

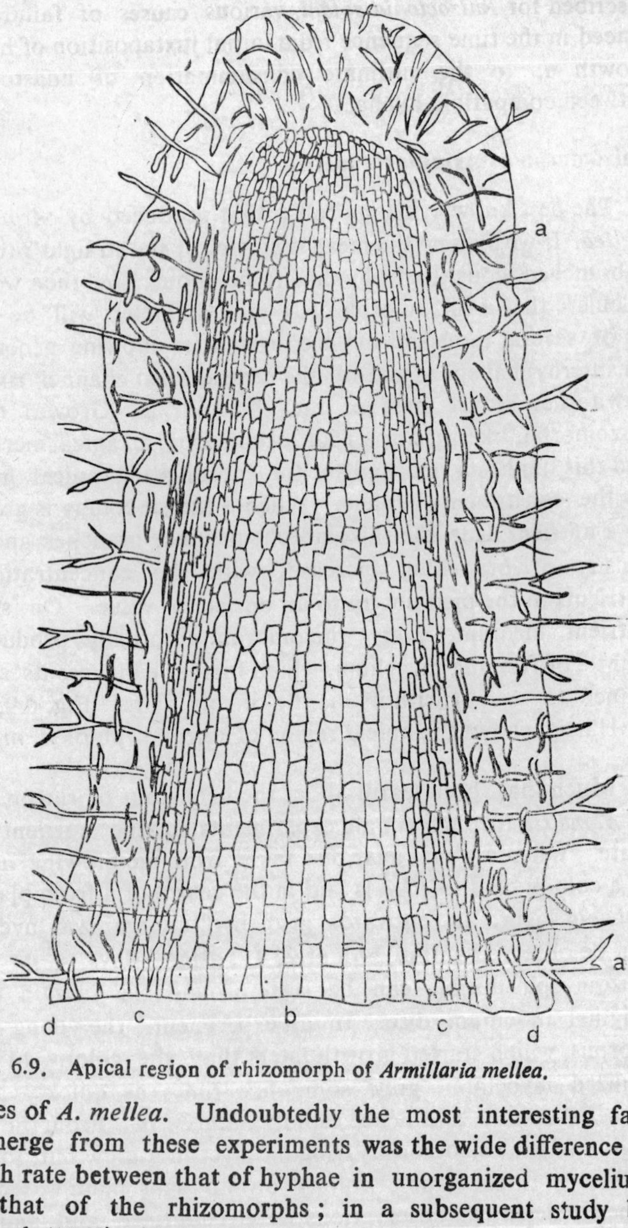

Fig. 6.9. Apical region of rhizomorph of *Armillaria mellea.*

isolates of *A. mellea.* Undoubtedly the most interesting fact to emerge from these experiments was the wide difference in growth rate between that of hyphae in unorganized mycelium and that of the rhizomorphs; in a subsequent study by Rishbeth (1968), rhizomorphs of one isolate grew some thirteen times as quickly as the margin of the mycelial colony. A possible explanation for this difference is that the constituent hyphae making up a rhizomorph are unbranched, and so the hyphal apices do not suffer from nutrient competition from the subordinate apices of branches. In a further study on growth of rhizomorphs from small woody inocula through glass tubes of soil, Scientist observed a progressive deceleration with time in the growth rate of rhizomorphs, and obtained evidence that this was due to increasing competition for nutrients from the subordinate apices of rhizomorph branches. But whatever the reason for this very high growth rate of rizomorphs, as compared with that of individual hyphae in unorganized mycelium, it must have one important consequence for the mechanism of ectotrophic infection. If our interpretation of this mechanism, as illustrated in Fig. 6.4 be accepted, then the faster the rate of epiphytic spread of the rhizomorphs, the more effective will be the ectotrophic infection-habit on overcoming active host resistance.

Another outstanding difference in morphogenesis of mycelial strands and rhizomorphs has also become apparent ; whereas strand-building is delayed by the presence in the growth medium of nutrients above a low concentration, the production of rhizomorph initials increases with glucose concentration at least up to the level of 2%. In an experiments on production of rhizomorphs on nutrient agar, vitamins were provided by addition of thiamin and also by a supplement of malt extract and/or peptone, to guard against any possible vitamin deficiency. Employing a synthetic growth medium with 1 mg thiamin/l as the only vitamin supplement, Weinhold and Garraway (1966) found that rhizomorphs where not produced either in the liquid medium or on the same solidified with agar unless the medium was further supplemented by ethanol, at the rate of 500 ppm. As the medium was provided with 6 g glucose/l as a carbon source, the ethanol supplement was evidently acting as a rhizomorph-inducing factor rather than as a carbon source, even though ethanol by itself at the rate of 4.6 g/l provided a satisfactory carbon source for btoh mycelial growth and rhizomorph production. Mycelial colonies grew well enough in the absence of ethanol, though growth was usually improved by its addition to the medium. *A. mellea* grew well enough in a synthetic liquid medium with thiamin as the only vitamin supplement, but rhizomorphs were virtually absent. These observations with this single isolate were confirmed by Azevedo (1963) with two isolates from England and two from the Azores, respectively. For some isolates, at least, of *A. mellea*, ethanol and other low molecular-weight alcohols thus seem to act specifically as a rhizomorph-inducing factor, though they can be replaced by some unknown factor or factors present in malt extract and in peptone. The results of further experiments by Garraway and Weinhold (1968) showed that for maximum mycelial growth and rhizomorph production their isolates of *A. mellea* required ethanol for the first 15 days of colony growth ; thereafter, lack of ethanol affected neither rhizomorph production nor total growth.

Production of rhizomorphs is not invariably associated with activity of *A. mellea* as a root-infecting pathogen. Their complete absence from areas of root disease in cacao plantations in Ghana was noted by Dade (1927) and their rarity in Malawi by R. Leach (1939). They are abundantly produced in the forests of the Kenya Highlands but are seldom found in the forests of Rhodesia surveyed by Swift (1968). Where rhizomorphs are thus absent, *A. mellea* spreads from infected to healthy roots by direct contact, and thereafter by mycelial sheets in the bark and over the plane of the cambium ; the mechanism of this spread as an alternative to ectotrophic infection by initially epiphytic rhizomorphs clearly calls for further investigation. In an endeavour to elucidate this variability in the production of rhizomorphs by *A. mellea* in different areas, Gibson (1961) studied 140 isolates of *A. mellea* from various parts of the world and found that they varied widely in the regularity, abundance and vigour of rhizomorph production under optimum conditions of pure culture on nutrient agar. A total of 116 isolates was tested for their temperature optima over the three ranges of 22-4°C, 25-7°C and 27-9°C, respectively. Both for mycelial growth and for initiation and growth of rhizomorphs, most isolates had their optima within the range 25-7°C ; the frequency distribution of temperature optima did not differ significantly between isolates from temperate and from tropical regions, respectively. These results thus seemed to give little support to the idea that production of rhizomorphs under natural conditions was limited by soil temperature, *i.e.*, that too high a temperature prevented their initiation at low altitudes in the tropics, but not at higher elevations in the tropics nor in temperate climates.

Fortunately, however, this proposition has recently been reinvestigated by Rishbeth (1968), who has thereby brought to light a critical difference between temperature optima for initiation and growth of rhizomorphs on nutrient agar and from woody inocula through soil, respectively. Growth rate both of purely mycelial colonies and of rhizomorphs on 3% malt extract agar was found to be maximal at a temperature of approx. 25°C ; growth rate of rhizomorphs from woody inoculum segments through soil tubes was maximal at approx. 22°C (Fig. 6.10).

The optimum temperature for growth rate of rhizomorphs from a food base through soil is thus lower, by some 3°C, than the optimum for growth of rhizomorphs on malt-extract agar. Scientists had earlier found that growth rate of rhizomorphs from woody inoculum segments through soil tubes is nutrient-dependent. Growth is faster from large inocula than from small ones, and weekly growth increments of rhizomorphs invariably decrease with time from the beginning of growth ; They suggested that this latter effect was duc partly to a decrease in rate of mobilization of nutrients from the woody food base, and partly to an increasing intensity of competition for nutrients between the main apices of the rhizomorphs and the growing number of their subordinate branch apices. Rishbeth has used this evidence in support of his suggestion that at the optimum temperature of approx. 22°C, rhizomorph growth rate is in balance with translocation from the food base ; at higher temperatures, rate of translocation is insufficient for the higher rate of metabolism in the rhizomorph apices. This suggestion was supported by the effects of transferring soil tubes from a lower to a higher temperature and vice versa. It also explains Rishbeth's observation that the optimum temperature for growth of rhizomorphs on malt-extract agar was 25°C and sometimes higher; under these growth conditions, rhizomorph growth rate would be unlikely to be nutrient-limited, because uprake of nutrients can take place along the whole length of the growing rhizomorph. The optimum temperature for extension of mycelial sheets between bark and wood in cut lengths of small stems of *Acer pseudoplatanus* was also found to be near 25°C (Fig. 6.10); under these conditions also, growth rate was unlikely to have been limited by translocation rate, because the mycelium was in direct contact with its substrate for the whole extent of its growth.

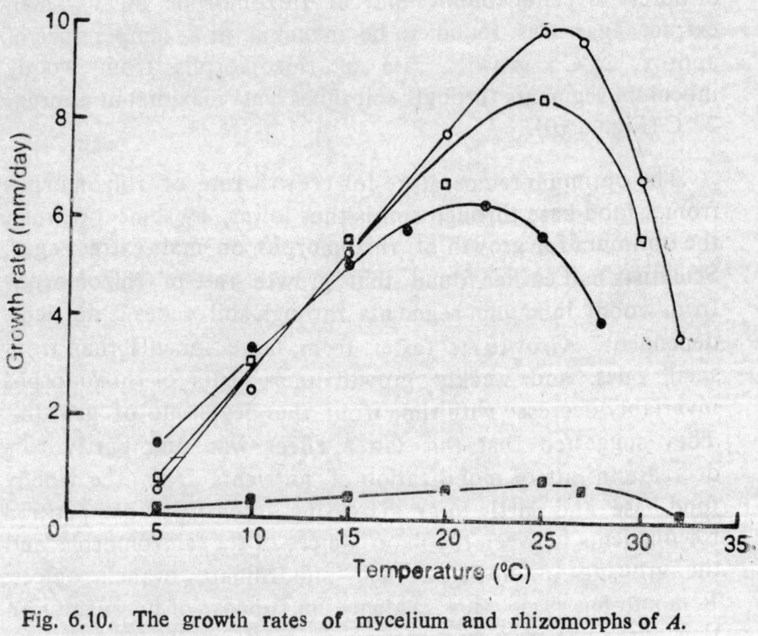

Fig. 6.10. The growth rates of mycelium and rhizomorphs of *A. mellea* in relation to temperature. O, Rihizomorphs in malt agar ; ● rhizomorphs in soil tubes ; □, mycelial sheets in wood lengths ; □, mycelium in malt agar.

In the preparation of his woody inoculum segments, Rishbeth employed a dose of basidiospores as inoculum so that his population of *A. mellea* must have been genetically variable and therefore a representative sample of the population in the area from which he obtained fructifications. Yet in no case did he observe rhizomorph initiation at a temperature above 25°C ; from his overall experience in this study, he selected a temperature of 18-20°C as the most reliable for rhizomorph initiation. His total experimental results thus clearly suggest that too high a soil temperature is likely to be one of the factors restricting production of rhizomorphs by *A. mellea* in the lowland tropics.

Rishbeth has, however, disclaimed any suggestion that supraoptimal soil temperature is the only factor limiting rhizomorph production in the lowland tropics, and has drawn attention to the work of Swift (1968) on the existence of a toxic factor in some Rhodesian forest soils. All the isolates of *A. mellea* employed by Swift produced abundant rhizomorphs on nutrient agar, so that an intrinsic inability to produce rhizomorphs could be eliminated as a possible factor in their behaviour. Moreover, although no rhizomorphs were produced from woody inoculum segments in growth tubes filled with a typical Rhodesian forest soil, they were initiated and grew well in the same soil after it had been autoclaved. Soil fumigation with propylene oxide did not remove the rhizomorph-inhibiting factor, so Swift was able to conclude that is was persistent in the soil in the absence of living microorganisms, at least for a time, though it was not thermostable because it was destroyed by autoclaving the soil. The persistence of this factor was also demonstrated in another way, by Seitz-filtration of a 1 : 1 soil-water extract and addition of the sterile filtrate to columns of autoclaved soil in the growth tubes ; this sterile filtrate caused total inhibition of rhizomorph growth.

Swift also considered another possibility, arising from the observation that rhizomorphs failed to make any appreciable growth into the short column of moist sand with which scientist covered inoculum segments, standing above soil columns in the growth tubes, in some of their experiments. They had thought that perhaps sand failed to provide some factor present in soil that stimulated outwards growth of rhizomorphs from the inoculum segments ; this possibility was later eliminated by Rishbeth's (1968) finding that rhizomorphs grew well enough from the inoculum segments when these were buried in 1 lb glass jars of sand. A satisfactory resolution of this discrepancy has now been provided by Griffin (1969), who has suggested that these observations can be explained by a drying-out of the short sand columns covering the inoculum segments at the top of the growth tubes. In this supposition he has been entirely correct ; they can indeed confirm that the short sand columns did dry out fairly quickly, in spite of the fact that the growth tubes were capped by moisture-proof, lacquered cellophane secured with a rubber band. The correct explanation did not occur to me because the drying sand still acted as an effective mulch to prevent drying out of the inoculum segments, and growth of rhizomorphs *downwards* into the soil columns was very satisfactory. But failure of rhizomorph initials to continue growth into the drying sand columns can now be explained by Griffin's discovery, in collaboration with A.M. Smith, that rhizomorphs will continue to grow only for so long as the apices are covered by an unbroken film of water. They have interpreted this requirement in terms of the reduced rate of melanin formation under the lower oxygen concentration around the rhizomorph apex when it is covered by a water film ; their explanation is compatible with the high values of the Michaelis constants for the enzymes laccase and tyrosinase. They postulate that the melanin sheath is mycostatic, protecting the mature part of the rhizomorph but inhibiting elongation of the apex if formed precociously.

SOME EFFECTS OF SOIL CONDITIONS ON ECTOTROPHIC SPREAD

Three ectotrophic root-infecting fungi are known to spread more quickly over the root systems of their hosts in alkaline soils than in neutral or acid ones. No explanation seems to have been advanced for this effect on the spread of *Phymatotrichum omnivorum*, but attempts have been made to explain a similar effect on the spread of *Ophiobolus graminis* and *Fomes annosus*, respectively.

During the investigation of the effect of soil conditions on rate of ectotrophic spread by *Ophiobolus graminis* along the roots of wheat seedlings, Scientists found that rate of spread was increased : (1) by improvement in soil aeration ; (2) by increase in pH value of the soil up to approx. pH 8. In a hypothesis attempting to unify these two sets of observations, they suggested that rate of advance of *O. graminis* was inversely proportional to the concentration of carbon dioxide at the surface of the respiring root ; thus at pH 8.3% undissociated carbon dioxide is in equilibrium with 97% as bicarbonate ion, whereas at pH 5.5, 90% free carbon dioxide is in equilibrium with 10% as bicarbonate. In further experiments to test this hypothesis, Scientists demonstrated that growth of *O. graminis* along the roots was greatly accelerated by forced aeration of the soil and that under these conditions soil pH value no longer affected the rate of fungal advance. This simple explanation has recently been challenged by the theoretical considerations and experimental results published by Greenwood and Nye (1968). They have pointed out that as diffusion through the gas phase is extremely rapid, oxygen and carbon dioxide partial pressures in any soil gas space that is directly connected with the atmosphere above the soil will be much the same as in air. In the soil-water phase, carbon dioxide is thirty times as soluble as oxygen and so will diffuse thirty times as fast. From this it could be concluded that a deficiency of oxygen at the root surface, due to both root and microbial respiration, is likely to limit the growth rate of *O. graminis*, and of similar ectotrophic root-infecting fungi, *before* the partial pressure of carbon dioxide has increased to a growth limiting value. Nevertheless, convincing as these arguments by Greenwood and Nye may seem, it is perhaps yet premature to come to any definite conclusion ; their arguments challenge not only my own hypothesis but also a substantial volume of other data hitherto accepted as evidence for a growth-limiting effect of carbon dioxide on the activities of various soil micro-organisms.

A different explanation was put forward by Rishbeth (1950) for the more rapid advance of *Fomes annosus* over the bark of pine roots in alkaline as compared with acid soils. He was able to reproduce this effect on 5 cm lengths cut from small pine roots growing in acid and alkaline soils, respectively, washed free of soil, and then incubated, after inoculation with *F. annosus* at one end, in a moist atmosphere. The design of this experiment eliminated any differential effect of carbon dioxide at the root surface, such as Scientists had postulated for *O. graminis.* It must also largely have eliminated any differential effect of active host resistance due to soil type, because as we have noted earlier, Wallis (1961) demonstrated that root-severing destroyed the effective defence-response of roots in either type of soil. The much retarded rate of advance by *F. annosus* over the root bark of roots taken from an acid soil in this experiment was ascribed by Rishbeth to antagonistic effects by root-surface fungi of acid soils, and particularly to those exerted by *Trichoderma viride* Pers. ex Fr. Gibbs (1967) has examined Rishbeth's suggestion of *T. viride* as an important antagonist of *F. annosus* on pine root bark in acid soils in the light of Rifai's (1969) subdivision of *T. viride*, as recognized by Bisby (1939), into nine species. Even within each of Rifai's narrowly defined species of *Trichoderma*, both Gibbs (1967) and Mughogho (1968) have found a wide variation amongst isolates in the degree of antagonistic that they exert against other fungi growing in close association with them. Gibbs was unable to find any clear evidence of a significant correlation between frequency of antagonistic isolates of *Trichoderma* and soil acidity. In place of this explanation, he has drawn attention to much published evidence in favour of the generalization that most antibiotics are inactivated much more quickly in neutral and alkaline soils than they are in acid soils.

On the light-textured, sandy soils of the Breckland area of East Anglia particularly studied by Rishbeth (1951), various other soil factors, and particularly organic content, can be closely correlated with soil pH. Thus the acid podsols are covered by a well marked litter layer, and so the moisture retaining and supplying power of these soils in dry seasons is higher than that of the alkaline sandy soils. Rishbeth adduced both observational and experimental evidence for the view that the more favourable moisture regime in the acid podsols enhanced the resistance of pine trees to *F. annosus*, especially in dry seasons. This opinion has been upheld by recent evidence produced by Gibbs (1968), showing that the volume of resin production following injury could be directly correlated with soil moisture availability, and was greater for pine trees on acid podsols than for those on the alkaline sandy soils. In the U.S.A., Towers and Stambaugh (1968) found that experimentally induced moisture stress increased the susceptibility of loblolly pine (*Pinus taeda*) to infection by *F. annosus.*

Competitive saprophytic colonization of substrates by root-infecting fungi

This reaction will be concerned with survival or pathogenic root-infecting fungi in the soil apart from living host plants. There are three possible modes of such survival : (1) by competitive saprophytic colonization of dead organic substrates, most commonly consisting of a corpus of dead plant tissue, and by saprophytic survival on these ; (2) by saprophytic survival of the pathogen in dead infected tissues of a host plant, first invaded by the pathogen during its parasitic phase ; (3) by dormant survival in the form of resting bodies (spores or sclerotia) which have been produced either in plant tissues colonized saprophytically, as under (1) above, or in tissues infected by the pathogen during its parasitic phase.

Because of the complexity of saprophytic competition for substrates amongst soil fungi, the circumstances under which some root-infecting fungi appear to have a phase of existence as competitive saprophytes in the soil are still incompletely known and imperfectly understood. In the early years of root-disase investigation, chance observations presented some valuable clues to a few gifted observers who were alert enough to be thinking about this problem already. There are the evidence for the conclusion that some tropical species of *Rosellinia* have a phase of active saprophytic colonization of leaf litter on the forest or plantation floor ; Scientist also described the chain of evidence that permits us to conclude that *Rhizoctonia solani* can exist in many cultivated soils as a free-living competitive saprophyte, for which it is well equipped by its ability to decompose cellulose quite vigorously. Since the beginning of the present century, mycologists from E.J. Butler (1907) onwards have been accustomed to investigate the occurrence of particular groups of fungi in the soil by burying therein 'baits' of particular types of dead plant or animal tissue so selected as to encourage competitive saprophytic colonization by the desired group of fungi, *e.g.*, boiled hemp seed for *Pythium* species, human hair for keratin-decomposing fungi, shrimp or lobster shell for chitin-decomposing fungi, and so on. Sadasivan (1939) baited various soils at the Rothamsted Experimental Station with pieces of fresh, mature wheat straw, and in this way was able to confirm a conclusion from earlier circmustantial evidence that *Fusarium roseum* f. *cerealis* (syn. *F. culmorum*) is a regular primary fungal colonizer of this substrate when it is introduced into cultivated soils.

Isolation of fungi from such baits of deal plant or animal tissue burried in natural and otherwise untreated soil provides the most direct evidence obtainable that such fungi can colonise these particular substrates in soil as competitive saprophytes ; for root infecting fungi, it constitutes evidence for the possibility of an independent saprophytic phase in the soil. This realization provided the basis for a technique for comparing degrees of competitive saprophytic ability amongst root-infecting fungi that was devised and developed by several scientists at the Cambridge Botany School from 1950 onwards ; it has been employed by workers elsewhere under the designation of the 'Cambridge Method'. Like many other experimented methods, this one purports to create a model of the natural situation or situations in the soil, and so invites questions as to its veracity. The questions in particular need answering. First, do substrates of the type employed experimentally occur at all frequently in the soil ? Secondly, when soil artificially inoculated with a particular fungus is employed, does the range of inoculum dosage correspond with the range of population level at which the fungus occurs naturally in cultivated soils ? This second question arises from the early discovery that the inoculum potential of a fungus at the surface of a substrate partly determines the share of the substrate that it is able to colonize and exploit.

For these reasons, interpretation of experimental findings on saprophytic behaviour of root-infecting fungi has been hampered by doubts concerning the validity of the experimental models of a kind that has scarcely vexed investigation of other phases of the life-cycle. Thus the parasitic phase, saprophytic mycelial survival in dead, infected host tissues, and survival of dormant propagules all present problems that are much less complicated than that of saprophytic competition, so that no one has really needed to ask whether the experimental models are satisfactory replicas of the natural situation.

The opportunities for purely saprophytic colonization of the root system are probably very restricted comparison with the volume of root tissue that is infected during the parasitic phase. Infection of living tissues continues into, or may even be initiated during, senescence whether natural or infection-induced ; it is finally halted only when tissue-resistance to invasion has declined to a level at which any obligate saprophyte can invade and occupy the remaining volume of root tissue. Living root tissues are reserved for the parasites, whereas dead tissues are open to invasion by all soil fungi ; both in time and in space, therefore, opportunities for parasitic invasion by root-infecting fungi greatly exceed their opportunities for saprophytic colonization, and this difference is greatest for the most specialized parasites, because these are the ones least effectively restricted by host resistance during their parasitic phase.

So far as occupation of the root system is concerned, therefore, we can conclude that saprophytic colonization by root-infecting fungi is of minor importance by comparison with infection of living tissues and that for the specialized pathogens it is of negligible importance. But shoot systems of plants eventually reach the soil, and are indeed regularly incorporated therein in the normal course of agriculture ; what opportunities does this give for saprophytic colonization by root-infecting fungi as an additional means of multiplication and survival ? Hudson (1968) has recently produced a very timely and comprehensive review of investigations conducted, particularly during the last two decades, into colonization of plant shoot systems by air-borne fungi. From this it appears that much of the plant shoot system is colonized by fungi whilst it is still erect ; specialised parasites invade the tissues whilst they are still in their prime, weak parasites follow during the phase of senescence, and even seprophytes can enter and colonize before the shoot system reaches the ground. This air-borne fungal colonization will begin earlier and extend further in a moist than in a dry climate, and its extent will increase with the time that the plant shoot system is allowed to remain standing. In a warm, dry climate or season, for instance, the straw of a wheat crop that is ploughed in before the first rains of autumn may consist largely of virgin, undecomposed plant tissue ; in a moist climate or reason and where ploughing has been delayed, the value of the straw as a substrate for soil fungi will have been correspondingly reduced by prior fungal colonization. It is important to keep in mind this limitation upon the availability of substrates for soil fungi, before we proceed to discuss what factors are likely to determine success in saprophytic competition. For any particular fungus, the share of a potential substrate that it can obtain will be determined : (1) directly by its *competitive saprophytic ability*, which is in intrinsic characteristic of the fungal species ; (2) directly by its inoculum potential at the surface of the substrate ; (3) inversely by the inoculum potential of its competitors.

Competitive saprophytic ability

Scientist first proposed this term in 1950 and later attempted definition of it as 'the summation of physiological characteristics that make for success in competitive colonization of dead organic substrates'. A definition for a new concept has something in common with a specific binomial for a newly discovered organism ; the less one knows about it, the more important it is to be able to identify it. This definition is chiefly revealing about our state of ignorance concerning competitive saprophytic ability at that time. Since then, research has clearly demonstrated that the possession of competitive saprophytic ability is not, as they had originally visualized it, the key to success in competitive colonization of substrates of all kinds ; instead, we are beginning to realize that the kinds of competitive saprophytic ability are almost as diverse as the types of substrate open to saprophytic colonization. An obvious paralled is the diversity of athletic prowess that makes for human success in athletic competitions ; a champion weight-lifter is unlikely to excell in short-distance sprinting and *vice versa.* A closer analogy for plant pathologists is to be found in the term 'pathogenicity', which is host-specific, and most of all so in specialised parasites, Just so, competitive saprophytic ability is substrate-specific. In the case of obligately saprophytic fungi, competitive saprophytic ability for the colonization of one or more types of substrate can be equated quited precisely with the general fitness of the fungal species for existence in one or more ecological niches, and the term thus appears to be redundant. But in the case of root-infecting parasites, for which scientist proposed this term in the first place, it may be conceded a certain usefulness for the answer of certain questions such as the following. Can this fungus exist as a free-living saprophyte in the absence of host plants ? If so, how well can it compete with obligate saprophytes ? Is the possession of competitive saprophytic ability incompatible with specialization to a parasitic existence ? The evidence to be presented later will show that we have made some progress towards an answer to these various questions.

To forestall possible confusion, it is now necessary to point out that the term "substrate' is widely used in two senses, through the particular sense is usually evident from the context: (1) in a wide sense, to denote a corpus of material, usually of plant tissue, which a fungus has colonized and on which it is substituting ; (2) in a narrow sense, to indicate to particular constituent of plant tissue that is being decomposed by a particular fungus, *e.g.*, a sugar, pectic substance, hemicellulose, cellulose or lignin. This latter sense is similar to that in which the term substrate for an enzyme is employed. When exploiting a corpus of plant tissue as a substrate (in the wide sense), a fungus may be utilizing several constituents of the tissue (*i.e.*, substrates in the narrow sense) simultaneously ; the biochemical versality of the whole organism is wider than that of any one of its enzyme-sequences. "Saprophytic sugar fungi' have been defined as those able to utilize only sugars and carbon constituents of plant tissues that are simpler than cellulose, *i.e.* they are incapable of degrading cellulose or lignin. Such a fungus may have a high degree of competitive saprophytic ability for colonization and exploitation of those carbon compounds that it can decompose but will clearly have none for cellulose or lignin, which it cannot degrade even in pure culture. Conversely, a cellulose-decomposing fungus with a high degree of competitive saprophytic ability for this substrate may succeed in securing only a small proportion of the sugars and simpler carbon compounds present in a corpus of fresh plant tissue buried in soil, because its growth rate is lower than that of the fast-growing saprophytic sugar fungi which are usually the poineers colonizers ; they get there first and may consume most of the simpler carbon compounds before slower growing cellulose-decomposers can establish themselves on the substrate.

It is certainly possible to compile a list, as scientists to do in 1950, of all the physiological and biochemical attributes that could conceivably contribute to competitive saprophytic ability of one kind or another ; few successful saprophytes will need, or are likely, to possess all these attributes in high degree, and the particular spectrum of attributes will be determined by the substrate. A short list of the more obvious attributes might be as follows :

(1) rapid germination of fungal propagules and speedy growth of young hyphae when stimulated by soluble nutrients diffusing from a substrate ;

(2) appropriate enzyme equipment for decomposition of the more resistant carbon constituents of plant tissues, such as cellulose and lignin ;

(3) excertion of fungistatic and bacteriostatic growth products, including antibiotics ;

(4) tolerance of fungistatic substances produced by other soil micro-organisms.

The term 'antibiotic' is generally reserved for microbial growth-products active against other micro-organisms at a concentration of 10 ppm or less, and somewhat specific in their growth-static effects. The term 'fungistatic' substance is usually used with reference to a microbial growth-product that is active only at a significantly higher concentration than the minimum for antibiotic action, and it is non-specific in its effects; examples are carbon dioxide, the bicarbonate ion, and substances generating extremes of pH in the ambient solution.

We will briefly indicate the way in which these various attributes of a fungal species contributing to its competitive saprophytic ability may be distributed amongst various substrate-groups of soil fungi, in the form of a tabulation as shown:

	Speedy germination and growth	*Enzymes for cellulose and/or lignin degradation*	*Production of antibiotics etc.*	*Tolerance of antibiotics etc.*
Primary saprophytic sugar fungi, as pioneer colonizers of fresh plant-tissue substrates	+	—	—	—
Secondary saprophytic sugar fungi, associated with cellulose and lignin decomposers	—	—	+	+
Root-surface and rhozisphere fungi	—	—	+	+
Cellulose and lignin decomposers	—	+	+	+

Some of the evidence for the conclusions summarized in this tabulation will be presented later in this chapter, but a brief commentary will be helpful at this stage. The primary saprophytic sugar fungi are, by definition, pioneer colonizers, and they occupy their substrate ahead of competitors so that neither production of nor tolerance to antibiotic substances is likely to give them any significant competitive advantage. The other three substrate-groups live in an environment in which they are exposed to the full intensity of microbial competition, so we can expect that both production of and tolerance to antibiotic substances will confer an additional degree of competitive saprophytic ability.

Secondary saprophytic sugar fungi are designated as those that live as commensals with cellulose-and lignin-decomposing fungi, *i.e.* on the cellobiose and glucose released by hydrolysis of cellulose, and on soluble products of lignin breakdown. Arising from the particular interest in the utilization of cellulose during saprophytic survival of cereal foot-rot fungi in infected wheat straw, scientist postulated the existence of this commensal group of scientists saprophytic sugar fungi. The occurrence of this type of commensalism has been most elegantly demonstrated by Tribe (1966). He was able to grow several species of *Pythium* (sugar fungi) on cellulose film in association with various species of cellulose-decomposing fungi ; the amount of growth made by a species of *Pythium* in presence/absence of an associated cellulose decomposer was estimated quantitatively by oospore counts.

Commensalism of this type has also been demonstrated by Chang (1967) for another non-cellulolytic fungus, *Humicola lanuginosa,* when grown in association with the strongly cellulolytic *Chaetomium thermophile* on filter-paper cellulose. When *C. thermophile* was grown by itself, soluble products of cellulolysis accumulated in the filter-paper pads, and the presence of reducing sugars was demonstrated ; when grown in association with *H. lanuginosa,* weight of solutes was reduced, and only a trace of reducing sugars could be detected. This and other evidence on the ecological niche occupied by secondary saprophytic sugar fungi has been recently discussed by Hudson (1968).

Inoculum potential

We know the difinition of inoculum potential as a determinant of infection. With a slight change in wording, this definition can be made to apply equally well to competitive saprophytic colonization : *inoculum potential is defined as the energy of growth of a fungus available for colonization of a substrate at the surface of the substrate to be colonized.* It is perfectly true that the population level of a particular fungus in natural soil, or its inoculum dosage in an experiment, is one of the most important factors determining the share of a substrate that will be obtained by that fungal species in competition with others. But it is not the only factor, because the success of a population of fungal propagules in competitive saprophytic colonization will also be affected by other factors, both endogenous and exogenous. The growth vigour of the propagules will be determined partly by their age and nutrient status (endogenous factors) and partly by nutrients diffusing from the surface of the substrate and by environmental conditions (exogenous factors). This whole complex of factors affecting the energy of growth of a fungus at the surface of a substrate, and similarly that of its competitors, is summarized by the concept of inoculum potential, and so this concept is just as apt for research on saprophytic colonization as for research on the process of infection. In the broad sense of the term 'competition', fungi compete for a share of substrate not only with other fungi, but also with bacteria and actinomycetes, and at least with members of the soil microfauna amongst soil animals. Nevertheless, it is axiomatic in the science of ecology that the keenest competition occurs between organisms occupying the same or closely similar ecological niches ; for this reason, no doubt, all the investigations to be discussed in this chapter have been restricted to competition between fungi, and therefore my discussion on microbial competition will be restricted to saprophytic competition between fungi.

Although we are using the concept of inoculum potential in order to elucidate the mechanism of saprophytic competition, we have to distinguish carefully between its role in this situation, and that as a determinant of root infection. In the process of infection, inoculum potential of the fungus is the force that opposes resistance of the host ; only if the inoculum potential of the fungus is sufficient to overcome resistance of the host does a successful, progressive infection ensue. A substrate, on the other hand, is inert and offers no active resistance to colonization ; the resistance to be overcome is constituted by the competition of other organism. For this reason, the propagule-population (per unit area of host or substrate surface) component of inoculum potential appears to be relatively more important in competitive saprophytic colonization than it is in host infection. Perhaps we can make this more clear by an analogy with human behaviour in a situation all of us must have experienced. In some countries at all times, and in all at some times, there is disorderly behaviour amongst a crowd of intending passengers trying to get into a public-transport vehicle. The door is open, so the vehicle offers no 'resistance' to invasion ; for any one intending passenger, the only resistance comes from the competition of others. But success does not necessarily go only to the strongest ; a person who happens to be immediately opposite the open door when the vehicle comes to a halt may be able to board it virtually without opposition, provided always that the crowd accepts a certain minimum level of civilized behaviour. Thus an 'advantage of position' may more than compensate for physical weakness ; just so, a high propagule-population of a particular fungal species at the surface of a substrate gives the species an advantage of position which can more than compensate for a low degree of competitive saprophytic ability. If thus favoured by a high population level, almost any fungus can obtain some share of a waiting substrate, as we shall see shortly.

In conclusion, we can epitomize the mechanism of saprophytic competition between fungi in the following statement : *the share of a substrate obtained by any particular fungal species will be determined partly by its intrinsic competitive saprophytic ability and partly by the balance between its inoculum potential and that of competing species.* Environmental conditions in the soil, as we shall see, can strongly affect the outcome of saprophytic competition ; they do this by altering the balance between inoculum potentials of the competing fungi. Saprophytic competition between fungi presents some of the most difficult problems in the whole field of root-disease investigation ; for this reason, we have thought it best to outline enough of what we have learnt about soprophytic competition as will be necessary to help comprehension of the experimental evidence, now to follow.

The Cambridge method

The Cambridge method for investigating competitive saprophytic colonization of substrates by root-infecting fungi, has been developed and employed by the following scientists F. C. Butler (1953), R. L. Lucas (1955), A. S. Rao (1959), R. C. F. Macer (1961) and R.L. Wastie (1961). Pure cultures of a fungus to be tested were grown upon a mixture of quartz sand +3% maizemeal ; when the culture was fully grown, it was easily miscible with soil in any desired preparation. In preparation for an experiment, the maizemealsand culture of the fungus was progressively diluted with an unsterilized soil of medium texture to give a dilution series ; a series that they commonly employed contained the following percentages of pure-culture inoculum : 100 (pure culture alone), 98, 90, 50, 10, 2 and 0 (unsterilized soil alone). Several glass jars were filled with each of these mixtures : in each jar were then buried fifty standard *substrate units* of dead plant tissue. The substrate units were incubated in the inoculum-soil mixtures for about 4 weeks, so as to give the inoculant fungus ample time to colonize as many of them, in competition with other soil fungi, as it was capable of doing. At the end of this period, the number of substrate units colonized by the inoculant fungus in each of the inoculum-soil mixtures was determined.

For investigation by the Cambridge method, scientists selected four (subsequently five) cereal foot-rot fungi, because at that time most of the previous research experience had been with this group of pathogens. These fungi were *Fusarium roseum* f. *cerealis* (syn. *F. culmorum*), *Ophiobolus graminis*, *Cochliobolus sativus* (Ito and Kuribay.) Drechsl. ex Dastur (stat. conid.=*Helminthosporium sativum* Pamm., King and Bakke), and *Curvularia ramosa* (Bain.) Boedijn. At that time, there was already good evidence for supposing that *F. roseum* was a vigorous competitive saprophyte and that *O. graminis* was a poor one; these two fungi thus provided reasonably reliable points of reference by which to judge the value of the method. The choice of this group of fungi had the further practical advantage that the host-plant tissue, which seemed the most appropriate to use in these tests, could be wheat straw, which was easily obtained and cut up into standard substrate units; short lengths (approx. 4 cm) were so cut from the long straw as to have a node at the lower end, and this served to hold the straw pieces together during subsequent handling. Before burial in the inoculum-soil mixtures, the straws were soaked in water or various nutrient solutions, and then autoclaved.

Various methods were tested for assessing the percentage of substrate units colonized by the inoculant fungus after the 4-week incubation period in the series of inoculum-soil mixtures. Washing, followed by surface-sterilization and plating out on nutrient agar, was found to be a reliable method only for fungi that have a high degree of competitive saprophytic ability for colonization of nutrient agar, like *F. roseium*; fungi with a low degree of competitive saprophytic ability for this substrate are suppressed by competitors, as will be described in more detail later in this chapter. A perfectly reliable, though laborious, method was the wheat seedling test scientists had originally devised for determining longevity of saprophytic survival by *O. graminis* in similar pieces of infected or colonized wheat straw. A wheat seed is placed inside the lumen of each piece of wheat straw to be tested for presence of viable *O. graminis*: a number of such inseminated wheat straws are then planted together in a box of sand, which provides a rooting medium optimum for infection by *O. graminis*. Viable *O. graminis* is subsequently demonstrated by the occurrence of runner hyphae and lesions on the roots of the wheat seedlings after approx. 3 weeks; the method is suitable also for detection of other cereal root-rot fungi. A second reliable method was the 'sand plate' devised by F. C. Butler (1953) by which wheat straws were recovered from the inoculum-soil mixtures, washed and plated out on a layer of moist sand in Petri dishes; Macer (1961) subsequently found that surface-sterilization after washing the straws was not only unnecessary but was actually harmful in partially suppressing the inoculant fungus, and so was best omitted. The sand plate method is merely a variant of the 'moist chamber' method long employed by mycologists to encourage fungi to sporulate and thus reveal their presence within infected or colonized plant tissue. Sporulation on the surface of a colonized substrate within a moist chamber does not involve any appreciable degree of competition between fungi *already* established on the substrate, and so the method does not have the disadvantage of strong selectivity, as does plating out on nutrient agar. Is is most satisfactory for fungi like *C. sativus* and *C. ramosa* that produce large, dark coloured spores that are easily visible under the stereoscopic microscope.

The results obtained by the Cambridge method for these four fungi can be illustrated by a selection from the data of F. C. Butler (1953). Butler tried all three methods described above for assessing percentage of straws colonized by his inoculant fungi; Scientists have selected for each fungus the data from an experiment in which assessment of colonization was made by the wheat seedling test. As they have pointed out elsewhere, the dilution end-point is likely to be lower for infection of wheat seedling roots than for sporulation in a moist chamber, so it is justifiable to compare only results obtained by the same technique, as it is done in Table 6.2.

Table 6.2. Saprophytic colonization of wheat straw by four cereal foot-rot fungi

	Percentage straws colonized						
Percentage inoculum	100	98	90	50	10	2	0
Fusarium roseum	90	87	80	67	65	55	22*
Curvularia ramosa	100	96	96	96	85	88	0†
Cochliobolus sativus	100	33	46	10	8	2	0
Ophiobolus graminis	98	46	8	10	2	0	0

* Colonized by *F. roseum* propagules occurring in the soil.
† *C. ramosa* does not occur naturally in English soils.

As we had expected to find, the percentage of straws colonized by each inoculant fungus declined progressively as the inoculum was diluted with unsterilized soil ; this not only diluted the inoculum but provided an increasing degree of competition from soil fungi and other micro-organisms with the increasing proportion of unsterilized soil. But as we had hoped to find, the *rate* of this decline with dilution of the inoculum provided an estimate of the degree of competitive saprophytic ability in the four inoculant fungi under comparison. Decline in percentage straws colonized by *F. roseum* and *C. ramosa*, respectively, was quite shallow down the dilution series ; in contrast, decline by *O. graminis* and *C. sativus*, respectively, was comparatively steep. We thus felt justified in concluding that the first group possessed a high degree of competitive saprophytic ability for colonization of wheat straw, whereas the second group possessed a much lower degree, at least under the conditions of these tests. These results for *F. roseum* and *O. graminis*, respectively, agreed with expectation from earlier evidence on their competitive saprophytic abilities.

In a second paper, Butler (1953) gave the results of various experiments designed to elucidate the difference in competitive saprophytic ability between *C. sativus* and *C.ramosa.* These two fungi had a growth rate over the surface of nutrient agar of a comparable order. Tests for antibiotic production by the stronger saprophyte, *C. ramosa*, were negative, whereas both isolates of *C. sativus* produced antifungal culture filtrates ; this difference was thus the reverse of that required to explain the difference in their respective competitive saprophytic abilities. But a possible explanation was furnished by the finding that *C. sativus* was much more sensitive to antifungal antibiotics, of which ten were tested, than was *C. ramosa* ; *C. sativus* was also more sensitive to antagonism by some common soil bacteria in culture.

A fortunate discovery by Lucas (1955) led to an important new insight into the nature of saprophytic competition during the invasion of wheat straw tissue. Quite by chance, he discovered that, in straws from all the inoculum-soil mixtures he tested, *O. graminis* was more abundant in the outer than in the inner tissues of the straw. He interpreted this observation by supposing that the struggle for colonization of the straw between *O. graminis* and other soil micro-organisms in the soil component of the mixture was not settled once and for all in the outer layer of straw tissue ; competition was postulated as continuing just as intensely whilst the organisms progressed inwards with the result that *O. graminis,* as a fungus of low competitive saprophytic ability, got left behind. If this supposition were correct, then we should predict that a vigorous competitive saprophyte like *F. roseum* should not get left behind in this way but should be as abundant within the inner as in the outer tissues. This point was verified by Macer (1961) and so these observations have confirmed the difference in saprophytic ability between these two fungi in another way. This finding led Macer to determine respective rates of straw tissue penetration in pure culture by these four fungi, together with a fifth, *Cercosporella herpotrichoides* Fron, which causes the eyespot lodging disease of cereals. The difference in straw penetration rates by *O. graminis* and *F. roseum,* respectively, turned out to greater than that between their respective surface growth rates over nutrient agar.

Sufficient evidence has now accumulated for an attempt to interpret the marked difference in competitive saprophytic ability for straw tissue colonization between these two pairs of fungi : the vigorous saprophytes, *F. roseum* and *C. ramosa,* and the weak saprophytes, *O. graminis* and *C. sativus.* The data assembled for this purpose in Table 6.3 are : (*a*) the approximate percentage of unsterilized soil in the inoculum-soil mixture that permits 50% of the straws to be colonized by each inoculant fungus (Table 6.2) ; this gives a single-figure estimate (the ED50) for competitive saprophytic colonization ; (*b*) superficial growth rates of the four fungi over nutrient agar, taken from Macer (1961) ; (*c*) figures for straw penetration rate in arbitrary units, taken from Macer (1961) ; (*d*) a *reduction factor* for each fungus, expressing the proportion by which growth is likely to be reduced by antagonistic effect from competing micro-organisms. These values of the reduction factor each of the four fungi have been taken from results obtained with a cellophane-covered agar plate test devised by

Wastie (1961), which will be described in the next section of this chapter.

It can be seen in Table 6.3 that there is no correlation between saprophytic success (*a*) and growth rate over nutrient agar (*b*). Straw penetration rate (*c*) is virtually the same for three of these fungi, but *O. graminis* is handicapped by a straw penetration rate only about 0.6 that of the others. But if values for straw penetration rate (*c*) are adjusted by multiplying by the reduction factor (*d*), to allow for the respective proportions by which growth rate through the straw tissue is likely to be reduced by fungistatic effects from other soil micro-organisms, then we get the product (*cd*) shown in the fifth column of Table 6.3. Mean values of *cd* for the two groups of vigorous and weak saprophytes, respectively, are correlated well enough with mean values for saprophytic performance.

Table 6.3. Distribution of characteristics possibly determining competitive saprophytic ability for colonization of wheat straw tissue amongst four cereal foot-rot fungi

	(*a*) *Percene, unster, soil permitting colonization of 50% straws by inoc. fungus as lab. temp. (16-22° C)*	(*b*) *Growth rate over agar at 22.5° C (mm/34 h)*	(*c*) *Straw penetration rate at 22.5° C (arbitrary units)*	(*d*) *Reduction factor due to competitors*	(*cd*)
Fusarium roseum	approx. 98	10.8	10.0	0.87	8.7
Cuvularia ramosa	>98	4.8	10.4	0.51	5.3
Cochliobolus sativus	approx. 10	3.8	10.3	0.16	1.6
Ophiobolus graminis	approx. 2	8 0	6.4	0.24	1.5

We can conclude that the poor performance of *C. sativus* under these experimental conditions can be ascribed chiefly to its low tolerance of fungistatic growth produced by soil micro-organisms : this affords independent confirmation, from Wastie's values for the reduction factor, of Butler's (1953) original conclusion. The poorest saprophyte, *O. graminis*, is handicapped both by low tolerance of fungistats and by a straw penetration rate lower than that of any other three fungi.

These experiments at Cambridge were carried out under standardized conditions, so as to be comparnble with one another, and the jars of inoculum-soil mixtures were incubated at laboratory temperature (range 16-22°C). But in the discussion of inoculum potential earlier in this chapter, it was pointed out that the share of substrate obtained by any particular fungal species would be determined by the balance between its inoculum potential and that of its competitors ; we should further expect that this balance will be altered by a sufficient change in environmental conditions, which are one of the determinants of inoculum potential. This expectation has been confirmed by the work of Burgess and Griffin (1967) at the University of Sydney, employing the Cambridge method with two of our fungi, *Cochliobolus sativus* and *Fusarium roseum* f. *cerealis*, and two others, *Gibberella zeae* and *Cochliobolus spicifer* Nelson. Their experiments were designed to investigate the effect of temperature upon competitive saprophy tic colonization by these four cereal foot-rot fungi ; the temperatures they employed were 10, 20 and 30°C, together with two programmed regimes of fluctuating temperature, 10±5°C and 30±8° C. All four fungi colonized the maximum percentage of wheat straws at the lowest temperature, 10° C. Their increased success at 10° C cannot be ascribed to a direct effect of temperature on their own inoculum potential, because for all of them the optimum for growth in pure culture was 25° C or above ; instead, as Burgess and Griffin have suggested, we must conclude that a soil temperature of 10° C depressed the overall inoculum potential of competing saprophytes more than it depressed that of any of the four fungi under test. This situation thus closely parallels the effect of temperature upon pre-emergence killing of seedlings, elucidated by L. D. Leach (1947) ; although the optimum temperature for growth of the fungal pathogens is around 25° C, percentage pre-emergence killing in some host-pathogen combinations may be greatest at a much lower temperature, because the decrease in temperature retards velocity of seedling emergence more than it retards fungal activity.

In addition to reporting this important and hitherto undescribed effect to temperature upon competitive saprophytic colonization, Burgess & Griffin have also pointed out that at 20° C, approximating to the mean laboratory temperature at which Butler's (1953) experiments were incubated, they obtained higher percentages of saprophytic colonization by *Cochiobolus sativus* than Butler did. They suggest that this can be explained by supposing that the intensity of microbial competition was lower in their soil than in that employed by Butler. In a further paper, Burgess and Griffin (1968) reported natural colonization by *C. sativus* of wheat straws buried in two uninoculated soils and incubated at 10° C, although the spore population of *C. sativus* in both soils was too low to cause infection of wheat seedlings. They concluded that, at this temperature, *C. sativus* might be able to increase its population by saprophytic activity, and thereby attain an inoculum potential high enough for infection of wheat seedlings ; their suggestion that wheat stubble should be ploughed into the soil in summer, whilst soil temperatures are still high enough to suppress saprophytic activity of *C. sativus*, seems a logical application of their findings to farm practice. A similar effect of temperature upon saprophytic activity of *Ophiobolus graminis* has been found by Gerlagh (1968), also employing the Cambridge method ; percentage saprophytic colonization of wheat straws was found to increase down the temperature series 23, 17 and 10° C.

Whereas the results of Burgess and Griffin and of Gerlagh show that our original studies at Cambridge, made over a temperature range of 16-22° C, underestimated the competitive saprophytic activity of cereal foot-rot fungi under a lower temperature regime in the field, the work of Cook & Bruhel (1968) in the Pacific Northwest region of the U.S.A. suggests that another characteristic of the Cambridge method has probably overestimated it ; these two opposite effects may to some extent cancer each other out. Selecting clean, bright unsterilized wheat straw per burial in containers of soil held at optimum moisture content and at 20° C, Cook & Bruehl found that 94.8% of straws were colonized by *Fusarium roseum* f. *cerealis* from propagules occurring naturally in the soil, which was not inoculated ; determination of *F. roseum* was made by plating out straws on nutrient agar a method quite appropriate for this particular fungus. Their results from this laboratory test thus reapeated ours in Cambridge. But when they made a survey of the fungal population in and straw pieces recovered 2-12 months after ploughing or cultivating into soil in the Pacific Northwest region of the U.S.A., they found that the pattern of colonization of the straw pieces by *C. roseum* closely resembled the pattern of *infection* of the wheat plant by this fungus around harvest time, *i.e. F. roseum* was isolated most frequently from the foot or collar region of the straw and frequency of isolation declined with increasing height above soil level. Cook and Bruehl have therefore concluded that, in their wheat-growing area at least, multiplication of *F. roseum* in its parasitic phase much exceeds its increase in population by competitive saprophytic colonization of wheat straw ; this conclusion accords with the more general view we expressed in the preamble to this chapter. These authors have discussed possible reasons for this apparent discrepancy between the results from the laboratory test and the field survey : (1) use of clean, bright straw for the laboratory test, whereas much of the straw ploughed into the soil after harvest may be 'weathered', *i.e.* precolonized by air-borne fungi ; (2) plating of straws within a month of burial in the laboratory test but sometimes longer after ploughing in the field survey ; thus Walker (1941) obtained a decreasing percentage isolation of *F. roseum* with time from first burial of straws in soil over the period 4-20 weeks tested ; (3) employment of optimum soil moisture content in the laboratory containers, as compared with a drier surface soil after harvest in the field. In a survey of eighty randomly selected fields of wheat, Cook and Bruehl were unable to detect *F. roseum* by plating on a selective agar in fifty-eight fields, and in another thirteen fields of population was < 100 propagules/soil ; according to Cook (1968) these propagules are mainly chlamydospores. They argue that if *F. roseum* were a regular saprophytic colonizer of wheat straw in their areas, then a higher and more consistent population level of this fungus in the soil would be found. From the survey made by Snyder and Nash (1968) of populations of *Fusarium* species in the soil of various fields at the Rothamsted Experimental Station, their is no doubt that frequent cropping with cereals does favour the multiplication of *F. roseum* f. *cerealis*. Thus populations of 2-3 $\times 10^3$ propagules/g soil were characteristic of Broadbalk Field cropped continuously with wheat since 1843, whereas from the soil of Barnfield, under root crops alone since 1876, *F. roseum* was rarely isolated. But what proportion of the population on Broadbalk has come from saprophytic colonization still remains a matter for conjecture.

Thus far we have endeavoured to interpret the available evidence on competitive saprophytic ability for colonization of a wheat straw substrate in terms of speed of propagule germination and fungal growth, production of fungistatic substances and tolerance of those produced by other soil micro-organisms. It is also possible that the rate at which a fungus can decompose cellulose and other more resistant components of wheat straw tissue may be a component of competitive saprophytic ability for this substrate, though this possibility has not been explored ; the connection between cellulolysis rate of various cereal foot-rot fungi. Another possible component of competitive saprophytic ability has been suggesteed by Bytser (1965) and Lindsey (1965) at the University of Colorado. Thus Byther found that *Fusarium roseum* was superior to *F. solani* in its utilization of inorganic nitrogen, and especially of nitrate nitrogen, in pure culture under sub-optimal conditions for growth, such as a low level of carbon nutrient, a high pH value of the growth medium and a low ambient temperature. Lindsey employed this information to interpret his data on saprophytic competition between *F. roseum* and *F. solani* for colonization of autoclaved soil and then for establishment in soil microbiological sampling tubes. *F. roseum* was the stronger saprophyte of the two in this experimental situation, but the degree of its dominance over *F. solani* was enhanced by shortage of available nitrogen, and reduced when nitrogen supply was not limiting growth. These results thus suggest that the speed and efficiency with which a fungus can absorb nutrients from a limited supply may be a factor contributing to its competitive saprophytic ability.

Colonization of the nutrient agar plate

This subject has been of interest to two groups in particular amongst mycologists. First, mycologists studying the fungal flora of the soil have employed various types of nutrient agar as a substrate for the isolation of soil fungi. The techniques fall into two groups. In the first group, a soil suspension at a suitable dilution is incorporated with the melted agar to give a soil dilution plate, or finely sieved soil is mixed with melted agar in the Petri dish to give a 'soil plate'. In the second group, nutrient agar is exposed to colonization by soil fungi *in situ* in the soil, as in Chester's (1940) soil immersion tube and its modifications, the screened immersion plate of Thornton (1952) and the soil microbiological sampling tube of Mueller & Durrell (1957). Secondly, all root-disease pathologists studying root-infecting fungi have frequent occasion to attempt isolation of their primary parasite, either for checking a disease diagnosis or for procuring a fresh isolate of the fungus for study. This isolation is often difficult of accomplishment without special precautions ; specialized parasites more often than not have a low competitive saprophytic ability for colonization of the agar plate, and so they are suppressed by unspecialized parasites and saprophytes that have followed them into root tissues. The special precautions that have to be taken for success include selection of recently invaded tissue and its surface-sterilization before plating out, both of which reduce the inoculum potential of competing fungi, and employment of a selective medium and/or temperature of incubation to favor the primary parasite over its fungal competitors. Competing bacteria are more easily suppressed, formerly by acidifying the agar and more recently by incorporation of antibacterial antibiotics with the plating medium.

It therefore occurred to us to modify the Cambridge method so as to make a more systematic study of competitive saprophytic colonization of the nutrient agar plate than had hitherto been attempted, and this was done in turn by Rao (1959) and Wastie (1961), employing a range of root-infecting fungi as subjects. In this agar-plate variant of the Cambridge method the inoculum soil mixtures were made up as before and sufficient aliquots of each to provide replicates were spread out as a layer 1.0-1.5 mm deep over the bottom of a Petri dish and then impregnated by and covered with water agar, first melted and then cooled down to approx. 40° C. After the agar had solidified, discs 4 mm diam, were cut out with a cork borer and set out (four to a plate), on five replicate plates of modified Czapek-Dox+0.05% yeast-extract agar. The number of pure colonies of the inoculant fungus arising from a total of twenty discs for each inoculum-soil mixture was then recorded. Partial colonies, *i.e.* a sector of the inoculant fungus in a composite colony with one or more other fungi, were recorded as fractions, the sum of which was then added to the number of pure colonies established by the inoculant fungus.

An improvement to the original Cambridge method in standardization of the pure-culture inoculum was introduced by Rao and employed by Wastie in all his final assessments of saprophytic performance on the agar plate by eleven of the root-infecting fungi that he tested. By the original method, pure cultures of all inoculant fungi on sand +3% maizemeal were incubated at 25° C for about 4 weeks. We had realized that this standard period of about 4 weeks might not coincide with 'peak vigour' of all the fungi grown on the medium, and this period of incubation was fixed thus so as to be long enough for the slower-growing fungi. For all his eleven fungi, therefore, Wastie performed separate colonization experiments with pure cultures ranging in age (at 25° C) from 5 days by steps of 5 to at least 30 days. For his final results with each fungus, he then selected that age of the inoculum that had given maximum colonization of the agar plates by the inoculant fungus. In this way, it was possible to compare saprophytic performance by each fungus at its peak inoculum, vigour, and so to eliminate one source of experimental error.

Wastie confined his experimental consideration of competitive saprophytic ability to the two characteristics of a fungus that then seemed most likely to determine its success on the agar plate : (1) growth rate in pure culture on the virgin nutrient agar ; (2) tolerance of fungistatic microbial growth-products as expressed by the ratio 'growth, rate on stated agar/ growth rate on vigin agar'. For the purpose of determining these two values for each fungus, Wastie' successfully developed an ancillary technique, the cellophane-agar-plate test. First of all, he inoculated an agar plate with soil at four equidistant loci, and then covered the whole plate with a thin sheet of wet cellophane. He then inoculated the surface of the cellophane, immediately above each site of soil inoculum, with an agar inoculum disc (4 mm diam.) of the fungus to be tested. In four series of replicate plates, the fungus was inoculated onto the cellophane at 0, 12, 24 and 48 h after the original inoculation of the plate with soil. By comparing rate of growth of the inoculant fungus over the cellophane on plates that had received no soil inoculum with those over plates receiving soil 0, 12, 24, and 48 h previously, Wastie was able to measure the depression in growth rate due to growth of soil fungi from soil inocula under the cellophane. This depression was ascribed to production of fungi static growth-products by these fungi, accumulating in the agar and diffusing through the cellophane. Wastie has listed for fourteen root-infecting fungi the values of the ratio 'growth rate on 48 h stated agar/growth rate on virgin agar'. Under the name of Wastie's 'reduction factor', we have already made use of this ratio in Table 6.3 ; as Wastie's values for all fourteen fungi are of general interest, we have reproduced them here in Table 6.4.

Wastie obtained a complete set of the required data for eleven out of the fourteen fungi listed in Table 6.4. This comprised : (1) a colonization rating for each fungus at peak vigour of its inoculum ; the rating was the mean percentage colonization for the whole range of inoculum-soil mixtures, which

Table 6.4. Tolerance of fourteen root-infecting fungi towards fungi-static microbial growth-products, expressed by Wastie's 'reduction factor'.

Rhizoctomia solani	0.94	Macrophomina phaseoli	0 33
Fusarium roseum f. cerealis	0.87	Pythium mamillatum	0·25
F. oxysporum f. vasinfectum	0.80	Ophiobolus graminis	0·24
F. oxysporum f. cubense	0.66	Cochiobolus sativus	0·16
F. avenaceum	0.53	Verticillium dahliae	0·11
Curvularia ramosa	0.51	Helicobasidium purpureum	0·07
Fusarium caeruleum	0.35	Fomes annosus	0·06

contained 98, 90, 75, 50 and 10% inoculum, respectively ; (2) growth rate of each fungus in pure culture on cellophane overlying virgin agar, and growth rates across the cellophane when the underlying agar had been inoculated with soil 0, 12, 24 and 48 h previously. Wastie set out his final conclusions in the form of four correlation diagrams ; we have given a summary of them in Table 6.5, more concisely and with a test of statistical significance, by means of correlation coefficients.

We can see in Table 6.5 that values of *r*, the correlation coefficient, fail to show any significant carrelation between the colonization rating and either growth rate of the various fungi

Table 6.5. Success of eleven root-infecting fungi in competitive saprophytic colonization of virgin nutrient agar from a series of inoculum-soil mixtures.

Between colonization rating (arcsin-transformed) and the following	*Value of r*	*Value of p*	*Significance of r*
Tolerance of fungistatic growth-products (=reductson factor)	0.2547	> 0.1	N.S.
Growth rate on 48-h stated agar	0.5000	0.1	N.S.
Growth rate on virgin agar	0.6631	0.02	At 2% level

on 48 h staled agar or their tolerance of fungistatic growth-products. But the value of *r* for the correlation between saprophytic success and fungal growth rate on virgin agar is significant at the 2% level. This suggests that the outcome of the struggle between the inoculant fungus and other soil fungi in the inoculum-soil agar disc is decided before any fungistatic concentration of fungal growth-products has had time to accumulate in the agar ; the issue may even be decided within the first 24 h after plating, because outwards hyphal growth from the inoculum-soil discs can usually be seen by the end of this period, at least under the microscope. If this assumption is correct, then Wastie would have obtained a still better correlation between saprophytic success and fungal growth rate if he had measured growth of his fungi in pure culture after the first 24 h. For nearly on the fungi that he tested, the first 24 h would have come within the lag phase of growth, *i.e.*, the phase before the steady-state growth rate is attained. The lag phase is made up of the period before the fungus makes any microscopically visible growth together with the ensuing period over which growth rate is gradually increasing to the steady-state phase. For all his fungi except two very fast and one very slow grower, Wastie estimated growth rate by measuring colony diameters after 59 h. This value does not tell us enough about growth over the first 24 h, because the shape of the lag-phase part of the growth curve may vary widely between a number of fungi that have all attained the same colony diameter after 59 h.

We can thus conclude from Wastie's data that for a primary saprophytic sugar fungus acting as a pioneer colonizer of a sugarbased nutrient agar, or of some kinds of dead but fresh plant tissue, the most important component of its competitive saprophytic ability consists in rapid germination and early growth of its propagules when activated by a nutrient stimulus from the substrate. This constitutes part of the evidence for the schema on competitive saprophytic ability of various substrate-groups of soil fungi set out earlier in this chapter. This conclusion will not come as a surprise to any mycologist who has studied the soil fungus flora by one of the cultural isolation methods to which it is referred earlier in this section. In recent years, fungistatic substances like rose bengal have been added to nutrient agars employed for isolation of soil fungi, in order to slow down fastgrowing fungi so that others get a better chance of developing colonies ; this function of rose bengal is additional to its valuable bacteriostatic effect. Wastie's conclusion is also supported by the findings of Lindsey (1965), who inoculated pairs of fungi simultaneously into autoclaved soil, which was later assayed by the soil microbiological sampling tubes of Mueller and Durrell (1957) ; the faster-growing fungus of each pair was the one predominantly isolated by this procedure.

Fungal growth rate is also likely to be the paramount determinant in intraspecific fungal competition on the nutrient agar plate or slant tube. In a mass isolation of a fungal pathogen from infected tissue containing more than one wild-type strain of the fungus, the fastest-growing one is likely eventually to suppress the others in prolonged culture over a succession of transfers. This is well illustrated by Stover's (1950) study of the brown and gray wild types of *Thielaviopsis basicola*, causing black root-rot of tobacco. On potato-dextrose agar, the growth rate of the brown wild type was greater by approximately one-third than that of the gray, which tended to be dominated and eventually suppressed by the brown. Virulence in unspecialized pathogens tends to be associated with a high mycelial growth rate ; this association is found also in *T. basicola*, inasmuch as the faster-growing brown wild type is a more virulent pathogen than the gray. Prolonged culture on nutrient agar, with a succession of transfers to fresh media, does not usually result in loss of pathogenicity by stock cultures of unspecialized fungal pathogens ; the procedure is selective for the fastest-growing strains in a mixture and therefore also for the most virulent ones. Unfortunately this mechanism operates in the reverse direction for the more highly specialized parasites, in which growth rate tends to be *inversely* correlated with the degree of parasitic specialization. It was found that stock cultures of *Ophiobolus graminis* are particularly liable to loss of pathogenicity in this way ; no doubt many other mycologists have similarly lost both time and temper with other fungi. This difficulty in the maintenance of pathogenic wild-types has been much reduced by the modern practice of maintaining stock cultures in a dormant condition ; agar cultures once grown can be covered with mineral oil or freeze-dried, or cultures can be grown on autoclaved soil that is then allowed to become air-dry.

Competition on staled nutrient agar, and effects of surface sterilants

The conclusions concerning saprophytic colonization of virgin nutrient agar do not apply to colonization of the same agar when it has become staled by fungal growth, as we have demonstrated in a recent study at Cambridge. We found that the species-spectrum of fungi colonizing staled agar changed progressively with the time and degree of staling due to fungal growth ; the most highly staled agar was colonized only by species of *Penicillium* and *Trichoderma*, which showed a high degree of tolerance to fungistatic growth-products of other fungi.

When plating out pieces of plant tissue for assay of their fungal colonizers, the tope of surface-sterilizing agent employed and the duration of the treatment exert a selective effect on the species-spectrum of fungi forming colonies on the isolation medium. This was shown by Walker's (1941) study of fungi colonizing wheat straw baits buried in various soils collected from fields at the Rothamsted Experimental Station. *Fusarium roseum* f. *cerealis* was the dominant fungus on the isolation plates after treatment of the straws with calcium hypochlorite, a mild sterilant, or after prolonged washing with sterile water. But after treatment of the straws with mereuric chloride or silver nitrate, the place of *F. roseum* was taken by species of *Penicillium* and *Trichoderma* which, after the longer periods of surface sterilization, were often the only fungi developing on the isolation plates.

This effect has been put to good use in the search for surface sterilants that are selective for particular fungal pathogens. In an early application of this principle, Davies (1935) found that a higher percentage isolation of *Ophiobolus graminis* from infected wheat tissue could be obtained by employing silver nitrate instead of the more usual mercuric chloride, and it was found that his method gives better results than any other. Since then, this principle has been widely employed in the search for selective surface sterilants, and for selective fungistatic agents to be incorporated in the isolation media designed for picking up particular pathogens.

COLONIZATION BY RHIZOCTONIA SOLANI

We must first mention, about some most elegant work by Bateman (1964) on the decomposition of cellulose by *R. solani* in pure culture. He demonstrated that *R. solani* could decompose cotton fibres, which are generally agreed to be the form of native cellulose most resistant of any to decomposition, and so this fungus must be regarded as a highly efficient cellulose-decomposer. Bateman's paper is accompanied by some excellent photomicrographs showing enzymic erosion of cotton and filter-paper fibres taken from his pure cultures of *R. solani* on these two substrates. He identified areas of cell-wall hydrolysis by their loss of birefringence under polarized light, thus recalling the earlier studies by F. Baker (1939) on cell-wall degradation in comminuted fragments of plant tissue by the rumen microfloras of various herbivores.

Much attention has been devoted to the saprophytic life of *R. solani* by G.C. Papavizas and C B. Davey of the U.S.D.A. at Beltsville, Maryland. In their first papers on this subject (1961), they reported obtaining high percentages of *R. solani* from mature, dry stem segments of buckwheat (*Fagopyrum esculentum*) when buried in various soils for 4 days. Colonization of the baits by *R. solani* was determined by plating them out on water agar fortified with aureomycin, neomycin and streptomycin. If the baits were left in the soil for longer than 4 days, then percentage isolation of *R. solani* declined, presumably because of increasing competition for colonization of the agar from later fungal invaders of the bait tissues. Water agar is often employed instead of nutrient agar for attempted isolation of particular fungi from infected or colonized plant tissue, for the reason that both density of mycelial growth and production of fungistatic growth-products are restricted by the absence of nutrients, and particularly of carbon nutrients. But in view of Wastie's (1961) conclusion that success in fungal colonization of the virgin nutrient agar plate is determined chiefly by fungal growth rate, it is doubtful whether the precaution of using water agar in place of nutrient agar was of much help to Papavizas and Davey in isolating *R. solani* from their bait segments. For the particular case of *R. solani*, Wastie's data strongly suggest that its lack of success in colonization of the nutrient agar plate is due to the fact that it is not a fast grower, and even more to the fact that it is probably a late starter in the race. Thus out of eleven fungi for which full data were available, *R. solani* was ninth in percentage of colonies established on the plates, eighth in its growth rate on virgin agar, and first in its degree of tolerance to fungistatic microbial growth-products. It could be further confirmed from earlier experience in attempting to isolate *R. solani* from roots of young cereal plants suffering from the 'no growth' disease in South Australia ; it was rarely possible to isolate the pathogen from diseased roots unless they were in the early stage of infection.

Papavizas and Davey obtained higher percentages of isolation of *R. solani* when their baits were buried at soil moisture contents below rather than above 50% saturation. Rather surprisingly, the dead buckwheat stem segments gave higher percentages of *R. solani* on plating out than did infected hypocotyls of snap bean (*Phaseolus vulgaris*). In a subsequent paper, Papavizas and Davey (1962) described some modifications in their plating out procedure for isolation of *R. solani* from mature stem segments (5-8 mm long) of twelve plant species tested as baits ; the best results were given by buckwheat, cotton and lima bean and the poorest by maize, oats and soybean. Twenty clones of *R. solani* obtained from ten different soils were tested for pathogenicity against seedlings of snap bean, radish, soybean, sugar beet and wheat, and these saprophytically isolated clones were found to range from non-pathogenic to highly pathogenic ; three clones were highly pathogenic to all five seedling species. Percentage isolation of *R. solani* from these baits ranged from 1% from a soil under grass for › years up to 97% from a soil cropped intensively with susceptible species for at least 6 years. Following upon a demonstration by Papavizas and Davey (1962) that percentage saprophytic colonization of buckwheat baits by *R. solani* was progressively inhibited in four different soils by concentrations of carbon dioxide ranging from to through 20 to 30%. Papavizas (1964) studied carbon dioxide tolerance of twenty isolates each of *R. solani* and *Rhizoctonia praticola* Kotila, respectively. With a few exceptions, a high degree of competitive saprophytic ability amongst these forty isolates was found to be correlated with good tolerance of the higher concentrations of carbon dioxide. The results of Papavizas and Davey with stem segments as baits for *R. solani* have been confirmed by Zarka (1963), employing stem segments of *Corchorus olitorius* as substrate units in conjunction with the full Cambridge method ; in his tests, the saprophytic performance of *R. solani* was as good as that we obtained with *Fusarium roseum* and *Curvularia ramosa* in colonizing wheat straw at Cambridge. Nevertheless. it is unsafe to generalize about the relative competitive saprophytic abilities of different fungal species from a comparison between single isolates of each ; this has been demonstrated by the variation in saprophytic performance amongst the forty isolates of *R. solani* and *R. praticola* in the experiments by Papavizas (1964) just quoted, and has been further emphasized for *R. solani* by Baker, Flentje, Olsen and Stretton (1967).

The findings of Papavizas and Davey in the U.S.A. have been very thoroughly confirmed by Sneh, Katan, Henis and Wahl (1966) in Israel, in a paper that provides some comparisons of particular relevance to this discussion. Sneh *et al.* employed the method of Papavizas and Davey, and their isolation medium for plating the baits was water agar+250 ppm chloramphenicol. As baits they employed stem segments taken from plants of bean (*Phaseolus vulgaris*), cotton, buckwheat, oats, wheat and barley ; the bean and cotton plants were grown to an age of 6 weeks, wnereas stems of the other four species of crop plant were described as 'mature'. No one acquainted with these six plant species would doubt that the greatest contrast in condition of the plant tissues would have bean provided by the living green tissues of the 6-weeks-old bean and cotton plants, on the one hand, and the mature, dry and nearly dead tissues of wheat and barley on the other. We should therefore expect that the green tissues of bean and cotton would possess at least some residual resistance to invasion, thus favouring parasitic fungi like *R. solani* over obligate saprophytes, whereas the nearly dead mature tissues of wheat and barley would have lost any significant resistance to invasion by obligate saprophytes. Sneh *et al.* have obligingly furnished us with a means of testing this assumption, by killing half their bait segments in the autoclave before burial in the soil. A relevant selection from their data is given in Table 6.6.

We can see in Table 6.6 that percentage isolation of *Rhizoctonia* species (*sic*) from the bait segments has reached a higher maximum, and has continued at a higher level, in the

Table 6.6. Percentages of untreated and autoclaved stem segments giving colonies of Rhizoctonia species on water-chloramphenicol agar after burial in a naturally infested soil for the periods shown.

	No. days burial in soil					
	2	*3*	*4*	*7*	*14*	*21*
Bean, untreated	27	57	75	42	45	36
autoclaved	46	57	56	27	18	25
Cotton, untreated	24	64	70	66	45	45
autoclaved	70	60	61	41	36	29
Barley, untreated	51	47	38	7	0	0
autoclaved	51	51	37	2	0	0
Wheat, untreated	41	49	30	7	2	0
autoclaved	50	51	30	6	0	0

untreated *living* stem segments of bean and cotton than in those *killed* by autoclaving before burial in the soil. This is exactly the result we should have expected ; the living segments retain enough tissue-resistance to delay colonization by competing obligate saprophytes but not sufficient to delay infection by the parasitic *Rhizoctonia* species. Killing the stem segments by autoclaving has destroyed the advantage enjoyed by *Rhizoctonia* species of being able to invade the plant tissues in advance of obligate saprophytes. When we look at the corresponding figures for stem segments of wheat and barley, we see that killing the tissues by autoclaving has made no significant difference to percentage recovery of *Rhizoctonia* species on the isolation plates ; it is clear that these mature tissues must have been virtually dead before autoclaving, so far as any resistance to colonization by obligate saprophytes was concerned.

This experiment by Sneh *et al.* has been most admirably designed for a demonstration of this important distinction between infection of still living tissues by parasites and saprophytic colonization of dead tissues by obligate saprophytes. The results of this experiment, more convincingly than any others scientists underline the importance of the practical recommendation. In any situation of disease risk, it is safer to employ dry, dead crop residues as an organic manure rather than the still living green tissues of a cover crop.

COLONIZATION OF WOODY SUBSTRATES

The problems we have just been discussing have not been easy ones to investigate ; nevertheless, if we now pass on to consider competitive saprophytic colonization of woody tissues, as provided by the old roots, branches and trunks of bushes and trees, we encounter situations of still greater difficulty. We have just been considering the importance of distinguishing between infection of still living tissues by parasites, and colonization of dead tissues by obligate saprophytes ; this can be a difficult distinction to make in practice for the tissues of herbaceous plants, but with woody perennials the difficulty is formidable indeed. Yet some way of making this distinction has to be found, because reliable root-disease control in horticultural and sylvicultural practice can be securely based only on a thorough understanding of the fundamental issues involved.

By applying the metnod of ring-barking control introduced by R. Leach (1939) in East Africa against *Armillaria mellea,* old trees that have to be removed in preparation for establishment of a young plantation crop on the site are ring-barked a year or more before felling ; the relatively early death oi the root system that follows ring-barking permits earlier invasion of the tissues by saprophytic wood-decomposing fungi, and their occupancy brings to a halt further invasion by *A. mellea.* Prior saprophytic colonization thus provides a very useful control measure against the further spread of infection by specialized parasites like *A. mellea,* and so we need more information about this subject.

In an attempt to provide some information on this general question, Scientists performed some laboratory experiments with *A. mellea* of a similar type to that by Sneh *et al.* (1966) just described. As a substrate for *A. mellea,* they employed segments of willow (*Salix alba* var. *coerulea*) shoots (approx. 1.75 cm diam.), which were buried in two types of soil in laboratory containers. Before burial, half of the living shoot segments were killed by autoclaving. Segments were buried in soil for various periods by to an effective maximum of 49 day before inoculation with *A. mellea* rhizomorphs growing from woody inocula. The relative volume of each substrate segment colonized by the fungus after an incubation period of 4 weeks was then assessed by measuring weekly growth increments of rhizomorphs put out from the segment in soil growth tubes over an assessment period of 5 weeks, employing a technique devised in an earlier study. The substrate value of the living segments for *A. mellea* was maintained for a period of 7 weeks previous burial in soil, which was the maximum tested. But the substrate value of dead segments declined with increasing length of burial in soil from 3 weeks onwards ; this Scientists attributed to partial occupation of the segments by saprophytic fungi before they were inoculated with *A. mellea.* The results thus agreed with prediction from Leach's original hypothesis, though they had reduced the time-scale as well as the space dimensions of the experiment from those of the plantatio n cropto those of the laboratory.

The results of this experiment thus suggest that *A mellea* can colonize dead woody tissues in the soil, or above it, if favourably placed to do so. Findlay (1951) has reported extensive damage by *A. mellea* to pit props in very damp coal mines. Other fungi infecting tree roots may similarly have opportunities for purely saprophytic invasion of woody tissues. Wallis and Reynolds (1965) have demonstrated that this is at least a possibility for *Poria weirii* Murr., causing a root disease of Douglas fir. Stakes of Douglas fir heartwood, air dried for 6 months and thus 'presumed dead', were buried beneath a 50-year-old stand of Douglas fir in British Columbia for periods of 0,3,6 and 12 months, respectively. Ten replicate stakes at each sampling period were removed from the soil, inoculated with wood-block inocula of *P. weirii* and then reburied. Saprophytic colonization of these stakes by *P. weirii* from the inocula did not decline with increasing periods of burial in the soil up to 6 months; after 12 months, however, there was a significant reduction in the extent of saprophytic colonization by *P. weirii.*

SAPROPHYTIC SURVIVAL OF ROOT-INFECTING FUNGI IN INFECTED OR COLONIZED HOST TISSUES

Saprophytic survival is defined as mycelial survival of a root-infecting fungus in infected host tissues invaded during its parasitic phase. We can also include in this term the mycelial survival of a parasite in plant tissues of any kind that it has been able to colonize competitively in its saprophytic phase, if it has one. In respect of saprophytic survival, no valid distinction can be made between the mechanism of mycelial survival in infected and saprophytically colonized tissues, respectively; in fact, most of the investigations into saprophytic survival in infected tissues have actually been carried out with colonized tissues. For such experiments, large numbers of infected or colonized tissue-units are required; infected material of uniform quality is difficult to collect from diseased crops in the field in sufficient quantity for a large experiment, and laborious and time-consuming to produce from inoculated plants in the glasshouse or in small field plots. But by inoculating prepared tissue-units after autoclaving in culture flasks, large quantities of material uniformly colonized by the desired pathogen can be prepared with the minimum of labour. Confirmatory tests have shown that such colonized tissue-units satisfactorily reproduce the behaviour of naturally infected tissue-units so far as saprophytic survival of the pathogen is concerned.

During infection of living tissues or saprophytic colonization of dead tissues, the fungus establishes a mycelial network throughout the tissues. Thereafter, whether the dead tissues remain in soil or are cultivated into it, the mycelial network of the pathogen survives saprophytically by a slow but continuous decomposition and exploitation of the tissues ; sugars and the more easily assimilated compounds are quickly exhausted but the cellular framework of hemicellulose, cellulose and lignin is more resistant to decomposition, and provides for a considerable longevity of the pathogen ; this longevity extends up to a year and sometimes more for pathogens of annual crops and for periods of years for pathogens of woody perennials. As we have already remarked above survival of a pathogen in inoculated and colonized tissue-units seems to reproduce satisfactorily, for experimental purposes, its survival in similar units of naturally infected tissue. In either case, the pathogen establishes its mycelial network in the plant tissue well in advance of other invaders ; during infection, the pathogen invades living plant tissue, and host resistance keeps out secondary fungi, at least until the mycelium of the pathogen has completely occupied most, if not all, of the host tissue. Inoculation of sterilized tissue-units with the pathogen reproduces, reasonably well according to all available evidence, infection of living tissues for the production of survival-units. But when an unspecialized parasite with sufficient competitive saprophytic ability colonizes competitively dead crop residues that have been cultivated into the soil, it may not achieve so complete an occupation of the plant tissues as if it had infected them during its parasite phase ; such incomplete occupation of the plant tissues is likely to be reflected in a decreased longevity of saprophytic survival.

Saprophytic survival of the mycelial network in plant tissues needs to be sharply distinguished from *dormant survival* of a pathogen, within the tissues or set free in the soil, in the form of resting spores or sclerotia. Such a distinction is easy to make in theory, but special precautions may have to be devised in experimental practice to separate these two types of survival, when they occur concurrently during the survival of a single fungus in plant tissues.

Survival in tissues of herbaceous hosts

Most of the investigations so far conducted into saprophytic survival have been made with the five cereal foot-rot fungi *viz. Ophiobolus graminis*, *Fusarium roseum* f. *cerealis* (syn. *F. culmorum*), *Cochliobolus sativus* (stat. conid. =*Helminthosporium sativum*), *Curvularia ramosa*, and *Cercosporella herpotrichoides*. The specialized ectotrophic parasite, *O. graminis*, infects first the root system, and thence the stem (tiller) bases of its cereal hosts ; the root-rot thus develops into a foot-rot. There is less certainty about the usual sequence of infection in the foot-rots caused by *F. roseum*, *C. sativus* and *C. ramosa* but it seems probable that stem-base tissues are often invaded by these less specialized pathogens direct from the surrounding soil, and not necessarily via infection of the roots. The specialized pathogen *C. herpotrichoides*, causing the eyespot lodging disease of cereals, is not a root-infecting fungus at all. Primary infection of autumn-sown cereal seedling occurs during the winter and early spring from rain-splashed spores ; these spores are abundantly produced, when night temperatures are around 0°C, on infected stubble from a previous diseased cereal crop that has been incompletely buried during ploughing Although *C. herpotrichoides* does not infect the roots of its cereal hosts, the fungus is a soil-borne pathogen inasmuch as it survives saprophytically in infected cereal stubble and so its survival has been studied by the same techniques as have been employed for the other four fungi listed above. To these five fungi has recently been added a sixth ; this is *Cephalosporium gramineum* Nisikado and Ikata (perfect state=*Hymenula cerealis* Ellis and Everh.), which causes a vascular wilt, known as scripe disease, of winter wheat in Washington State, U.S.A.

In an early studies of the saprophytic survival of *Ophiobolus graminis*, Scientists employed a standardized survival-unit of wheat straw tissue colonized in pure culture by the fungus, though they took the precaution of confirming the more important conclusions resulting from such experiments by comparable tests with pieces of naturally infected wheat straw collected from diseased crops around harvest time. These survival-units were cut from lengths of mature wheat straw to give pieces about 4 cm long, each with a node at the lower end, so that the internodal part above it consisted of the culm surrounded by the subtending leaf sheath ; inclusion of a node with each straw piece was essential to secure coherence during decomposition in the soil for prolonged periods, and during subsequent manipulation in the test for survival of viable. *O. graminis.* After the straws had been colonized in pure culture by *O. graminis*, during incubation for about one month at 25°C, they were buried in lots of fifty in approx. 200 ml soil in glass tumblers or jars. At intervals, a sample of not less than 100 straws in two soil containers was taken for the survival test ; a wheat seed was planted, germ end downwards, within the lumen of each straw, and fifty straws thus inseminated were planted in a seedling box filled with sand, which is a rooting medium optimum for infection and spread by *O. graminis.* This wheat seedling test is the only reliable one for *O. graminis,* which is a poor competitor on the nutrient agar plate , the microconidia of this fungus are unreliable for diagnosis, and production of perithecia in a moist chamber cannot be induced either quickly or with the complete assurance required for a survival test. This difficulty does not arise with *Cochliobolus sativus* or with *Curvularia ramosa*, which produce their large and distinctive, septate, dark-coloured conidia as soon as straws are incubated in a moist chamber of the sand-plate type devised by Butler (1953) and employed by him in subsequent studies of saprophytic survival (Butler, 1959). When straws have been incubated on a sand plate for about 1 week at 25°C, sporulating conidiophores on the straw surfaces can be recorded during rapid scanning under the stereoscopic microscope ; if a detailed record is required, the complete length of the straw can be divided up into successive fields of view, each of 1 mm diam., by employing a mechanical stage. Survival of *Fusarium roseum* can be reliably assessed by plating out surface-sterilized straws on nutrient agar, because *F. roseum* has a high competitive saprophytic ability for this substrate. As for *Ophiobolus graminis*, survival of *F. roseum*, *C. sativus* and *C. ramosa* can be assessed by the wheat seedling test ; the test is sufficiently reliable for experiments in which straws have been completely colonized by the original inoculant fungus, and there is no substantial probability that the test wheat seedlings will be infected by other soil-borne fungal pathogens. This proviso has to be made, because roots infected by all three pathogens show a generalized discoloration due to cortical necrosis, and symptoms of root infection by one pathogen cannot be distinguished from those produced by the other two. This difficult of recognition does not apply to roots infected by *Ophiobolus graminis*, which show two distinctive features : (1) an intense necrosis and discoloration of the vascular cylinder ; (2) external growth of the characteristic, dark-coloured runner hyphae, which are easily visible under the stereoscopic microscope. The wheat seedling test for survival of *F. roseum* can be replaced either by plating out surface-sterilized straws on nutrient agar, as noted above, or by a modification of Butler's sand-plate test introduced by Macer (1961). The hyaline macroconidia of *F. roseum* are neither large nor distinctive enough to be identified with certainty under the stereoscopic microscope but they can be scraped off the straw surface into a drop of water, which is then searched under a sufficiently high power of the compound microscope. Macer employed this modified sand-plate method also for assessing survival in straw of *Cercosporella herpotrichoides* ; to induce sporulation by this fungus, sand plates have to be subjected to a daily alternation of temperature with night temperature going down to near 0°C, as was first demonstrated by Glynne (1953). For *C. herpotrichoides*, Macer also successfully employed a wheat seedling test ; nodes were cut off from straws after recovery from the soil, and the internodal segments were slipped over the coleoptiles of just emerged wheat seedlings as a collar. The inoculum segments were then covered with a layer of moist sand to ensure infection, the occurrence of which was demonstrated after some 8 weeks by appearance of the characteristic eyespot lessons.

During the first part of the investigation into the saprophytic survival of *O. graminis*, it became possible to correlate the rapidity of disappearance of the fungus from the straws in a general way with the maintenance of soil conditions optimum for microbial activity ; thus loss of viability was most rapid under conditions of medium to high temperature, suitable moisture content, and good soil aeration. Viability of *O. graminis* was preserved, on the other hand, at a low temperature (2-3°C) and in air-dry soil, and was actually maintained for longer in a water-logged soil than in one held at a moisture content of 50% saturation. The general interpretation placed upon these results is that under conditions conducive to maximum microbiological activity, the available food reserves of the straw substrate are most quickly exhausted by *O. graminis* itself and by associated micro-organisms that have followed it into the tissue. Conditions restricting microbiological activity, such as low temperature, dryness and poor aeration, are thus likely to promote maximum longevity of the fungus in infected host tissuses, providing that the effect of such conditions is not so extreme as to be actually lethal. Thus *O. graminis* is likely to suffer less loss of viability during the low temperatures of winter, or during the extreme drought of such a summer as occurs in South Australia, than during periods of the year in which soil temperature and moisture content are more favourable for microbiological activity.

This conclusion has been fully confirmed by the work of Fellows (1941), using a technique approximating more closely to actual survival of *O. graminis* in the field. Fellows studied the rate of dicappearance of *O. graminis* in naturally infested soil taken straight from the field and incubated under the eight possible combinations of high and low temperature, loose and firm packing, and moist and air-dry conditions. The low-temperature samples were stored in a deep cave and the high-temperature ones were kept in the glasshouse. Looseness was maintained by occasional stirring, and a firm packing achieved by thorough tamping of the soil whilst most. The most soils were maintained at approximately 60% of saturation, whereas the others were allowed to airdry. Four experiments were made, and periods of incubation varied from 230 to 777 days ; at the end of the incubation period, all soil samples were brought to favourable tilth and moisture content, and a test crop of wheat grown in them to maturity, in the glass-house. The degree of infection on the plants at harvest was expressed by means of a 'severity rating'. In almost every comparison provided by this series of extensive factorial experiments, the fungus disappeared more quickly in a warm soil than in a cold one, and more quickly in a loose soil than in a compact soil. The influence of moisture was less straightforward; under cool conditions survival appeared to be rather better in a moist than in a dry soil, whereas under warm conditions the reverse was the case. In all four series of experiments, without exception, disappearance of the fungus was most rapid in the warm, moist, loose soil, *i.e.* under just those conditions of high temperature, adequate moisture content and good aeration postulated above as most conducive to rapid loss of viability.

Effect of nitrogen on saprophytic survival

The C/N ratio of mature wheat straw varies around a value of 80; mature stem tissues of other annual crop plants similarly have a fairly high C/N ratio. It was first realized many years ago, as recounted by Waksman (1932), that such mature plant tissues when composted were decomposed much more quickly if nitrogen, in some readily available form, was added to the compost; similarly, tissues of this type disappeared more rapidly in nitrogen-rich than in nitrogen-poor soils. The mechanism of this effect depends upon the fact that microbial degradation of pectins, hemicelluloses, celluloses and lignins of the plant cell wall is brought about by the action of enzymes. These enzymes are both bound to the microbial cell wall and set free in soil moisture films; the action of bound enzymes is necessarily very local and that of the free enzyme molecules produced and by the effective radius of their diffusion. The largest constituent of mature plant tissue is cellulose and it is this, together with the superimposed lignin, that survives for longest in the decomposing tissue. From an active length of fungal hypha lying alongside a plane of plant cell wall, cellulase enzymes diffuse outwards and the products of cellulose hydrolysis, *i.e.* cellobiose and eventually glucose, diffuse inwards so that a proportion is absorbed by the fungal hypha. So the radius of cellulase action around the enzyme-producing fungal hypha must be limited both by the intensity and duration of its production, and also by any factors limiting its outwards diffusion. For these various reasons, no individual fungal hypha can produce more than a limited zone of enzymic erosion around itself; when this zone is exhausted, then this particular fungal hypha will die from carbon starvation.

Zones of enzymic erosion around fungal hyphae and bacterial cells have been demonstrated under the microscope by the use of differential staining methods and also by loss of birefringence of degraded cellulose when viewed by plane-polarized light. Both methods were used by F. Baker (1939), who has figured very clearly these zones of erosion around bacterial cells lying alongside cell walls in fragments of plant tissue taken from the alimentary canal of various herbivorous animals. Bateman (1964) has similarly figured them on cotton and filter-paper fibres taken from his pure cultures of *Rhizoctonia solani* on these types of native cellulose. Enzymic erosion is also demonstrated particularly well by the so-called 'boreholes' around hyphae of wood-destroying fungi; Bailey and Vestal (1937) have used the geometrical configuration of these boreholes as an aid to interpreting the fine structure of cellulose fibrils. Proctor (1941) has provided microscopical evidence to show that penetration of the thickened secondary walls of plant xylem elements by fungal hyphae is through enzymic action and that mechanical pressure plays only a subsidiary role at most; he has suggested that the term 'borehole' is thus a misnomer. The observed fact that these boreholes do not usually attain a diameter exceeding three times that of the fungal hypha that has made them (Cowling, 1963) suggests that the functional life of any particular length of fungal hypha in cellulolysis is of limited duration. In decomposing cellulosic tissues of herbaceous plants, which are of a construction much looser than that of secondary wood in woody perennials, there must be a effective barrier to outwards diffusion of both cellulase at the sugars derived from its action. The sugars will be quickly absorbed by associated micro-organisms living as commen sales of the cellulose-decomposer in the plant tissue. We can therefore conclude that production of cellulose by a fungal hypha is likely to have become limited by the operation of natural selection to an optimum or economic quantity ; we can certainly visualize the possibility that production of cellulose in excess of this optimum quantity would be uneconomic, because further increases in production would yield a progressively diminishing return of sugar by inwards diffusion towards the cellulase-producing hypha in the presence of sugar-absorbing commensal organisms. A way in which this limitation of cellulase-production may be brought about is suggested by the fact that the cellulases are *adaptive* (as distinct from *constitutive*) enzymes and so contact with cellulose is required for their induction hydrolysis of the immediately surrounding cellulose will remove a fungal hypha from contact with its cellulose substrate, and so induction of cellulase will cease.

Nevertheless the mycellium of which our particular length of hypha under consideration forms an part can survive granted one essential condition ; this is that a continued slow growth and extension of the mycelium within the substrate tissue is not limited by lake of nitrogen or of any other mineral nutrient. So we can now understand that, because both the working life and the survival of individual hyphae are limited by the radius of enzyme action, it must follow that a continued extension of the mycelium by branching and growth within the substrate tissue is an essential condition for continued cellulolysis, and also for survival of the mycelium effecting that cellulolysis. Rate of decomposition of plant tissue is thus limited by the supply of mineral nutrients essential for the synthesis of fungal and other microbial protoplasm. Nitrogen is the mineral element required in largest amount, followed by phosphate and potash. Decomposition-rate of mature plant tissue in compost or in nitrogen-poor soil is thus limited by the supply of available nitrogen ; phosphate may sometimes limited decomposition-rate in compost, but much more rarely in the soil.

An understanding of the mechanism of microbial decomposition of plant tissue, as outlined above, is an absolute prerequisite for comprehension of the factors determining longevity of saprophytic survival by root-infecting fungi. At the outset of the work on saprophytic survival of *Ophiobiolus graminis*, scientist had anticipated that rate of its disappearance from the wheat straws would be correlated with the rate of overall microbial activity in the soil. In general terms this turned out to be true, as they have outlined above, but with respect to the effect of nitrogen the original expectation was exactly reversed. *O. graminis* was wound to do die out early from the rigid and little-decomposed straws buried in nitrogen-poor soils, whereas it was found to survive in almost 100% of straws in nitrogen-rich soils that had become so soft, attenuated and disintegrated by decomposition that they were difficult to handle for the wheat-seedling test. At first scientists adopted a wrong hypothesis to explain this quite unexpected finding, but further thought upon the problem followed by microscopical examination of the straws after burial in the soil showed that the mycelium of *O. graminis* continued to develop in the straw tissue lying within the soil, and that this development was greater, and continued for longer, in nitrogen-rich soil. This observations has recently been corfirmed by Chambers and Flentje. We can therefore conclude that *O. graminis* dies out prematurely from carbon starvation in wheat straw buried in nitrogen-poor soil, because it cannot obtain the nitrogen needed for continued mycelial growth and hydrolysis of fresh areas of cellulose within the straw tissue. The fungus is thus starving in the middle of a well-stocked larder, because it can no longer get at its food.

After this work on *O. graminis* scientist had fully expected that they should find a similar effect of nitrogen in prolonging saprophytic survival of other cereal foot-rot fungi, and perhaps also of other root-infecting fungi similarly surviving in other mature plant tissues with a high C/N ratio. They were therefore surprised when Butler (1953) discovered that excess of soil nitrogen actually reduced longevity of *Cochliobolus sativus* in wheat straw. Subsequently Butler (1959), continuing this work in New South Wales, confirmed this result by comparing longevity of saprophytic survival of *C. sativus* and three other cereal foot-rot fungi in a ley soil of high fertility (total N= 0.13%) and in an arable soil of low fertility (total N=0.07%). Survival of both *O. graminis* and *Fusarium roseum* f. *cerealis* was longer in the high-nitrogen soil, whereas that of *C. sativus* was longer in the low-nitrogen soil; survival of *Curvularia ramosa* appeared to be indifferent to the level of soil fertility and nitrogen content (Fig. 6.11).

In discussing these results, Butler suggested that this effects of nitrogen in actually reducing longevity of *Cochliobolus*

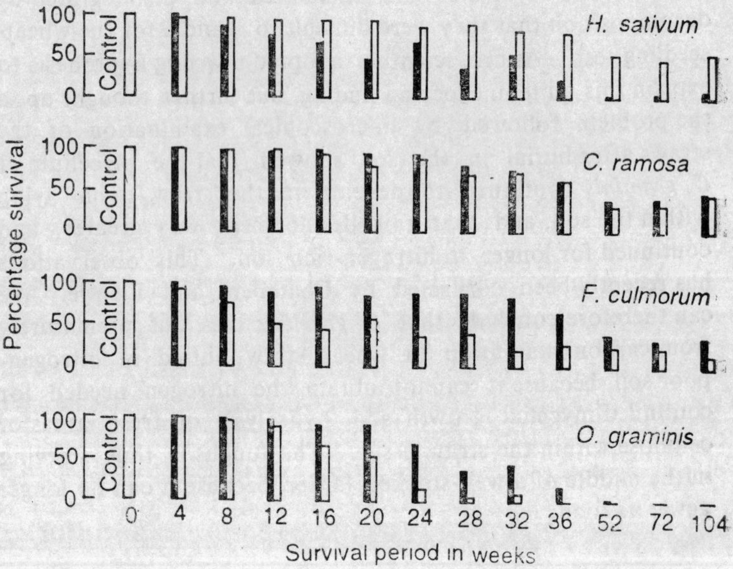

Fig. 6.11. Saprophytic survival of four cereal foot-root fungi in wheat straw. The left-hand column in each pair represents percentage survival in a soil of high fertility, and the right-hand one represents survival in a soil of low fertility.

sativus might be connected with its low tolerance of fungistatic growth-products excreted by competing soil micro-organisms. He supposed that this might limit the slow but continued mycelial devolopment such as it was observed for *O. graminis* in the wheat straw tissue ; if further mycelial development of *C. sativus*, were thus limited, then diffusion of soluble nitrogen into the straw tissue from the surrounding soil would not benefit *C. sativus*, but might well help other soil micro-organisms. and especially soil fungi, to invade the subtrate, and thus to increase the population of competing micro-organisms within the straw tissue. At this stage of the discussion, it will be helpful to summarize evidence on effects of nitrogen on longevity of saprophytic survival by five cereal foot-rot fungi ; in Table 6.7 a 'positive effect' of nitrogen indicates that this nutrient prolongs survival, whereas a 'negative effect' denotes the reverse.

Table 6.7. Effects of nitrogen on longevity of saprophytic survival by five cereal foot-rot fungi.

Pathogen	*Disease*	*Effect of nitrogen*
Ophiobolus graminis	Take-all	Strongly positive
Fusarium roseum f. *cerealis*	Foot-rot	Strongly positive
Cercosporella herpotrichoides	Eyespot	Weakly positive
Curvularia ramosa	Foot-rot	Weakly positive Indifferent
Cochliobolus sativus	Foot-rot	Strongly negative

We can see in Table 6.7 that a strong positive effect of nitrogen on longevity of *O. graminis* has been repeatedly confirmed and so has a strong negative effect on longevity of *C. sativus*. The strong positive effect on *F. roseum* rests on the results of a single experiment by Butler (1959). Responses to nitrogen by *C. Ramosa* and *C. herpotrichoides*, respectively, has been either weak or indifferent.

Scientists suspected that these differing responses to nitrogen in longevity of saprophytic survival by these five fungi might be connected in some way with their respective cellulolytic abilities. The cellulose cotent of mature wheat straw comprises about 40% of total dry matter, yet no information on cellulolytic ability of these five species appeared to be available. To procure this, mycologists measured loss in dry weight by filter-paper cultures of the five fungi over a standard incubation period of 7 weeks at 22.5 °C. *C. sativus* was found to have the highest cellulolysis rate and it seemed possible that this was somehow connected with its strong negative response to nitrogen longevity of survival. Nevertheless, cellulolysis rate by itself could not be correlated at all clearly with the nitrogen effects ; they then came to realize that the adequate of the cellulolysis rate for continued saprophytic survival of a fungus in wheat straw tissue is not determined by this rate alone, but by the relation between this rate and the general metabolic rate of the fungus. The general metabolic rate determines the rate at which carbohydrates are consumed by the processes of respiration and growth ; the cellulolysis rate determines the rate at which soluble carbohydrates are made available for the fungus. Only if the celluloysis rate is *adequate* to maintain the general metabolic rate will a fungus be able to continue surviving in the straw tissue. Thus a fast-growing fungus with a high metabolic rate will require a higher rate of cellulolysis to supply its needs than will a show-growing fungus with a slow metabolic rate. To express these relationships, they proposed the *Cellulolysis Adequacy Index*, which is obtained by dividing cellulolysis rate by linear growth, rate (as a parameter of general metabolic rate), each expressed in suitable units. If cellulolysis rate is expressed by percentage loss in dry weight of filter-paper cultures over the standard incubation period of 7 weeks at 22.5 °C and linear growth rate as mm/24 h at 22.5°C, then the cellulolysis adequacy index is found to vary around unity for the five cereal foot-rot fungi under investigation. In Table 6.8, mean values of the index for *F. roseum*, *C. herpotrichoides* and *C. ramosa* have been taken from Garrett (1966) and the mean value for *O. Gramlnis* from Garett (1967).

Table 6.8. Cellulolysis adequacy indices for five cereal foot-rot fungi.

	No. isolated tested	*Percent. loss in filter-paper dry wt. after 7 wks at 22.5°C* Range	Mean	*Linear growth rate of fungus (mm/24 h) at 22.5°C*	*Mean value for cellulolysis adequacy index*
Ophiobolus graminis	5	2.1-3.6	2.8	7.3	0.38
Fusarium roseum f. *cerealis*	4	6.5-8.5	7.5	11 0	0.68
Cercoporella herpotrichides	4	0.8-1.1	0.9	1.3	0 69
Curvularia ramosa	1	—	6.6	5.3	1.25
Cochliobolus sativus	6	5.5-8.4	7.5	3.6	2.10

A comparison of the data given in Table 6.8 with those in Table 6.7 suggest that a low value of the cellulolysis adequacy index (hereafter referred to as the CAI) is associated with a strongly positive response to nitrogen in longevity of survival and a high value of the index with a strongly negative response. The concept of the CAI can thus explain why a fungus such as *Cochliobolus sativus* does not need extra nitrogen in order to survive in wheat straw tissue ; a further extension of this hypothesis can also explain why extra nitrogen actually *shortens* its period of survival. This is because a supply of nitrogen much is excess of the level needed to keep cellulolysis going at a rate just adequate for survival is going to promote an *uneconomic* rate of cellulose consumption ; this will result in a complete exhaustion of cellulose in the substrate in a shorter time than if nitrogen had been supplied at a rate just sufficient to keep the fungal colony alive and metabolizing the substrate.

This hypothesis thus postulates that the primary factor determining the response of a fungus to nitrogen is its own behaviour in metabolizing the cellulose of the substrate, as expressed by its CAI value, and that other fungi and other soil micro-organisms play a subsidiary role in decomposition and eventual exhaustion of the substrate. The *direct effect* of nitrogen upon the cellulytic activity of the fungus itself thus differs from the *indirect effect* proposed by Butler (1953), who thought that nitrogen acted by assisting other micro-organisms to invate and colonize the substrate alongside its primary colonizer. These two effects are not mutually exclusive but an experiment has been designed with *Cochliobolus sativus* to determine which effect was the more important. A total amount of nitrate nitrogen sufficient to curtail saprophytic survival of *C. sativus* was provided in two ways : (1) to the straws in the pure culture flasks before inoculation ; (2) to the soil at time of burying the colonized straws. Scientist argued that if the effect of nitrogen was mainly on the cellulolytic activity of *C. sativus* itself (the *direct effect* hypothesis), then nitrogen would reduce longevity more if added to the pure culture flasks, because it would have a month longer in which to act on the fungus, and because most of it would be absorbed and retained by the mycelium of *C. sativus*. But if the main effect of nitrogen was to assist other fungi and micro-organisms to invade and colonize the substrate (the *indirect effect* or *replacement* hypothesis), then nitrogen would reduce longevity of *C. sativus* more if it was added to the soil at time of straw burial. The longevity of *C. sativus* was estimated on a sample of fifty straws from one soil jar for each of the six experimental treatments (two fungal isolates times three nitrogen treatment) over the period 4-16 weeks. After recovery from the soil, the encasing leaf-sheath around each straw was removed and the surface of the culm segments was well rubbed under the running tap before sand-plating. The object of this was to remove any conidia, which act as survival spores in *C. sativus* ; if this precaution had not been taken, dormant survival by conidia attached to the straws might have been wrongly recorded as saprophytic survival. After 9 days incubation on the sand plates at 22.5°C, each straw was thoroughly scanned under the stereoscopic microscope (50× magnification) ; any straw bearing one or more effect, sporulating conidiophores was recorded as containing still viable *C. sativus* (Table 6.9).

The results of this experiment, displayed in Table 6.9, have returned a definite answer in favour of the *direct effect* as the

Table 6.9. Effect of nitrogen and its time of application on longevity of saprophytic survival by Cochliobolus sativus in wheat straws buried in soil.

	Percentage straws with C. sativus still viable			
	Period in soil (weaks)			
	4	8	12	16
Isolate P 1, no nitrogen	100	100	92	82
soil nitrogen	100	100	60	14
pure-culture nitrogen	100	72	48	-
Isolate BS 790, no nitrogen	100	100	100	74
soil nitrogen	100	100	92	42
pure-culture nitrogen	100	100	78	2

more important of the two. After this, scientists realised that the most convincing evidence for the direct effect hypothesis would be to reproduce the whole spectrum of nitrogen responses in a fungus growing in pure culture upon a cellulose substrate. Although, over the range of nitrogen levels tested, a significantly negative response had been shown only by *C. sativus*, it seemed likely that the response of any these five fungi would switch from positive to negative once the critical nitrogen level had been exceeded. The optimum nitrogen level for survival of *Ophiobolus graminis* in straw buried in soil is around 0.5 g N/100 g air-dry straw ; results of an earlier experiment suggested that 0.9 g N was supra-optimal for maximum longevity of survival. *O. graminis* was selected as the subject for this experiment : (1) because its saprophytic survival is shorter than that of the other four fungi ; (2) because symptoms of root infection in the wheat-seedling test are completely reliable for diagnosis. As most fungi survive in pure culture for a long time, they took all possible precautions to reduce the survival period in this experiment, and the substrate was reduced to a single circle of Whatman no. 3 filter paper lying on a reservoir of inorganic nutrient salts in pure quartz sand. At each sampling time, survival of *O. graminis* on the filter-paper circle was estimated by percentage infection of the roots of twelve wheat seedlings planted thereupon ; percentage figures given in Table 6.10 are based on examination of some

120 roots for each treatment-series at each sampling time (Time 6.10).

Table 6.10. Effect of nitrogen level on longevity of survival by Ophiobolus graminis in pure culture on filter-paper cellulose.

Porcentage roots infected by O. graminis				
		Time of sampling (weeks)		
		10	*15*	*23½*
Nitrogen series	S/4	80	80	16
	S/2	99	97	55
	S	98	99	92
	2S	100	85	82
	4S	100	73	73

Standard error of means =3.79 : least difference for significance at 5% level=11.

Table 6.10 shows that the optimum level of nitrate nitrogen for maximum longevity of *O. graminis* on filter-paper cellulose was that provided by series S, which was 0.235 g N/100g airdry filter paper. At the final sampling after 23½ weeks, the effect of the highest nitrogen treatment (0.94 g N/100 g air-dry filter paper in the 4 S series) in reducing longevity of *O. graminis* was highly significant (at the 0.1% level). The results of this experiment were thus in complete accord with prediction from the *direct effect* hypothesis concerning the various effects of nitrogen on longevity of saprophytic survival.

Other factors affecting survival

We have already mentioned the possibility that saprophytic survival of mycelium may be augmented by the production of dormant propagules in the host tissues. *Ophiobolus graminis* is not known to produce any type of dormant propagule ; in the experiment with *Cochliobolus sativus* to which Table 6.9 refers, Scientists had taken due precautions to separate saprophytic mycelial survival, which they were studying, from dormant survival by conidia. This precaution was necessary, because conidia of *C. sativus* function as dormant propagules in soil ; their effectiveness as such is more readily understood since the recent announcement by Meronuck and Pepper (1968) that chlamydospores are produced within the cells of conidia over-wintering in soil. Since this complication of saprophytic mycelial survival by dormant survival was eliminated from the comparison between *O. graminis* and *C. sativus* in the previous section, there seems no reason to doubt that the contrasting effects of nitrogen on the longevity of these two fungi in wheat straws were indeed produced by the influence of nitrogen on saprophytic mycelial survival. But this conclusion cannot be extended without qualification to the nitrogen response of *Fusarium roseum* f. *cerealis,* the only other one of these five cereal foot-rot fungi showing a strong response (positive) to nitrogen in its longevity of survival in wheat straws (Table 6.7). Cook (1968) has observed the regular production of mycelial chlamydospores by *F. roseum* in infected wheat tissues. They therefore have to ask themselves whether nitrogen, which increases longevity of *F. roseum* in wheat straws, does so by promoting an increased production of chlamydospores in the infected or colonized tissues. To this question we can return a cautions negative ; the available evidence suggests that production of chlamydospores by *Fusarium* species, as well as that of other types of dormant propagule, is favoured by a high rather than by a low C/N ratio of plant tissues.

Although we can thus conclude that the effect of nitrogen on longevity of these cereal foot-rot fungi in wheat straw operates mainly through its influence on saprophytic mycelial behaviour, it seems most likely that overall longevity is determined by whether dormant propagules are produced and if so by their efficiency in promoting longevity. *Ophiobolus graminis* is the only one of these fungi that is not known to produce any type of dormant propagule, and it survives in wheat straw for not more than a year ; Butler (1959) obtained survival of the other three fungi shown in Fig. 6.11 for a period of up to 2 years. The fifth fungus, *Cercosporella herpotrichoides*, was found by Macer (1961) to survive under some soil conditions in approx. 75% of colonized straws buried outside in the field for 3 years. This difference in longevity between *O. graminis* and *C. herpotrichoides* agrees with the experience of English farming practice ; the take-all disease can usually be controlled by a one-year break in the rotation between susceptible crops, whereas the eyespot lodging disease requires a break of several years. *C. herpotrichoides* was reported by Sprague (1931) and Dickens (1964), to produce micro-sclerotia on nutrient agar, but their function and efficiency as dormant propagules in soil needs to be investigated.

In view of the intensity of microbial competition for substrates in the soil one of the most interesting facts brought to light by these and other studies on saprophytic survival is the remarkable longevity of pioneer fungal colonizers of plant tissue substrates lying in the soil. As we have just noted, *Ophiobolus graminis* survives under favourable soil conditions for up to a year in wheat straw, and its survival is not assisted by production of any dormant propagules ; on the nutrient agar plate, at least, it is particularly sensitive to fungistatic growth products of other soil micro-organisms (Table 6.4). How does this come about ? Earlier in this chapter, we have emphasized the fact that when a primary parasite invades living plant tissues, host resistance to infection protects it, for a time at least, from following fungi, and so it establishes itself virtually in pure culture in the host tissues ; in experiments with colonized plant tissues, this situation has been reproduced by establishing pure cultures of the inoculant fungi on sterilized host tissue. Establishment of the mycelium as a pure culture throughout the plant tissue must give the inoculant fungus a substantial initial advantage, but how is this advantage maintained when the tissue is buried in soil and open to invasion by all the surrounding soil micro-organisms ? It has been suggested that production of fungistatic substances by the primary colonizer enables it to obstruct invasion of its substrate by following fungi, but this explanation could fit only a minority of the situations already investigated. Amongst the five cereal foot-rot fungi so far discussed, only *Cochliobolus sativus* is known to produce strongly fungistatic culture filtrates to which it is itself sensitive, as demonstrated by strong staling of cultures on most nutrient agars. Such a suggestion has been advanced by Tyner (1966), on the basis of his demonstration that *C. sativus* produces a fungistatic culture filtrate when grown on ground-up wheat straw. But demonstration of such an effect during early growth of the fungus on this substrate does not tell us whether the fungus is still producing a significant concentration of fungistatic substances in the later stages of its saprophytic survival, when it is subsisting on the relatively low concentrations of sugars released by cellulolysis, and when it is in greater danger than earlier of being replaced by other fungi from the surrounding soil. Antibiotics, as they are produced in cultures of micro-organisms both experimentally and industrially, are products of 'shunt metabolism', to employ an apt description coined by Foster (1949). Antibiotics are produced by fungi in pure culture on a sufficiently rich nutrient medium when mycelial growth has been halted by some limiting factor, such as a nutrient deficiency. In this situation, the mycelium is unable to stop absorbing sugar from the culture medium, and the excess sugar is excreted as a shunt metabolite, which may have antibiotic properties, as Woodruff (1966) has explained in an admirably lucid review of antibiotic production. These conditions for antibiotic production may sometimes be satisfied when a fungus first invades a substrate of fresh plant tissue for instance, Brian, Elson and Lowe (1956) have demonstrated production of patulin by *Penicillium expansum* Link em. Thom during its invasion of apple fruits, which have a fairly high sugar content. But it seems unlikely that the sugars released by cellulolysis of residual cellulose during the later stages of saprophytic-mycelial survival in a mature plant tissue will reach a high enough concentration to permit a significant production of antibiotic. The rate of cellulolysis is self-limiting, because accumulation of sugar above a certain concentration temporarily inhibits further production of cellulases, as *adaptive* enzymes. In a study with *Verticillium albo-atrum*, Talboys (1958) found that production of cellulase in the presence of cellulose was reduced by 0.1% and inhibited by 1% glucose. Similarly, Horton and Keen (1966) reported that production of cellulase by *Pyrenochaeta terrestris* (Hansen) Gorenz, Walker and Larson was repressed by 0.09% glucose.

On present evidence, therefore, we are bound to conclude that production of an antifungal antibiotic is not essential to saprophytic survival by a fungus for its full term in infected or colonized plant tissue, even though it may assist survival if conditions for its production are satisfied. Bruehl, Lai and Huisman (1964) demonstrated production of an antifungal antibiotic by *Cephalosporium gramineum* in 2% cornmeal extract acidified to pH 4.8, and thought that production of this antibiotic helped *C. gramineum* to establish colonies from infected-tissue transplants on the agar medium made from this nutrient solution, Lai and Bruehl (1966) later suggested that continued production of this wide-spectrum antifungal antibiotic by *C. gramineum* in the infected wheat tissue, even at a low concentration, might prolong its saprophytic survival. But a subsequent paper by Bruehl and Lai (1966), on saprophytic survival in colonized wheat straws by four cereal foot-rot fungi and six saprophytic fungi, is entitled 'Prior-colonization as a factor in the saprophytic survival of several fungi in wheat straw' ; the conclusion about this situation implied by their title coincides with my own, and with a more general one enshrined in the English proverb 'Possession is nine points of the law'.

Survival in tissues of woody perennial hosts

A change of scene to problems of saprophytic survival by root-infecting fungi in root systems and collar regions of tree species does not necessitate an immediate change from the main topic of the preceding section, because here again the problem of nitrogen as the main mineral nutrient has attracted recent interest, as exemplified by the work of E.B. Cowling, W. Merrill and their associates in the U.S.A. As Cowling and Merrill (1966) have pointed out, the C/N ratio of wood in different species of tree mostly varies within the range 350-500, though in some species it is as high as 1250. It is apparent, therefore, that wood-inhabiting fungi must exercise a quite remarkable economy in the deployment and conservation of so meagre a supply of nitrogen, to account for the known rapidity with which they can decompose timber. Levi, Merrill and Cowling (1968) have provided evidence for the hypothesis that these fungi conserve nitrogen by recycling it from the older and autolysing regions of the mycelium to the younger and actively metabolizing hyphae. In the presence of glucose, various fungi were found able to utilize several fractions of their own mycelia as sole source of nitrogen in pure culture. Cell-wall nitrogen was taken up less quickly than cytoplasmic nitrogen ; experiments both with mycelial fractions and with model nitrogenous compounds showed that proteins, peptides and amino acids were more quickly utilized than nucleic acids, nucleotides or cell-wall fractions.

More recently, Levi and Cowling (1969) have turned their attention to the means whereby a limited supply of nitrogen may be deployed with maximum economy in the mycelia of these wood-destroying fungi. They found that when *Polyporus versicolor* L. ex Fr. was grown on media varying widely in C/N ratio, the nitrogen content of its mycelium ranged from 8.2% on a medium of C/N 4 to one of 0.1% on a medium of C/N 1600. Comparative analyses of these nitrogen rich and nitrogen-poor mycelia showed that under conditions of nitrogen starvation, nitrogen was allocated to metabolically active cell constituents such as the nucleic acids, in preference to structural components of the cytoplasm and cell walls. A representative group of white-rot fungi was found able to produce cellulases on a cellulose medium of C/N 2000 ; other groups of cellulose-decomposing fungi were unable to produee cellulases at this C/N ratio of the culture medium. The white-rot group of fungi are the most highly specialized amongst wood-decomposers, inasmuch as they can degrade lignin as well as cellulose ; from these data of Levi and Cowling, it seems that their specialization has extended to include adaptation to the nitrogen-poverty of their substrate.

The massive woody root systems and collar regions of tree species differ so much from the tissues of herbaceous plants as to make the problem of saprophytic survival quite different in most of its aspects. The much greater size of tree roots and stumps, as well as their mechanical construction, greatly reduces the proportion of extraneous nitrogen that can diffuse into the inner tissues from the surrounding soil, at least during the earlier stages of decomposition whilst roots and stumps are still intact. In the absence of wounds, the bark offers a mechanical barrier of cork tissue to fungal invasion; even after this barrier has been penetrated, initial invasion of the wood is limited to cellulose-decomposing fungi. Sylvicultural wounds that expose transverse sections of wood, however, will enable non-cellulose-decomposers, such as the blue-staining species of *Ceratocystis*, to invade the lumina of exposed vessels; another route for more rapid fungal invasion is provided by the tunnels of beetles and other wood-boring insects. Nevertheless, fungal invasion even of dead wood is slower than that of softer plant tissues; mechanical structure, chemical composition (*e.g.*, shortage of nitrogen) and sometimes the presence of fungi-static substances, as in the heartwood of some tree species, all combine to impede the progress of fungal invasion, and to exercise a selective effect upon the species of colonizing fungi.

Most studies so far made of wood colonization by fungi have shown that considerable volumes of wood are often occupied by a single fungal colony. The situation thus often arises in dying or dead tree roots of a meeting between two fungal colonies advancing along the root in opposite directions. This situation has been experimentally studied by Rishbeth (1950, 1951) in the course of his work on *Fomes annosus*. Root lengths occupied by *F. annosus* were juxtaposed with root lengths occupied by another fungus; competing fungi thus investigated were various isolates of *Trichoderma viride* (*sensu* Bisby, 1939), species of *Ceratocystis* causing blue-stain of timber, and some common basidiomycete competitors of *F. annosus* found in pine stumps, including *Peniophora gigantea* (Fr.) Massee, *Stereum sanguinolentum* (Fr.) Fr. and *Polystictus abietinus* (Dickson) Fr. Amongst the pairings tested by Rishbeth, all three possible results were obtained, *i.e. F. annosus* replacing its opponent or being replaced by it, or a stalemate. Over all pairings, the competitive advantage of *F. annosus* was greater at 10° than at 15° C; this recalls a similar effect of temperature found for several cereal foot-rot fungi. Although both protagonists and conditions of competition were quite different in these two situations, they had one thing in common, *i.e.* that a parasite was competing with other fungi, some of which were obligate saprophytes; this effect of temperature is worth further exploration. Rishbeth also found, as might have been anticipated, that *F. annosus* was displaced less frequently from recently infected roots than from old infected roots in an advanced stage of decay. Lastly, *F. annosus* was found to survive for longer in infected sinker roots lying at some depth in the soil than in the shallower-growing lateral roots, and for longer in alkaline than in acid soild.

Replacement of one fungus by another implies the possibility of a fungal succession in such woody tissues, and such a succession of fungi in unprotected stumps was studied by Meredith (1960) in East Anglican plantations of Scots and Corsican pines. Stump surfaces exposed by tree-felling and left without any of the various protection-treatments now employed were at first infected by air-borne spores either of *F. annosus* or of one or other of two common basidomycete competitors, *Peniophora gigantea* and *Stereum sanguinolentum*. These two latter fungi are weak parasites, inasmuch as they can invade still living stump tissues in which the defence-response to infection has been much weakened by felling of the trunk, but they are not sylvicultural pathogens like *F. annosus*. The basidiomycetes were found to be accompanied in their early invasion of the stump tissues by blue-staining species of *Ceratocystis*, which are non-cellulolytic, and thus occupy an ecological niche distinct from that of the basidiomycetes. These primary basidiomycete invaders were often found to have been replaced in 3 to 4-year-old stumps by *Hypholoma fasciculare* (Hudson) Fr., and after 5 years *Tricholoma rutilans* (Schaeffer) Fr. had become the commonest basidiomycete occupant of these stumps (Table 6.11).

Despite the liability of *F. annosus* eventually to be replaced by other fungi in stump root systems, its longevity in such tissues appears to vary widely according to size of stump and soil type. On an alkaline soil, Rishbeth (1951) found *F. annosus* to be still alive in 85% of stumps of 60-year-old pines no less than 25 years after the trees had been felled.

Table 6.11. Succession of basidiomycetes in pine stumps.

Percentage stumps occupied by fungi listed, at various ages from tree-felling	Age of stumps (years)					
	1	*2*	*3*	*4*	*5*	*8*
*Fomes annosus**	43	37	16	7	17	14
*Peniophora gigantea**	85	89	24	3	—	—
*Stereum sanguinolentum**	32	29	26	15	6	5
*Polystictus abietinus**	—	4	35	13	33	—
*Hypholoma fasciculare**	—	—	45	55	24	12
Tricholoma rutilans	—	—	15	43	73	66
Polyporus amorphus	—	—	5	8	4	—
Paxillus atrotomentosus	—	—	—	—	5	4
P. involutus	—	—	—	—	—	1
Collybia maculata	—	—	—	—	4	1
Clitocybe aurantiaca	—	—	—	—	—	3
Flammula sapinea	—	—	—	3	5	2
Pholiota flammans	—	—	—	—	—	1

*Species identified by mycelial characters ; remaining species identified by presence of sporophores.

Colonies of wood-rotting basidiomycetes are often bounded in the woody tissue by a distinctive, dark-coloured envelope. In a section across the wood, this envelope is seen as a dark-brown or black, narrow zone, and is therefore called a 'zone-line' because it is viewed in only two out of its three dimensions (and even then this term violates the geometrical definition of a line as 'having length but no breadth'). One of the earliest descriptions of zonelines was that by Brooks (1915) of the conspicuous black zonelines produced by *Ustulina zonata* (Lev.) Sacc. in wood of the rubber tree. Campbell (1934) has described the zone-lines formed by *Xylaria polymorpha* (Pers.) Grev. and *Armillaria mellea*. The boundary envelope of the fungal colony in the wood is formed by a tissue of dark-coloured, sclerotoid and inflated fungal cells. This boundary layer thus resembles the rind of a fungal sclerotium, and so Campbell proposed the name of *pseudosclerotium* for the whole fungal colony in the wood. It cannot be denied that Campbell's analogy is a useful one, provided that the parallel is not pushed too far and allowed to obscure the real physiological difference between these two types of fungal tissue. It is true that, by definition, a sclerotium can incorporate host tissue and even soil ; nevertheless, the host tissue if present is usually ghost-like in the mature sclerotium, and its nutrients have been transferred to the fungal cells, as in sclerotia of *Helicobasidium purpureum*, formed around small lateral roots on the tuberous tap roots of sugar beet and carrot. So the mature sclerotium of *H. purpureum* and similar fungi represents *dormant survival* of an aggregation of hyphal cells, well stocked with reserves of carbon and other nutrients and comparable to a well filled larder. But in the pseudosclerotium, as defined by Campbell, the fungal larder is still in process of being filled up by active fungal metabolization of the woody host tissue ; the pseudosclerotium thus seems to have more in common with saprophytic survival by a colony of still active mycelium than with dormant survival by a mature sclerotium.

To say this, however, is not to declare that the boundary envelope (or zone-line) of the pseudosclerotium lacks the protective function commonly assigned to the rind of a sclerotium ; indeed, there is good evidence for the protective function of zonelines. In a study with several wood-rotting fungi, Hopp (1938) concluded that zone-lines were formed only when the air-moisture regime in wood lay within certain limits ; because air and water contents of wood are complementary, as they are in soil, water-saturated wood is oxygen-deficient, and dry wood is water-deficient. Hopp's findings were confirmed by Rishbeth (1951), who concluded that gradual desiccation promoted formation of zonelines by *Fomes annosus* in pine roots. He found that *F. annosus* could be replaced by *Trichoderma viride* and by *Torula ligniperda* (Willkom) Sacc. in root lengths in which zonelines had either not been formed by *F. annosus* or had been damaged ; *F. annosus* was not replaced by these or other fungi when protected by intact zone-lines. Rishbeth has suggested that *F. annosus* surviving in a stump and its roots is protected from replacement by other fungi as a result of zonelines induced by progressive desiccation, from the stump surface downwards and from the distal region of the roots upwards. Fluctuation in soil moisture content during the summer causes expansion and contraction of the wood, which is likely to result in breaching of the zone-lines, especially if wood decay due to *F. annosus* is in an advanced stage. Continuing desiccation of the decaying wood under a low enough soil moisture content then enables *F. annosus* to form another zone-line nearer to the stump centre, and so the fungus gradually retreats in a succession of rearguard actions, marked by successive series of zone-lines.

These conclusions about the protective function of zone-lines have been upheld by Nelson's (1964) study of survival by *Poria weirii* in naturally infected wood blocks (2 in^3), cut from the heartwood of a Dougals fir, and buried in soil under a young stand of the same species. A sample of thirty-six blocks was taken after 6, 9, 12 and 20 months from burial; at each sampling, presence/absence of zone-lines in each block was recorded, and viability of *P. weirii* was determinded by tissue-plating. The percentage of blocks with zone-lines varied from 69 to 83% over the sampling period. *P. weirii* was recorded from only 10% blocks *without* zone-lines after 6 months in soil, and from none thereafter. But from blocks *with* zone-lines, it was recovered from 100, 90, 70 and 32% at successive sampling periods up to 20 months.

The protective function of the zone-line boundary envelope of the pseudosclerotium was an essential ingredient of the hypothesis proposed by Bliss (1951), to explain the mechanism whereby *Armilaria mellea* in infected citrus roots is killed by soil fumigation with carbon disulphide. As a background to his hypothesis, Bliss noted that *A. mellea* generally persisted for 6 years or more in infected citrus roots of not less than 3 cm diam, in Californian soils ; he attributed this considerable longevity to the boundary layer of the pseudosclerotium. In the course of testing for survival of *A. mellea* in infected roots after soil fumigation. Bliss observed that *Trichoderma viride* regularly grew out from tissue platings that failed to yield *A. mellea*, and this led him to suggest that *A. mellea* had been killed not by direct fungicidal action of carbon disulphide, but by antibiotics produced by *T. viride*. Dominance of *T. viride* in soil after fumigation with quite a wide variety of fumigants had already been reported by that time. Bliss therefore suggested that this much augmented soil population of *T. viride* resulting from fumigation was able to penetrate the boundary layer of the *A. mellea* pseudosclerotium, whereas the much smaller population in unfumigated soil was unable to do so. He adduced three kinds of evidence in support of his hypothesis : (1) that the death of *A. mellea* did not immediately follow soil fumigation, at least in the stouter infected roots, but was spread out over a period of some 3 weeks ; (2) that *A. mellea* in infected roots could survive a similar fumigation in sterile soil ; (3) that *A. mellea* was killed if infected roots were buried in a pure culture of *T viride.*

It was not incompatible with this very useful hypothesis to admit the possibility that carbon disulphide might kill *A. mellea* within roots of small diameter by direct fungicidal action, and that it might further damage the mycelium of this fungus in the outer part of stout roots within the limits of fumigant-diffusion, and so impair the resistance of *A. mellea* to replacement by *T. viride.* Employing woody inoculum segments of *A. mellea* (approx. 1·75 cm diam.) Scientists found that *A. mellea* could be killed by direct fungicidal action of carbon disulphide in sterile soil, under which conditions no alternative explanation was possible. But they were also able to demonstrate an indirect, microbiological effect of soil fumigation, as postulated by Bliss, upon the viability of *A. mellea.* They buried similar small inoculum segments in soil that had earlier been treated with carbon disulphide at a dosage concentration and period previously ascertained to ensure a maximum development of *T. viride* in the soil after fumigation. After fumigation at this optimum dosage, the soil containers were uncovered to permit has fumigant to disperse from the soil into the atmosphere over a period of 3 weeks ; during the third week, green plaques of sporulating *T. viride* could be seen developing between the soil and the glass wall of the containers. At the conclusion of this period, inoculum segments containing *A. mellea* were buried in the fumigated but now fumigant-free soil, and incubated for 23 days ; the inocula were then removed and set up in glass tubes filled with soil to assay their capacity for rhizomorph production, as a test for the viability of *A. mellea* within each inoculum segment. In a comparison between twenty inoculum segments buried in unfumigated soil and twenty buried in soil fumigated with an optimum dosage of carbon disulphide, Scientists found that thirty-nine of forty segment-ends from the control series produced rhizomorphs within 26 days in the growth tubes ; in the series from the fumigated soil, only twenty-eight of forty segment-ends had produced rhizomorphs after 75 days. Incubation of the inoculem segments for 23 days in the fumigated but fumigant-free soil had thus caused 30% of segment-ends to lose their ability to produce rhizomorphs. This effect must have been due to some change in the soil brought about by fumigation ; in conjunction with the visible development of *T. viride* at the soil-glass interfaces of the containers, it seemed to support at least some of the claims made by Bliss.

Bliss had also reported a complete kill of *A. mellea* in infected roots buried in a pure culture of *T. viride*, and this suggested exploration of the possibility that the inoculum potential of *T. viride* was a determinant of success in its lethal action against *A. mellea*, as was indeed explicitly postulated by Bliss. In an experiment to test this myologists prepared pure-culture inoculum of *T. viride* on autoclaved soil, and aliquots of this were diluted to one-quarter and one-eight strengths with the same soil unsterilized. Twenty inoculum segments were buried in each of these soil mixtures for 23 days, and all segments were then transferred to soil growth tubes to assay capacity of inocula to produce rhizomorphs, as before. *A. mellea* failed to produce rhizomorphs, and was therefore presumed dead, in 95% of inoculum segments buried in the pure culture of *T. viride*. Amongst the twenty inoculum segments buried in one-quarter strength inoculum of *T. viride*, 25% of segment-ends failed to produce rhizomorphs within 76 days ; this loss of rhizomorph-producing capacity was thus closely comparable to that following incubation of inoculum segments in a soil that had received a fumigant dosage optimum for subsequent development of *T. viride* in the soil. Taking

Bliss's results and the results of the above mentioned experiments together, it was concluded that some post-fumigation biological control of *A. mellea* by *T. viride* could occur under optimum conditions, though less than Bliss had originally claimed. More modestly on behalf of *T. viride*, we would say that *T. viride* could help to finish off what fumigation had begun. At tnat time, Scientists did not appreciate that populations of *T. viride* in fumigated soils could vary very widely in their antagonistic effects against *A. mellea* and other fungi, as will shortly transpire. The Scientists were lucky in obtaining positive results, though whether it is 'lucky' to obtain results than can be confirmed only with difficulty by subsequent workers is open to question.

Reasons for the development of dominance by *Trichoderma viride* after soil fumigation with a suitable dosage of carbon disulphide have been explored by Saksena (1960) and Moubasher (1963). Both these workers found in fumigation trials with a range of carbon disulphide doses that *T. viride* possesses only a moderate of tolerance to carbon disulphide ; various species of *Penicillium* and *Aspergillus*, mostly ascosporic, can survive a dose at least five times that tolerated by *T. viride* Saksena therefore concluded that *T. viride* owes its abiliiy to dominate the fungal population recolonizing fumigated soil to possession of a moderate but sufficient degree of fumigant tolerance allied with a high growth rate through fumigated soil ; some fungi with a still higher growth rate possess only a low fumigant tolerance whereas the highly tolerant fungi are mostly rather slow growers. This combination of these two characteristics thus possessed by *T. viride* appeared to give it a unique advantage for recolonization of fumigated soil, at least amongst the common soil fungi studied by Saksena and Moubasher. The fuller data collected by Moubasher are compatible with Saksena's explanation, and he also provided a useful quantitative expression for *recolonizing ability*, in the ratio obtained by dividing the soil population of a particular fungal species 36 days after fumigation by its population 1 day after fumigation. He estimated soil populations by soil-dilution-plate counts. This recolonizing ratio for *T. viride* was 9·2, and was higher than that for any of the other twenty-three common soil fungi that Moubasher investigated.

Analysis of this problem was carried further by Mughogho (1968), who began work in the hope that he could find a fumigant more reliable for promoting dominance of *T. viride* than we had found carbon disulphide to be. To obtain even a moderate development (as compared with that in sterile soil) of *T. viride* in fumigated soil, the dosage of carbon disulphide had to be kept within a narrow range and the prospects of being able to do this by soil fumigation in the field were not encouraging. Moreover, in the fumigation experiments with carbon disulphide, the indirect component of the fumigation effect on *A. mellea,* although significant, had not been good enough for a control measure in actual practice, and Scientists hoped to improve on this by finding another fumigant more effective for this particular purpose. Mughogho's first results in this search were encouraging, as the following fumigants were found superior to carbon disulphide for promoting dominance of *T. viride*, both in respect of breadth of suitable dosage range, and in the size of *T. viride* populations developing in the fumigated soils : allyl alcohol, chloropicrin, methyl bromide, D-D mixture, sorbic acid and acetylenedicarboxylic acid. It was therefore unexpectedly disappointing for Mughogho to find that these high populations of *T. viride* in the fumigated soils more often than not failed to extert any controlling effect either on production of rhizomorphs by *Armillaria mellea* or on mycelial growth through soil and infection of swede seedlings by *Rhizoctonia solani.* Most fortunately, however, it became possible to explain these unexpected results through the current research by J. Webster and M.A. Rifai at the University of Sheffield on a revision of the taxonomy of the genus. *Trichoderma.* In 1939, G.R. Bisby revised the taxonomy of the genus *Trichoderma* Pers. ex Fr. and incorporated all described species into *Trichoderma viride* Pers. ex Fr., thus making the genus a monotypic one. But Webster and Rifai have now found good taxonomic reasons for putting asunder what Bisby had joined together, and Rifai (1969) has published a conspectus of the genus *Trichoderma* in which nine species are recognized and keyed for identification.

The species, as now recognized by Rifai (1969), that Mughogho found to make up his populations of *Trichoderma* in these variously treated soils were *T. hamatum* (Bon.) Bain., *T. harzianum* Rifai, *T. Roningii* Oud., and *T. viride* Pers. ex S.F. Gray ; amongst these four species, *T. harzianum* usually predominated. He further found that different isolates of *Trichoderma*, even within a single species, varied widely in their degree of antagonism towards *Armillaria mellea* and *Rhizoctonia solani*. He concluded that dominant populations of *Trichoderma* in his fumigated soils must have been made up largely of isolates that were not effectively antagonistic towards either *A. mellea* or *R. solani*. The highest proportion of effective antagonists amongst isolates of *Trichoderma* tested by Mughogho belonged to *T. viride* Pers. ex S.F. Gray, as now delimited by Rifai (1969).

It cannot be denied that all these studies of the taxonomy and ecology of *Trichoderma* species in soil have greatly increased our comprehension of the sequence of mycological events occurring after soil fumigation ; equally, however, we have to admit that our hopes of mediating biological control of root-disease pathogens by *Trichoderma* species through soil fumigation, originally aroused by D.E. Bliss in 1951, have received a set-back.

DORMANT SURVIVAL BY RESTING PROPAGULES OF ROOT-INFECTING FUNGI

Dormant propagules constitute the most effective mode of survival by root-infecting fungi between one host plant and the next ; infective propagules with a period of dormancy are of the following kinds : (1) *dispersal spores*, which are produced either sexually or asexually by the great majority of fungi in all major taxonomic groups ; (2) *resting spores* resulting from sexual reproduction, such as the oospores of the Oomycetes and ascospores of some species of the Ascomycetes ; (3) thick-walled *chlamydospores* formed within hyphal and conidial cells, especially by species of *Fusarium* ; (4) *sclerotia*, which are multicellular resting bodies produced by fungal species in all taxonomic groups except the Phycomycetes. In a review of fungal spore germination, Gottlieb (1950) produced the following definition of *dormancy* : 'A spore is dormant when it does not germinate under the same nutritive and environmental influences which later allow germination'. After morphological demarcation of the spore, a further period of maturation must ensue before it is capable of germination under suitable environmental conditions. As Gottlieb has pointed out, a period of maturation is essential for spores of all types, and it is only the length of this period that distinguishes dispersal spores from resting spores. In some resting spores, such as the oospores of *Phytophthora cactorum*. Schrot., this maturation period may be both prolonged and of very variable duration in different individuals.

The subject of spore germination has more recently been reviewed by Gottlieb (1966), Sussman (1966) and Cochrane (1966). Dormancy as defined by Gottlieb (1950) has been qualified by Sussman as *constitutive dormancy*, to distinguish it from *exogenous dormancy* imposed upon the spore by environmental conditions that inhibit germination. As an example of constitutive dormancy, Sussman has cited the ascospores of *Neurospora* species, which require a heat-shock before they will germinate during the dormant period. Constitutively dormant spores often or even usually can germinate when ripe without an exogenous energy source, whereas for other spores this is often an essential. Cochrane (1966) has noted that the relatively short-lived asexual spores of the Ascomycetes and Fungi Imperfecti usually require exogenous sources of both carbon and nitrogen nutrients. He has termed such spores *heterophagous*, in contrast to *autophagous* spores that will germinate without exogenous nutrients. Cochrane has also emphasized the importance of certain precautions when experimentally determining a possible need for exogenous nutrients in spore germination. Washing spores off a nutrient agar surface will dissolve some of the substrate nutrients ; if then spores are centrifuge-washed to get rid of these, endogenous nutrients may be reduced by leaching below the critical level for germination. It is therefore best to remove spores from the parent culture with a vacuum collector, or by brushing, shaking, etc. ; the choice of method will depend upon whether the fungal sporophores produce dry spores (xerospores) or slime spores (gloiospores).

Spore germination in the soil

An exogenous dormancy is imposed upon fungal spores lying in the soil by a fungistatic factor, or factors, that is widespread in most soils, both natural and cultivated. General interest in this problem of soil fungistasis was first aroused by Dobbs and Hinson (1953), who reported that spores of various soil fungi that germinated freely in distilled water would not do so in soil solution diffusing through a cellophane membrane. Simmonds, Sallans and Ledingham (1950) had earlier observed a similar effect on spore germination, but with the conidia of just a single fungus, *Cochliobolus sativus*, so that their report failed to arouse the widespread interest that greeted the announcement by Dobbs and Hinson, who also demonstrated that a glucose solution of concentration only 0.1% was sufficient to override the fungistasis and cause spores to germinate. In spite of all the work that has been reported since that time, the overcoming of soil fungistasis by glucose and other nutrients can still be best interpreted through the profoundly important generalization made by William Brown (1922) as a result of his study of the effect of carbon dioxide upon spore germination and mycelial growth of various mould fungi ; Brown epitomized his findings by saying that the inhibiting effect of carbon dioxide was greatest where the fungal energy of growth was least. These observations by Dobbs and Hinson were soon confirmed by those of Jackson (1958) for a number of Nigerian and English soils, and with some other species of soil fungi, some of which were unaffected by the general soil fungistasis. In two soil profiles studied, fungistatic effects were confined to the upper 40 cm. The soil fungistatic effect was found to decrease with increasing acidity, suggesting to Jackson that soil bacteria might be important as producing agents. It is important to remember, however, that a sufficient degree of acidity in the soil solution can of itself reduce the percentage of spores germinating in any population. Dobbs and Gash (1965) have opportunely distinguished between *microbial fungistasis,* as originally reported by Dobbs and Hinson, and *residual fungistasis.* Microbial fungistasis is considered to result from the products microbal metabolism, because it can be removed by autoclaving and by all sterilizing agents studied, and does not return so long as a sterilized soil remains sterile ; it is usually absent, or virtually so, from subsoils, which have a low microbial population. Residual fungistasis, on the other hand, was found by Dobbs and Gash to be thermostable, and was not removed by soil sterilization. Scientists have already suggested that a residual fungistasis of this type could be caused by a sufficiently extreme degree of soil acidity, but that studied by Dobbs and Gash was observed in calcareous soils. They detected this residual fungistasis in sand from the foreshore and mobile dunes of the Anglesey coast, but not in that from the fixed dunes, with a higher organic content. Their experiments eliminated calcium carbonate as a direct cause of residual fungistasis but various results agreed in suggesting that the toxicity was due to unchelated iron at the high pH value (8.0 or above) of the sand ; chelation of the iron by the organic matter that had accumulated in the fixed dunes could thus provide an explanation for the disappearance of this residual fungistasis age of the dunes.

Much has been contributed towards solution of this problem of microbial fungistasis by the intensive studies of J. L. Lockwood and his associates in the U.S.A. Lingappa and Lockwood (1961) failed to extract any *persistent* soil toxin and therefore suggested that a fungistatic level must be maintained by continuous microbial production, especially by spore surface bacteria subsisting on spore-exudate nutrients. The same authors later (1964) reported a several-fold increase both in bacterial numbers and in soil respiration when soil was supplemented either with living fungal spores or with aqueous washings of the same. When washed cells of bacteria or of *Streptomyces* species were incubated with fungal spores in absence of any exogenous nutrients, spore germination was thereby inhibited. In a contemporaneous review, Lockwood (1964) noted that *washed* fungal spores germinated poorly or not at all in distilled water and suggested that bacteria and actinomycetes m ght prevent fungal spore germination in soil by consuming spore-exudate nutrients as quickly as they were released through the spore membrane. This hypothesis clearly implies that the germination of similar but *unwashed* spores in distilled water is due to a back reaction of the spore exudate upon the spore that has produced it ; when the exudate has attained a certain threshold concentration in the water immediately surrounding the spore, it must trigger off germination, according to this hypothesis. As we have mentioned already, and as we shall see in more detail later, many types of spore require exogenous carbon nutrients and some need exogenous as well before they will germinate. Such observations constitute part of the evidence in support of Lockwood's hypothesis ; successful spore germination in the absence of exogenous nutrients thus seems likely to depend not only on the total content of endogenous nutrient reserves, but also on their rate of mobilization in soluble form to the critical threshold concentration.

Continuing to emphasize this nutrient-deficiency component of microbial soil fungistasis, Chacko and Lockwood (1966) criticized the cellophane film employed by Dobbs and Hinson (1953) and the water-agar discs used by Jackson (1958) as spore carriers, on the grounds that neither of these carriers was completely nutrient-free. They recommended instead a more direct method devised by Lingappa and Lockwood (1963) whereby the fungal spores were set to germinate on a thoroughly smoothed surface of finely sieved soil, and after an incubation period were stained and then fixed by impregnating a very thin surface layer of the soil with collodion ; this was then peeled off when dry as a thin film for microscopical examination. Employing this technique, Chacko and Lockwood assessed the strength of soil fungistasis by incubating spores on the surface of a graded dilution series of a natural soil with the same soil sterilized by ethylene oxide vapour. Increasing dilution of the natural with the sterilized soil caused two progressive changes : (1) by provision of an increasing amount of nutrient substrate in the form of sterilized soil ; (2) by a concomitant reduction in the density of the microbial population in the mixture ; each of these effects must have contributed to a gradual decline in the intensity of fungistasis. This technique thus provided a method for quantitatively comparing both different soils for intensity of the fungistatic effect, and spores of different species of soil fungi for their susceptibility to this effect, through the ED 50 value, *i.e.* the proportion of sterile soil in the mixture just sufficient to permit 50% spore germination. The ED50 varied for different species of fungi in the same soil, as earlier work had shown. Intensity of fungistasis was found to decrease rapidly down the soil profile, from a value of 0 7 at 10 cm down to one of 0.05 at 61 cm. In a trial with eighteen species of fungi, Ko and Lockwood (1967) correlated ability of spores to germinate in natural soil with their ability to germinate in absence of exogenous nutrients ($r=0.94$; $P < 0.5\%$). But the spores of four species that were nutritionally independent in germination on glass slides were found not to germinate in the soil ; this was ascribed by Ko and Lockwood to a steep concentration gradient of spore-exudate nutrient from spore surface to soil caused by microbial assimilation at the spore surface, as originally suggested by Lookwood (1964). When spores are standing in a drop of sterile water on a glass slide, the nutrient concentration gradient due to diffusion is much more shallow. But when Ko and Lockwood increased the steepness of this concentration gradient in another way, by placing spores on filter paper continuously leached with distilled water, germination failed to occur. When they provided a substrate by incorporating ground dried alfalfa (*Medicoge sutica*) tissues with the soil, then all spores, including those of nutrient-dependent species, germinated. Chinn (1953) had earlier reported a similar effect of an amendment of soybean meal in inducing spore germination of *Cochliobolus sativus*, *Fusarium roseum* f. *cerealis* and seven other species in soil.

The use of water agar discs as a carrier for the spores to be tested against soil fungistasis by Jackson (1958) has already been noted ; in this technique, fungistatic substances that are water-soluble diffuse through the agar, as they do through the cellophane film employed by Dobbs and Hinson (1953) Dix.

(1967) improved upon Jackson's technique by slicing agar blocks into a graded series of thicknesses (200, 400, 600 and 800) with a microtome, so that the concentration of water-soluble fungistatic substances was graded by diffusion according to the thickness of the agar. His addition of 1% peptone to the agar is likely further to have widened the spectrum of germination-respone amongst the twelve species he studied, which were chosen from amongst a number isolated from the root surface and rhizosphere of *Phaseolus vulgaris.* A wide spread in the effect of soil fungistasis upon spore germination was found amongst these twelve species; *Penicillium janthinellum* Biourge and *Trichoderma koningii*, which were pioneer root-surface colonizers (apical 1 cm of roots) were amongst the most tolerant. In a comparable study at Cambridge, we have confirmed Dix's findings on the tolerance of these two species to fungistatic effects, and also his observation that *Mucor plumbeus* Bon. and *Penicillium nigricans* (Bain.) Thom are rather sensitive to fungistasis. Dix's experiments were carried out with washed spores and he was thus able to determine that three of his species, *Penicillium nigricans*, *Mucor plumbeus* and *Doratomyces purpureofuscus* (Fr.) Morton and Smith, required an exogenous supply of both carbon and nitrogen nutrients for spore germination; *M. plumbeus* further required one or more vitamins present in yeast extract.

The agar spore-carriers employed both by Jackson and by Dix make it probable that the fungistatic factors they studied were water-soluble but not necessarily volatile. Balis and Kouyeas (1968) have made use of the ability of silver and mercury salts to form complexes with volatile hydrocarbons to demonstrate the occurrence of a volatile fungistatic factor, possibly of this type, in the soils they studied. The fungistatic property was lost when soil was air-dried, but was restored when the air-dried soil was remoistened and incubated; restoration was most rapid and complete at a moisture content optimum for microbial activity, and did not occur in water-saturated soils. Fungistasis was annulled by passing an airstream over the sporiferous soil surface. Disappearance of fungistasis on air-drying of soil was ascribed to loss of the adsorbtive properties of clay mineral particles when air-dried.

We can now attempt to put this recent work on spore germination in the soil into perspective against the background of earlier work on nutritional factors affecting spore germination, which has been reviewed by Gottlieb (1950) and Cochrane (1958). It now seems likely that most thin-walled dispersal spores are nutrient-dependent in germination, at least partially. Some are absolutely dependent upon one or more exogenous nutrients; most others are partially nutrient-dependent, in the sense that percentage germination is usually higher, as often much higher, in a nutrient solution than in pure water. The greatest response is usually to sugars. The laboratory situation in which spores are set to germinate on a glass slide is not so far removed from that in a film of water overlying a leaf; as Brown (1922) showed with spores of *Botrytis cinerea*, germination-promoting substances diffuse through the cuticle, so that percentage germination is likely to be higher than that in water on a glass slide, though lower than that in a nutrient solution of optimum composition and concentration. But the situation of a spore germinating on a root surface is further removed from that on a glass slide; the more rapid exudation of nutrients from the root surface as compared with that through the cuticle, the often higher relative humidity of the soil atmosphere and the absence of extreme temperature fluctuations all combine to promote a degree of microbial activity at the root surface that is much higher than that on the leaf surface in most types of climate.

We can now understand the biological implications of this probably fairly general difference in behaviour between spores cf leaf-infecting and root-infecting fungi, respectively. As Kerr and Flentje (1957) were the first to point out, the spore of a leaf-infecting fungus once deposited on a leaf and stuck there has only two alternatives; to infect or to perish. There is thus no visible advantage in exogenous dormancy imposed by a critical need for exogenous nutrients in germination. But with spores of root-infecting fungi, nutrient-imposed exogenous dormancy prevents suicidal spontaneous germination in the absence of a substrate stimulus; because roots grow through the soil, waiting spores are almost certain to come into contact with a substrate sooner or later.

The behaviour of thick-walled resting spores is determined by a more complex set of variables. Some of them can certainly germinate quite well in pure water, once they are ripe to do so. We do not know how important dormancy-breaking treatments may be in the natural environment ; in the laboratory they provide a convenient means of saving time and imposing a greater uniformity in germination on a spore population. Such treatments as heat-shock, exposure to low temperature and to certain chemicals may affect either the permeability of the spore wall or the state of the protoplast, or both. Permeability is not necessarily related to wall thickness ; a thin wall may be as impermeable as a thick one. Germination-stimulating chemicals that are not nutrients and act at a low concentration raise further problems. Various volatile substances are effective activators of spore germination, and are responsible for the germination-promoting effect of some fresh plant tissues, as first demonstrated by William Brown (1922). Noble (1924) found that suitably low concentrations of some volatile chemicals, such as benzaldehyde, salicylaldehyde and acetone, stimulated germination of the chlamydospores of *Urocystis tritici* Korn., causing flag smut of wheat. Hooker, Walker and Link (1945) found a similar effect of a mustard oil, allyl isothiocyanate, on germination of the resting spores of *Plasmodiophora brassicae*, and suggested that this and similar volatile substances might be responsible for the germination-promoting effect of cruciferous host roots. King and Coley-Smith (1968) have reported that a volatile germination-activator for the sclerotia of *Sclerotium cepivorum* Berk. is given off by growing seedlings of onion, leek and garlic ; also stimulatory was the pure chemical diallyl sulphide.

A valuable insight into the complex of factors that determine the nature and duration of constitutive dormancy in thick-walled resting spores was early provided by Blackwell's (1943) intensive study of the oospores of *Phytophthora cactorum*; she has described their development, dormancy and eventual preparation for germination as follows. After fertilization by the antheridium, the young binucleate oospore, lying loosely within the oogonial wall, forms an *exospore* upon the inside of the oospore membrane. This exospore, composed of pectic substances and proteins, is thin, smooth and transparent ; it seems to be virtually impermeable, as few stains were found able to penetrate it. Within the exospore is then laid down the *endospore*, which eventually attains a thickness of 2 μ ; it is composed of hemicellulose, cellulose and protein and seems to function as a food reserve, because it is later eroded during preparation for germination.

Three stages have been distinguished by Blackwell in the development of this oospore :

(1) cytological development ; the period from fertilization to fusion of the two nuclei takes about 1 month at laboratory temperature (15-20° C), but longer at lower temperatures ;

(2) after-ripening period ; this occupies 6-8 months at 15-20° C, but only 1-2 months at a temperature just above 0° C.

(3) preparation for germination ; nuclear changes are accompanied by digestion of the endospore, and by desaturation and assimilation of the oil globule, and these events take at least a month (Fig. 6.12).

The most important conclusion drawn by Blackwell from this thorough and prolonged study was that individual oospores varied widely in their period of dormancy and in their germination-response to environmental conditions. As she has stressed, this wide physiological heterogeneity amongst a population of oospores promotes maximum longevity of the population, because under no single set of environmental conditions will more than a minor proportion of the population germinate at any one time. When hosts and substrates are absent, summer may be a more hazardous season for survival of the species than is winter, but the oospore population provides against all environmental contingencies.

Survival of spores

In this section we shall give an account of the behaviour in soil of various types of spore, selected as a representative range from amongst the most studied root-infecting fungi, as follows : ascospores of *Ophiobolus graminis*, basidiospores of *Fomes annosus*, conidia of *Cochliobolus sativus*, macroconidia and chlamydospores of various *Fusarium* species, ascospores of *Rhizina undulata* Fr. and resting spores of *Plasmodiophora brassicae* and *Spongospora subterranea* (Wallr.) Lagerh. Upon all these spores when lying in soil, the general soil fungistasis imposes an exogenous dormancy, which can be broken by a suitable host or substrate stimulus ; all of them can thus be said to function as dormant propagules, even though the total life of the spore population in soil may be short, as in the ascospores of *Ophiobolus graminis*, which are primarily dispersal spores of delicate structure and ephemeral existence. At the other end of the scale come the ascospores of *Rhizina undulata*, which are characterized by a well marked constitutive dormancy that can be terminated precociously by heat-shock. Similarly the resting spores of the Plasmodiophorales pass through a period of constitutive dormancy, though this appears to be of brief duration in some 95% or more of the population ; it is the longevity of the remaining 5% or less of the resting spore population that constitutes the problem for plant pathologists.

Ascospores of Ophiobolus graminis

Whether these ascospores could function as infective propagules remained an unsolved problem for many years. In many inoculations of the soil around wheat seedling roots, Scientists failed completely to secure infection, though it was easy to infect roots of seedlings growing in completely sterile soil or sand. They therefore concluded that these ascospores were so completely dependent upon exogenous nutrients in the root exudate that they could cause infection only when the exudate was wholly available to them, *i.e.* in the absence of other soil micro-organisms as in sterile sand. The corollary that the ascospores were most unlikely to infect roots growing in natural soil was later questioned by the occurrence of widespread outbreaks of take-all amongst early wheat crops on the recently drained Noord Oost Polder of the former Zuider Zee in the Netherlands. The soil of this Polder was virgin in the most literal sense of this word, because it had been drained from under the fresh water of the Ysselmeer, and so *Ophiobolus graminis* must have been transported thither from outside the

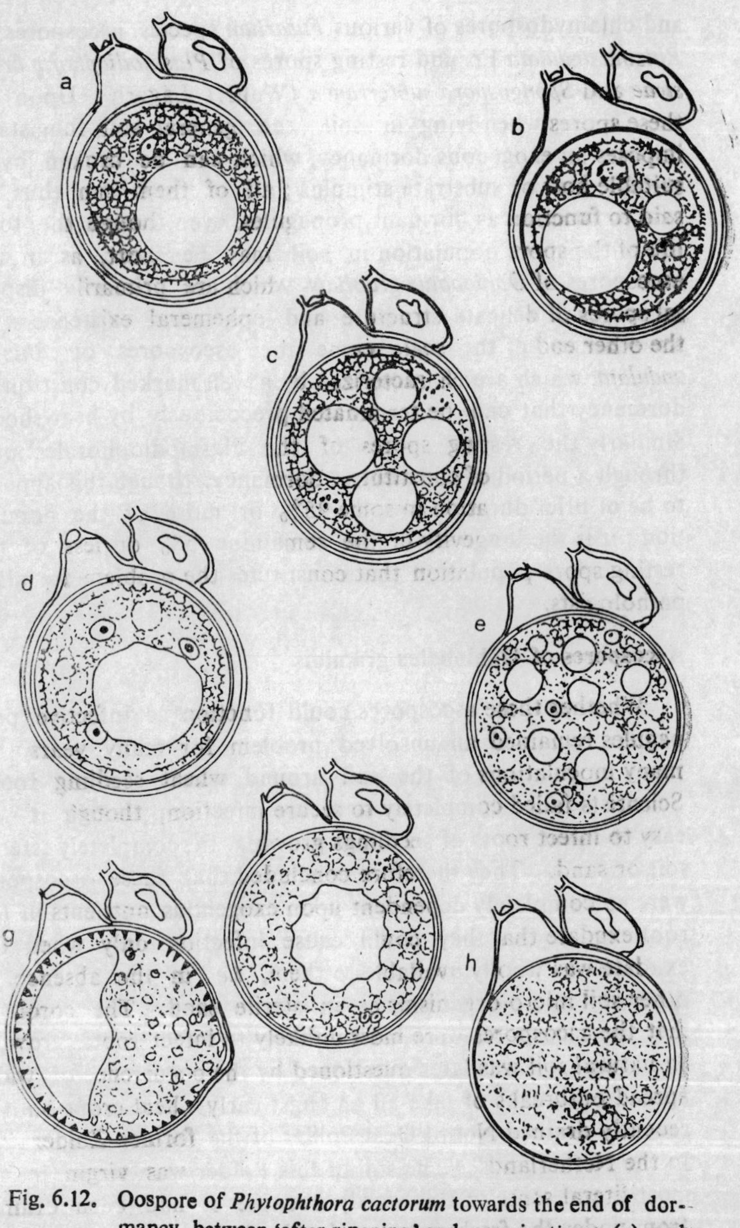

Fig. 6.12. Oospore of *Phytophthora cactorum* towards the end of dormancy, between 'after-ripening' and germination : (*a*) and (*b*) fusion nucleus in division, oil globule breaking up,

Polder. Several attempts to secure infection of roots of susceptible grasses by soil inoculation with ascospores ended equally in failure, and the problem was finally solved by D. H. Brooks (1965). He demonstrated that ascospores could readily infect the proximal part of the seminal roots of wheat seedlings that had been produced by allowing wheat seeds to germinate *on the surface* of wet soil. This proximal part of the seminal roots, between the seed and the soil, is not in direct contact with the soil and so is at first virtually free of a root-surface microflora. If ascospores are inoculated (or naturally alight) upon this infection site, then seed and root exudates are at first available to them in virtual absence of microbial competition. Compatible with this interpretation was Brook's finding that if seedling roots were left for a period after emergence before they were inoculated with ascospores, then the longer the delay the greater was the reduction in proportion of successful inoculations.

We concluded from Brooks discovery that ascospore infection by *O. graminis* could occur in the field when wheat seeds were germinating, at a time of ascospore dispersal, on the surface of sufficiently moist soil ; these conditions are fulfilled by rainy weather around harvest time, when some wheat seeds are shed from overripe ears, as witnessed by the not infrequent occurrence of 'volunteer' wheat plants in a following crop of another species. We further appreciated the possibility that the low microbial population of the virgin soils of the Nether-

endospore becoming eroded ; (*c*) second nuclear division, oil globule broken up ; (*d*) four nucleate stage, oil globule dispersing gradually ; (*e*) eight-nucleate stage, oil globule further subdivided, endospore almost completely absorbed; (*f*) fourth nuclear division, oil globule still single, endospore almost completely absorbed ; (*g*) about thirty-two nuclei, oil globule still single, endospore not yet absorbed, germ tube through exospore but not yet through oogonium wall ; (*h*) sixth nuclear division, nuclei very small, all reserves absorbed into protoplasm but germ tube not yet formed. (*a*)—(*h*) are composite drawings of median longitudinal sections (× 1000) not representing a sequence of stages.

lands polders immediately after draining might be insufficient to prevent ascospore infection of wheat or susceptible grass roots *within the soil*. Since then, this possibility has been confirmed in some most ingenious experiments by Gerlagh (1968) on the East Flevoland Polder, drained in 1957. For these experiments, he grew wheat seedlings in pots of subsoil clay taken from under the floor of the Polder, with due precautions to avoid contamination by the air spora ; in this way, he obtained a 'soil' closely comparable to that emerging when the water had first been drained off the Polder. He soon noted that spontaneous infection by *O. graminis* frequently developed on root systems of wheat plants growing in pots of subsoil clay that had *not* been inoculated with the pathogen, and suspected that these infections had arisen from ascospores present in the air spora of his glasshouse. He followed this up by suspending perithecia-bearing wheat stubble over plants of wheat, *Lolium perenne* and *Poa annua* growing in pots of subsoil clay. All three species became infected by the ascospores. The results of these glasshouse experiments were confirmed by a small field trial, in which a plot was excavated to a depth of 1 m, lined with plastic sheeting and filled up with fresh subsoil clay. The second season's wheat crop on this plot developed numerous whiteheads due to *O. graminis*. In conjunction with these experiments, Gerlagh made a microscopical examination of ascospore-inoculated roots of wheat seedlings growing in Perspex observation boxes. In the subsoil clay, roots were resularly infected ; in an old cultivated soil, only few plants became infected and lysis of ascospore germ tubes was observed.

Gerlagh has concluded that *O. graminis* first establishes itself by ascospore infection of the roots of susceptible grass species, which are amongst the pioneer colonizers of the freshly exposed floor of a polder after draining. He verified this supposition by finding widespread infection of grass root systems all over an artificial island in one of the polder lakes ; this island was made by pumping bottom sludge into a walled enclosure, and its mode of construction eliminated the possibility of contamination by any infective propagules of *O. graminis* other than the ascospores.

Basidiospores of Fomes annosus

In the course of his early work that demonstrated widespread infection of coniferous stump surfaces by the air-borne basidio spores of *F. annosus,* Rishbeth (1951) incidentally discovered that these dispersal spores can survive as dormant propagules in the soil, a discovery that certainly surprised the scientists at that time. This finding arose from a trial of covering freshly exposed stump surfaces with soil as a possible protection against basidiospore infection, as had been recommended against other air-borne fungal pathogens in the tropics ; in this trial, some 50% of the soil-covered stumps became infected by *F. annosus,* in the absence of any natural inoculum other than basidiospores, and this actually exceeded the percentage infection of the untreated control stumps. Rishbeth verified this conclusion by an experiment planned for the purpose. and also discovered that soil from under a coniferous plantation was infective to a depth of more than 50 cm, which was explicable only on the assumption that basidiospores get washed down through the soil during and after periods of heavy rainfall. This again is an example of exogenous dormancy, presumably imposed by soil fungistasis.

Conidia of Cochliobolus sativus

Studies of these conidia in Canada have been particularly concerned with their exogenous dormancy in soil and how far this varies according to the particular isolate of the fungus. When conidia of a fungus, whether lying in soil or after introduction into it, germinate in the absence of any discernible stimulus from a host root or from any other substrate, their behaviour is described as *spontaneous germination.* Chinn and Tinline (1963) have coined the phrase 'inherent germinability' to express the tendency towards spontaneous germination amongst isolates of *C. sativus* ; the degree of inherent germinability is to be equated with the percentage of spores spontaneously germinating. To determine the range of inherent germinability amongst a collection of 255 wild-type isolates, glass slides coated with a suspension of conidia in water agar were set in the soil, and the percentage of spores subsequently germinating was then recorded. Spontaneous germination amongst these isolates ranged from 0 to 87%, though the majority showed none at all. When reviewing this subject earlier. Scientist suggested that spontaneous germination was a departure from the conception of ideal behaviour amongst a population of resting propagules, and must result in considerable wastage. This speculation now rests on the firmer ground of evidence provided by Chinn and Tinline (1964), who estimated respective longevities in soil of spore populations varying in inherent germinability from 0 to 65%, longevity was estimated in two ways : (1) by recovering spores from soil by the oil-flotation method of Ledingham and Chinn (19:5); (2) by a wheat seedling test for viability of spores *in situ* in the soil. An inverse correlation was found between inherent germinability of an isolate and longevity of its spore populations in soil, and so expectations were duly confirmed. Continuing these studies, Chinn (1967) devised a dilution technique to provide a more sensitive assay for the degree of fungistasis in different soils, and for the sensitivity to fungistasis of various isolates of *C. sativus*. The method was similar to that of Chacko and Lockwood already described, except that dilution of the natural soil was made with inert silica particles (grade 70 mesh/in) instead of with the same soil sterilized. Amongst eleven soils tested, strength of fungistasis was greatest in peat soil, intermediate in clays and loams, and least in sandy soils. With the most sensitive isolate of *C. sativus*, spore germination was completely inhibited by addition of only 0.25% of a loam, clay or peat soil to the silica medium, or by addition of 1% of a sandy soil.

Up to the time of the most recent of the studies cited above, dormancy of these conidia of *C. sativus* was assumed to be maintained by soil fungistasis. There is still no need to challenge this assumption, but the recent announcement by Meronuck and Pepper (1968) of their discovery of chlamydospores within conidial cells of *C. sativus* does make it easier to comprehend the considerable longevity of some of these spore populations. Meronuck and Pepper introduced conidia into pots of autoclaved soil, which were then sunk in soil outside and left to experience the North Dakota winter. About 70%

of the 800 conidia recovered from the potted soil by the oil flotation technique, during and after the winter, showed internal chlamydospores as evidenced by thickening of the inner walls of one or more cells.

Macroconidia and chlamydospores of Fusarium species

Early observations on the behaviour of macroconidia of *Fusarium roseum* f. *cerealis* in soil were made by Chinn (1953) in parallel with his study of the conidia of *Cochliobolus sativus*, just noted. When 2% soybean meal was added to the soil, the macroconidia of *F. roseum* germinated and produced a limited mycelium on which chlamydospores were formed, thus ensuring continued survival of the fungus at an augmented population level. These observations were the precursor of many more.

The need of macroconidia of *Fusarium* species for exogenous nutrients in germination has been established for several species and may be fairly general. Sisler and Cox (1954) reported that washed macroconidia of *F. roseum* required both exogenous carbon and exogenous nitrogen, and a higher percentage of germination was obtained only in presence of some other constituents of the basal nutrient medium as well; similar results were obtained by Marchant and White (1966). Cochrane, Cochrane, Simon and Spaeth (1963) found that macroconidia of *F. solani* f. *phaseoli* required exogenous carbon and nitrogen and a growth factor present in yeast extract, which latter could be replaced completely by ethanol or acetoin, and partially by acetaldehyde or one of several amino acids. Rather surprisingly, perhaps, the requirement for an exogenous carbon nutrient was found to exceed the total dry weight of the spore, suggesting that endogenous reserves were inadequate for germination and that substrate supplementation, in addition to germination-activation, was an essential requirement supplied by exogenous carbon nutrients. This suggestion was reinforced by the finding that the requirement for exogenous carbon nutrients and ethanol could be reduced, though not eliminated, by producing the macroconidia on a high carbohydrate medium.

Nutrient requirements for infection are higher than those for spore germination and so the nutrients supplied by root exudate must generally be essential for infection. Toussoun, Nash and Snyder (1960), investigating infection of hypocotyls of snap bean (*Phaseolus vulgaris*) by macroconidia of *F. solani* f. *phaseoli*, determined that exogenous glucose was essential for the prepenetration stage of infection, *i.e.* germination of conidia and epiphytic develppment of mycelium on the hypocotyl surface ; nevertheless, glucose without nitrogen delayed host penetration and the onset of pathogenesis. Nitrogenous nutrients, on the other hand, favoured early penetration and pathogenesis ; organic forms of nitrogen were more effective than inorganic ones. This line of investigation has been further developed, but with chlamydospores instead of macroconidia, and so will come up for discussion again later.

In discussing the formae speciales of *F. oxysporum* causing vascular wilts the growing consensus of opinion are being noted that chlamydospores are the propagules responsible for the great longevity of some of these pathogens in soil free from host plants. According to Snyder and Toussoun (1965), chlamydospores are produced by most clones of *F. solani*, *F. oxysporum* and *F. roseum*, but not commonly by the other six species of this genus. Snyder and Hansen (1945) recognized nine species, which they distinguished according to shape of the macroconidia, presence/absence of microconidia and their position and kind, and presence/absence of chlamydospores.

The production of chlamydospores by macroconidia of *F. solani* f. *phaseoli* has been elucidated by Nash, Christou and Snyder (1961), who found that when macroconidia were seeded into field soil, they either germinated to produce a short germ tube which eventually produced one or more chlamydospores, or else they were converted directly into chlamydospores. Chlamydospores thus represent the normal survival propagule derived from macroconidia. Nash *et al.* employed the technique devised by Warcup (1955) for determining the kind of propagule that has produced a colony on a soil dilution plate ; by this method, they ascertained that most colonies of *F. solani* f. *phaseoli* arising on dilution plates made from soil of a naturally infested bean field originated from a chlamydospore, associated with or embedded in a fragment of plant tissue or of humus deriving from such a plant residue. Soil dilution plates made by Nash and Snyder (1962) from the soil of various infested bean fields, employing a selective peptone-PCNB agar, gave propagule counts of $1-3\times10^3$/g soil, and these propagules were found to be uniformly distributed through the soils down to the ploughing depth of 6-8 in. Ploughing and other cultivations quickly destroy any distribution pattern arising from that of the original production sites of pathogen propagules in the soil.

On account of the remarkable longevity of *F. oxysporum* f. *cubense* in soil in the absence of bananas or other host plants, much attention has recently been paid to the survival of this pathogen in the form of chlamydospores. The importance of these as survival propagules was clearly demonstrated by Newcombe (1960) in her study of the survival of this fungus on glass fibre tapes buried in soil. Failure of the fungus to survive in flooded soil was correlated by Newcombe with its failure to produce chlamydospores under anaerobic conditions. The oxygen requirement for production of chlamydospores by *F. roseum* f. *cerealis* has recently been demonstrated by Griffin (1968) by means of a new technique of exceptional elegance and promise. Very thin glass cells were filled up with glass microbeads (approx. 300 μ diam.) and the space between was partly filled up with nutrient solution, so as to give a growth medium simulating soil in its particulate nature and thus in its air-water regime but with the advantage of semi-transparency for direct microscopical observation. By this technique, Griffin was able to demonstrate an interesting difference between the air requirements for production of macroconidia and chlamydospores, respectively. Macroconidia were produced within water zones, though close to their air boundary ; chlamydospores, in contrast, were produced only in the air spaces of this particulate medium. Pigmentation of mycelium was similarly restricted to that occupying air spaces, and this recalls the general observation that epiphytic mycelia of root-infecting fungi are often coloured, whereas infection hyphae, at least in the deeper layers of host tissue, are rarely so.

Much information was collected by Rishbeth (1955) about the survival of *F. oxysporum* f. *cubense* in the banana soils of Jamaica. Although Rishbeth mentioned the possibility that longevity of this pathogen was due to chlamydospores, he did not attempt to prove it, nor does this uncertainty detract from the persisting value of his extensive data ; we can merely note that subsequent work makes it reasonably certain that his surviving propagules were chlamydospores. It is now accepted that *F. oxysporum* f. *cubense* cannot be identified with certainty from other forms of *F. oxysporum* on the agar plate, as Reinking and Manns (1933) had originally declared possible ; pathogenicity to *Musa* species must be established before identification can be assured, laborious as such host-testing may be. For his pathogenicity test, Rishbeth employed small Gros Michel banana suckers, taken from a disease-free area and set in glass-house pots filled with the soil to be tested. Rishbeth calibrated his technique by planting suckers in a series of disease-free soils inoculated with a suitable dosage range of macroconidia, and then counting mean number of rhizome infections per plant ; values ranged from 0 to 6 and the test was able to detect infective propagules down to the low population level of 2 propagules/g soil. *F. oxysporum* f. *cubense* was not detected by this technique in soils that had never carried bananas, nor usually on sites abandoned for bananas more than 8 years earlier. But a population below the detectable level for this technique, *i.e.* of less than 2 propagules/g soil, must usually have been present on such old banana sites, because when Gros Michel was planted, wilt usually developed before long. Indeed, Rishbeth quotes one instance in which only 2 banana plants out of 100 planted survived on a site that had not carried bananas for 20 years. In such cases, he was unable to detect any propagules of f. *cubense* until a few months after infected plants had wilted, when 50 propagules/g soil might be found, increasing in a year or two to a maximum of about 300 around diseased plants still standing. After removal of diseased plants, pathogen numbers declined after 2-3 years to a value around 10 propagules/g soil. But Rishbeth noted that if a diseased crop of Gros Michel was followed by a planting of the resistant Lacatan, instead of by fallow or a non-host crop, then this decline in population of *f. cubense* was slowed down ; in one such case, the pathogen was still present at a level of some 100 propagules/g soil 9 years after planting Lacatan. Non-progressive infections of this disease-resistant banana evidently maintain the population level of the pathogen ; in complete absence of bananas of any kind, Rishbeth observed that the pathogen population fell to approx. 3% of its peak level within 3 years, and to below 0.5% within 8 years.

Rishbeth's survey in Jamaica was followed by that made by Trujillo and Snyder (1963) of the banana soils of Honduras. The host for identification of f. *cubense* that they employed was *Musa balbisiana*, the advantage of which was brought to light by Stover and Waite (1960). Unlike Gros Michel, which sets seeds very rarely in its fruits (much to the detriment of dentists, as the seeds are large and excessively hard), *M. balbisiana* sets seed freely and so seedlings can be used for the host test ; this saves not only labour but also the difficulty of finding disease-free areas from which to obtain healthy suckers. By the time of Trujillo and Snyder's investigation, it had become much more likely that chlamydospores constituted the survival propagules of f. *cubense* and they reported finding very large numbers of these chlamydospores in infected banana tissues. They found a high population of pathogen propagules only around sites of recently dead banana plants ; this very local distribution of the pathogen was attributed to the absence of ploughing or other soil disturbance in banana plantations.

Following upon the original observations by Chinn (1953) noted above, it has now become well established that chlamydospores of various fusaria will germinate in response to a nutrient stimulus from substrates of fresh plant tissue and from root exudates from non-host as well as from host roots. This non-specific response of the chlamydospores, which might at first sight appear to be disadvantageous to survival of the population (as indeed it seems to be for the conidia of *Cochliobolus sativus*), actually enables them to augment their population. The fusarial chlamydospores germinate in the presence of the diffusing nutrients to form a brief mycelium on which more chlamydospores are produced ; this ephemeral saprophytic phase results in a real increase in the volume of the chlamydospore population. Thus conidial chlamydospores of *F. solani* f. *phaseoli* studied by Schroth and Hendrix (1962) responded both to chopped fresh plant material of various kinds and to root exudates of at least some non-host species ; in the rhizospheres of tomato, lettuce and maize, the chlamydospore population doubled itself. This saprophytic opportunism by f. *phaseoli* has been further confirmed by Toussoun, Patrick and Snyder (1963) and also by Papavizas, Adams and Lewis (1968).

Much of the recent work on these chiamydospores has been concerned with analysis of the nutrient composition of seed, root and hypocotyl exudates, and with the effects of these on chlamydospore behaviour. Toussoun and Snyder (1961) found that chlamydospores of *F. solani* f. *phaseoli* did not germinate spontaneously when air-dried soil was rewetted, but did germinate when placed in contact with bean seedling hypocotyls. Proximity to germinating seeds and to root tips was found by Schroth and Snyder (1961) to stimulate germination but the surface of mature roots was nonstimulating ; this difference could be correlated with differences in amino acid and sugar content of the exudates. The following were detected in seed and young root exudates, and found individually to stimulate chlamydospore germination : aspartic and glutamic acids, asparagine, glucose and sucrose. Hypocotyl exudates contained sugars, but only trace amounts of amino acids. Schroth, Toussoun and Snyder (1963) identified twenty-two amino acids together with glucose, fructose, sucrose and maltose in exudates from bean seeds and roots. When added at the rate of 0.2 ml of a 2% solution to 0·5 g chlamydospore-infested soil, all four sugars and twenty of the amino acids individually stimulated germination to some extent, maximum responses being to glucose (46% germination) and to asparagine (42%). A combination of three sugars and seven amino acids at a total concentration of only 0·005% absorbed into porous procelain seed models stimulated 12% chlamydospores to germinate, as compared with none in the water controls. When a combination of four sugars and four amino acids, in total amount equivalent to that in exudate from a single germinating bean seed, was dried down on an area of cover slip equal in area to the total surface of the seed, then 27% chlamydospores germinated; there was a 10% germination response to the sugar fraction alone, and one of 4% to the amino acid fraction alone. These responses were compared with 41% germination of chlamydospores in soil adjacent to germinating bean seeds, suggesting that sugars and amino acids together account for the major part of the germination stimulus from seed and root exudates. The relative germination response to sugars and amino acids, respectively, is likely to be affected by the available nitrogen content of the soil. In an alkaline loamy sand with low available nitrogen, Cook and Schroth (1965) obtained responses to single amino acids of three times the average response to single sugars; sugar and inorganic nitrogen in C/N ratio equivalent to that in asparagine gave a germination response equal to that elicited by asparagine. Penicillin and streptomycin much increased chlamydospore germination *in the absence of* exogenous nutrients, presumably by reducing either bacterial assimilation of spore-exudate nutrients or bacterial fungistatic excretions or both.

Reasons for failure of infestion by chlamydospores after a high percentage of germination have been investigated by Cook and Snyder (1965), who attempted to discover why germinating bean seeds are rarely infacted by f. *phaseoli*, whereas seedling hypocotyls are regularly infected in the same soil. This difference appeared to be due not to any difference in resistance to infection between seeds and hypocotyls, but rather to the effect of the respective exudates on chlamydospore behaviour. Thus 60% chlamydospores germinated within 16 h in soil in contact with germinating seeds, but growth of germ tubes then ceased and lysis shortly followed. In contrast, only 20%

chlamydospores germinated within 48 h in soil adjacent to hypocotyls, but the germ tubes continued to grow, forming mycelia appressed to the hypocotyl surface and infection duly developed. Cook and Snyder suspected that this difference was due to the difference in nutrient composition of seed and hypocotyl exudates, respectively ; in seed exudates, the sugars are balanced by amino acids whereas in hypocotyl exudates there are sugars but only a trace of amino acids. To produce a model for these two contrasting exudate effects, Cook and Snyder added carbon and nitrogen nutrients in various combinations to chlamydospore populations in unplanted soil. Glucose or sucrose alone stimulated only 20% of chlamydospores to germinate, but there was no lysis of the germ tubes : after a limited growth, the hyphae produced new chlamydospores. Both germination percentage and lysis of germ tubes increased when the nutrient solution contained nitrogen in balance with the sugar. Nitrogen as asparagine promoted earlier lysis than did inorganic nitrogen, and the lysis was maximal when yeast extract was added to a solution of glucose and asparagine.

Cook and Snyder have therefore concluded that ample provision of exogenous carbon and nitrogen nutrients together, either artificially or by way of seed exudates, promotes maximum germination of chlamydospores but simultaneously creates conditions that are fatal to the germlings, probably through the development of a large population of zymogenous bacteria. In support to this explanation, they quote a review by Brian (1960), who has carefully discriminated between various types of mycelial lysis. First, he has pointed out that autolysis in pure culture sets in only when carbon nutrients have become exhausted. Lysis caused by other soil microorganisms is distinguished as heterolysis. But effective competition by a large microbial population can deprive a fungus germling of nutrients even when these are continuously renewed by seed or root exudation, and so induce a carbon-starvation type of autolysis similar to that eventually occurring in pure culture through carbon-substrate exhaustion. Other mechanisms of heterolysis are more easily distinguished from pure-culture autolysis. Thus Brian has noted that a large microbial population can cause oxygen-lack in spore germlings, and also that some antibiotics act as respiration inhibitors and thus cause growth-stasis and death for the same ultimate reason. He further cites examples of bacterial and actinomycete enzymes that can disrupt fungal cell walls or protoplast membranes, so that heterolysis of fungal hyphae can act through a variety of mechanisms. Lloyd and Lockwood (1966) have also made observations on lysis of fungal sporelings and in their discussion have laid most emphasis on autolysis following unavailability or exhaustion of substrate nutrients. But the speed with which germ-tube lysis occurs in the situation examined by Cook and Snyder suggests that carbon starvation is not the only or even the most important cause.

These conclusions by Cook and Snyder were later employed by Cook and Flentje (1967) to interpret the behaviour of chlamydospores of *F. solani* f. *pisi* in a loamy sand soil (pH 7·5) in South Australia. Maximum germination (46%) of chlamydospores adjacent to germinating pea seeds occurred at the highest moisture content tested (81% saturation) but germling survival was better at lower moisture contents ; in subsequent trials, percentage of germ tubes lysed increased with soil moisture content rising above 81% saturation. These results were interpreted by Cook and Flentje not as a direct effect of soil moisture content upon chlamydospore and germling behaviour, but rather as an indirect effect from the influence of soil moisture level on exudation by germinating pea seeds ; exudation rate was shown (by dry weight loss of germinating pea seeds) to increase progressively with rising soil moisture content. A similar effect of rising soil moisture content upon rate of exudation by germinating pea seeds had earlier been reported by Kerr (1964) from the same laboratory. To test this interpretation, Cook and Flentje then added to their chlamydospore-containing soil a nutrient solution of 2.2% sucrose made up with ammonium sulphate to C/N 5 ; when any possible effect of soil moisture content on rate of nutrient accretion to the soil through seed exudation was thus blanketed, chlamydospore behaviour was not affected by soil moisture content over the range 58-90% saturation. After 17 h, some 53% chlamydospores had germinated and after 48 h some 35% were left with germ tubes still unlysed, irrespective of the particular soil moisture level. But addition of sucrose alone (2.2%) to the soil was not followed by any germ tube lysis, even in a soil at 100% saturation, so that all results were as predicted from the conclusion of Cook and Snyder about effects of exudate nutrients on chlamydospore behaviour.

Other observations on behaviour of chlamydospore populations in soils can possibly be interpreted by reference to these effects of carbon and nitrogen nutrients, and taking into account the modification of the C/N ratio of exudates by the nitrogen content of the soil solution. In general terms, we can expect that in a nutrient medium of low C/N ratio, carbon nutrients will be exhausted before available nitrogen, and so chlamydospore germlings will perish and autolyse through carbon starvation. But in a medium of high C/N ratio, exhaustion of nitrogen will terminate growth whilst carbon nutrients are still available, and so (in the absence of infectible host tissue) hyphae will continue to absorb surplus carbon nutrients and incorporate them into chlamydospore cell walls. Such a sequence of events could have determined the fate of macroconidia of *F. solani* f. *phaseoli* in two soils of the Columbia Basin region of the U.S.A. studied by Burke (1965). Both soils were of the same fine sandy loam type ; one represented virgin soils recently brought under irrigation and very liable to severe outbreaks of bean root-rot ; the other represented old irrigated soils of the same region, which had become 'resistant' to development of the disease. In his study of the behaviour of macroconidia in a film of water agar on glass slides buried in the soil, Burke found that the old, 'resistant' soil promoted extensive germ tube growth with a comparatively late development of small terminal chlamydospores on the mycelium. In the young, 'susceptible' soil, on the other hand, germ tube growth more quickly became arrested and numerous large chlamydospores were produced both within the macroconidia and also terminally on short germ tubes. Burke's observations can be explained on the assumption that the old soil had a higher available nitrogen content than the young one ; it is well known that in this type of virgin soil, both organic content and total nitrogen increase with continued cultivation of crops under irrigation. Thus in comparing nine virgin soils with three old irrigated soils in the same region of the U.S.A. as studied by Burke, Menzies (1959) found a mean increase in organic matter, attributable to irrigation and crop cultivation, of 53%.

This account of the chlamydospores of *Fusarium* species, which have been described and figured by Toussoun and Nelson (1968), would be incomplete without some mention of the structure of the chlamydospore wall, which may determine differences in longevity between populations of different species and formae speciales. Differences in ultrastructure of the wall have been described by Nash and Alexander (1965) between chlamydospores of *F. solani* f. *cucurbitae* and f. *phaseoli*, respectively, and they suggest that these differences may account for the much longer life of chlamy-dospore populations of f. *phaseoli* in soil. This posssibility has been further discussed by Nash Smith (syn. Nash, Shirley M.) in a review (1969).

Ascospores of Rhizina undulata

These ascospores are of exceptional interest, because they are the only ones in which the dormancy-breaking effect of heat-shock, so widely employed in laboratory practice, has been demonstrated as a factor of paramount importance for initiation of outbreaks of a root disease. *R. undulata* is the cause of a root disease of conifers known as 'group dying' and outbreaks have been known for some years to be associated with bonfire sites, both in Europe and in the U.S.A. Conditions governing initiation of these outbreaks have been elucidated by Jalaluddin (1967) for conifers in Great Britain. They arise only at the margin of bonfire sites, when branches from brashing, thinning or clear felling have been burned in close proximity to stumps of recently felled trees, or to standing trees of a very susceptible species. Although older trees can become infected, the greatest damage from the disease is suffered by young trees planted on a clearfelled site. Jalaluddin has demonstrated the sequence of events leading up to such an outbreak, as follows. During progress of a bonfire, the underlying soil and that on the perimeter is heated up ; peak soil temperatures during progress of the bonfire are zoned and decrease centrifugally. A zone thus occurs below and around every bonfine in which the peak temperature lies within the range 35-45° C, determined by Jalaluddin as suitable for dormancy-breaking in the ascospores by heat-shock ; the actual position of this zone, of course, will depend upon size and duration of the bonfire. Independently of its infection of tree seedlings, *R. undulata* manifests its presence by production of ascocarps on the soil surface. In the forest, conditions necessary for infection of tree seedlings and production of ascocarps are : (1) soil heating by a bonfire ; (2) presence of susceptible roots, infection of which provides *R. undulata* with a food base from which to spread as mycelial strands through the soil, and thence to infect the roots of seedling trees ; (3) an acid soil. Roots of recently felled trees are the most susceptible of all. If trunks of standing pine trees are close enough to the fire to be badly scorched, then root resistance may be reduced sufficiently to permit infection. Jalaluddin further observed that *R. undulata* could infect the roots of standing Nerway spruce even when trees had suffered no fire damage, and Sitka spruce appears to be equally susceptible. Douglas fir roots, in contrast, appeared more resistant than those of pines. Roots of deciduous trees seemed to be completely resistant to infection by *R. undulata* even after trees had been felled ; ascocarps did not appear around a bonfire site except in the presence of susceptible coniferous roots. The group-dying disease thus presents various features of exceptional interest to which Jalaluddin did full justice, particularly by providing quantitative estimates in the form of ascocarp counts. Table 6.12 taken from his second paper shows how he employed ascocarp counts to demonstrate : (*a*) that pine stump roots must be within reach of the mycelium of *R. undulata* arising from heat-shock-germinated ascospores so that it can infect them and thus establish itself in a first food base ; (*b*) that roots of freshly produced stumps provide the best food base for *R. undulata*; with elapse of time from tree-felling, stump roots become first infected and later saprophytically colonized by various soil fungi, and are then no longer available for infection by *R. undulata*. Figures given in Table 6.12 are means of five replicates.

Table 6.12. Ths effect on colonization of bonfire sites by Rhizina undulata of the position and age of pine stumps.

Mean no. ascocarps per fire site developing within 1 yr of bonfire			
A. Position of fresh pine stump in relation to fire site			
	Distance from margin of fire site (m)		
Within fire site	1.5	3	6
20	13	5	0
B. Age of pine stumps near fire site			
	Period since felling (yr)		
Freshly felled	1	2	3
16	10	6	0

Jalaluddin (1967) has also presented a variety of dada on germination of the ascospores of *R. undulata* with and without a heat-shock treatment. Germination of unheated ascospores on 3% malt agar did not occur below 15° C; over the range 18-30°C, rate at which germination occurred increased with rise in temperature. Stimulation by heat-shock occurred over the range 35-45° C; the best treatment was 72 h at 37° C, followed by incubation at 22°C. In a typical experiment, 50% ascospores germinated between the 6th and the 8th day after heat-treatment and no more thereafter up to 30 days; only 10% untreated ascospores germinated within 30 days and germination was spread over the period from 8th to 29th day. These figures illustrate very well the point that we had made early in this chapter, *viz.*, that ascospores of this type will germinate without heat-shock but they do so slowly and irregularly over an extended period; heat-shock terminates endogenous dormancy quite abruptly for a substantial proportion of the spore population, so that they germinate quickly and together when environmental conditions are favourable. The peculiar interest of Jalaluddin's observations is that here nature is seen as appearing to imitate scientific art rather than *vice versa.*

In view of the well-marked endogenous dormancy of these ascospores, it is not surprising that Jalaluddin found them to survive well when buried on glass slides in three different forest soils. After 1 year, the number of spores found viable when slides were coated with malt agar, heat-treated and then incubated at 22°C, was about one-quarter of those that germinated when similarly treated at the outset. After 2 years, the proportion so germinating was reduced to 7% of the original in one soil, but none germinated in the other two.

Resting spores of the Plasmodiophorales

Resting spores of *Plasmodiophora brassicae*, causing the clubroot disease of crucifers, can survive for a number of years in the soil. A longevity of 5 years in New Zealand was recorded by Gibbs (1939), and Fedorintchik (1935) in the U.S.S.R. found 27% of cabbage seedlings to become infected when potted in soil taken from a field that had not carried a cruciferous crop for 7 years. A 4-year rotation is often inadequate to control this disease ; on Agdell Field at the Rothamsted Experimental Station, which had been under a rotation of swedes with three non-cruciferous crops since 1848, severe outbreaks of clubroot eventually developed from about 1920 onwards. The disease is characterized by the development of large galls on the root system ; the gall tissue eventually becomes transformed into a mass of resting spores, which are later liberated into the soil through decomposition of the gall tissue (Fig. 6.13).

When such a high population of resting spores is released into the soil following upon a severely diseased crop, then longevity of the pathogen for some years is assured, even if only a minute fraction of the original population preserves its dormancy. Germination of these resting spores is very pH dependent, and exogenous dormancy is probably more widely and effectively maintained by a sufficient degree of soil alkalinity than by any effect of soil fungistasis of microbial origin ; this effect of soil alkalinity is thus of the type classified by

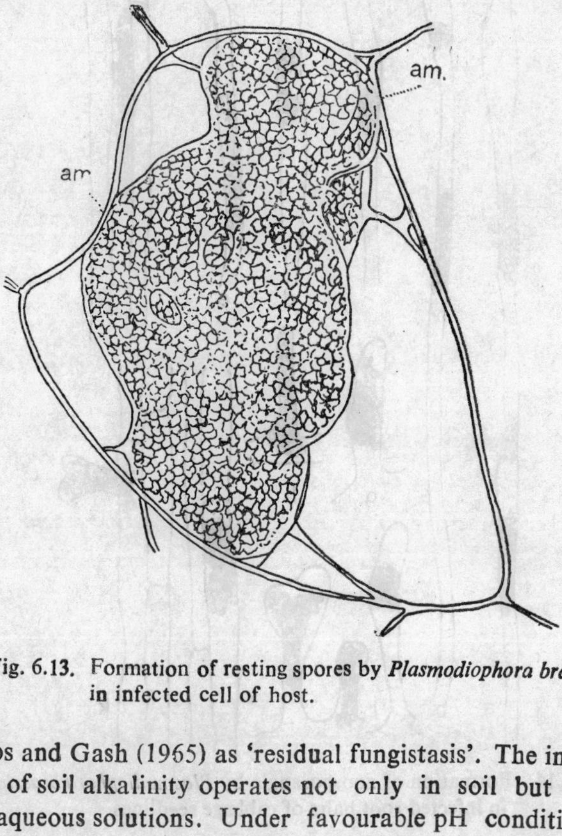

Fig. 6.13. Formation of resting spores by ***Plasmodiophora brassicae*** in infected cell of host.

Dobbs and Gash (1965) as 'residual fungistasis'. The inhibiting effect of soil alkalinity operates not only in soil but also in pure aqueous solutions. Under favourable pH conditions and in the presence of a cruciferous root exudate, the resting spore germinates to produce a single, biflagellate, primary zoospore, which has been described and electronmicrographed by Kole and Gielink (1962). Infection of the root hairs by these primary zoospores was described by Samuel and Garrett (1945), who have reviewed the earlier observation by Woronin (1878) and Cook and Schwartz (1930) ; the primary zoospore develops into a plasmodium within the root hair and within 7-8 days at 25°C this plasmodium becomes converted into a cluster of ripe zoosporangia, from which secondary zoospores emerge (Fig. 6.14).

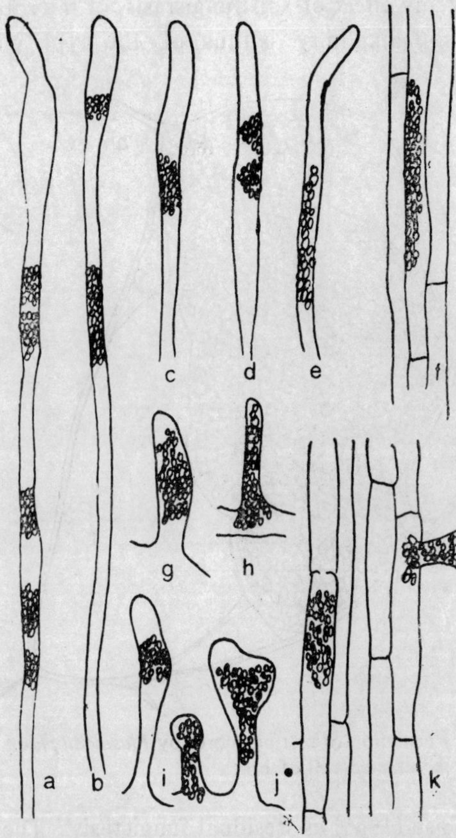

Fig. 6.14. Formation of zoosporangia by *Plasmodiophora brassicae* in infected root hairs of cabbage seedlings.

Employing an infected-root-hair-count method based on these observations, we found the optimum soil pH value for germination of resting spores and infection of root hairs was slightly above 6 ; an optimum pH range of 5-6 was found by Macfarlane (1958), employing the same root-hair-count method but in aqueous solutions instead of in soil or sand. Nevertheless, although the optimum pH value for root-hair infection lies within the pH range 5-6, alkalinity exercises a stronger inhibiting effect on spore germination than does a comparable degree of acidity, and this is one of the reasons for the general use of lime or ground chalk for control of clubroot in farm practice.

Of particular relevance to the longevity of resting-spore populations in soil are observations by various authors on wastage due to apparently spontaneous germination. To follow this downward drift with time in the resting-spore population, Bremer (1924) assessed by a plasmolysis technique the viability of spores in thin sections of galled cabbage roots buried under various soil conditions. A high proportion of spores were found to germinate spontaneously even in soil freshly treated with lime, although germination was more rapid and extensive in the acid, untreated soil. Similar results were reported by Macfarlane (1952), using the infected-root-hair-count method to estimate resting-spore populations ; he found a rapid decline in populations during the first few weeks after addition to unplanted soil. This decline was more rapid in a wet, slightly acid soil, providing optimum conditions for spontaneous germination, than in a drier, alkaline soil. Similarly Bochow (1961) determined rates of spontaneous germination in three different soils of pH value 6.6, 7.0 and 7.2, respectively ; in all three soils, more than 50% spores had germinated after 8-10 days, and more than 90% after 12 weeks.

In addition to this loss through spontaneous germination, resting-spore populations of *P. brassicae* in soil also suffer wastage through the germination-decoying effect of root exudates from some non-host species of crop plants and weeds. Webb (1949) was the first to report zoosporangia of *P. brassicae* in the root hairs of several grass species and other non-cruciferous plants ; such plants can be termed 'non-hosts' in the sense that infection proceeds no further than the zoosporangial stage in the root hairs, and so no resting spores are produced. Macfarlane (1952) found root-hair infection not only in species from families closely related to the Cruciferae, such as *Papaver rhoeas* and *Reseda odorata*, but also in roots of some grass species, *e.g. Lolium perenne*, *Agrostis alba stolonifera*, and *Dactylis glomerata*. In one of his experiments, a crop of perennial ryegrass appeared to reduce the resting-spore population, by comparison with that surviving in fallow soil. Kole and Philipsen (1956) have similarly described zoosporangia in the root hairs of *Trifolium pratense*. It further appears that diffusates from some kinds of organic material also stimulate resting-spore germination, as demonstrated by Bochow and Seidel (1964) both for farmyard manure and for rye-straw meal.

A similar germination-decoy effect has been demonstrated by White (1954), who tested the response of various species in the Solanaceae to inoculation with resting spores of *Spongospora subterranea,* causing powdery scab of potato tubers. Amongst the species tested, *Datura stramonium* was the only one that stimulated germination of the spore-balls of *S. subterranea* without subsequent development of disease galls on the root system. In a field trial, plots planted with potatoes as a test crop showed a reduction of powdery scab from 37% infected tubers after a fallow to 7% following a thick planting of *D. stramonium* as a decoy crop.

Survival of sclerotia

Sclerotia are multicellular resting propagules, composed of vegetative hyphal cells, and they sometimes incorporate some host plant tissue. Though often irregular in shape, the constituent cells tend to be more or less isodiametric, thus differing from the elongated cells of septate hyphae. Sometimes, as in the sclerotia of *Rhizoctonia solani,* there is little differentiation between inner and outer cells ; more often, however, the inner cells are thin-walled and colourless, and the selerotium is bounded by a rind of cells, often of smaller size, with thicker, dark-coloured walls. Chemical analyses of various types of sclerotia have shown that the content of reserve nutrients, and especially of carbon nutrients, is usually much higher than that of unorganized mycelium.

A fundamental distinction can be made between sclerotia produced by air-borne fungal pathogens and those produced by root-infecting fungi. Sclerotia of the former group usually produce either fruit bodies, such as the stalked apothecia of *Sclerotinia* species, or stromata in which spores are produced, such as the stalked, perithecia-producing stromata of *Claviceps* species (ergot fungi of grasses and cereals). Sclerotia of this type are often variable in shape, and usually rather variable in size within a single species ; they can attain a size of up to 2 cm in their largest dimension. In contrast with these are the sclerotia of most root-infecting fungi, which are both smaller and much more uniform in size and also more regular in shape, most frequently being subspherical. The sclerotia of *Phymatotrichum omnivorum*, causing Texas root-rot of cotton and many other dicotyledonous hosts, are mostly 1-2 mm in diameter, though occasionally reaching 1 cm ; they have been described as 'usually smaller than radish seed, by Taubenhaus and Ezekiel (1931). Both in size and shape ; the sclerotia of *Sclerotium rolfsii* again resemble radish seed, being mostly 1-2 mm in diameter (Leach and Davey, 1938). The sclerotia of *Sclerotium cepivorum*, causing the white rot disease of onions and related species, are subspherical but smaller than those of *S. rolfsii*, being mostly 0.2—0.5 mm diameter. The microsclerotia of *Verticillium dahliae* are aptly named, because their range in length is given as 50—200 μ.

As compared with sclerotia of air-borne fungal pathogens, sclerotia of root-infecting fungi are thus characterized by their smaller and more consistent size, and often by a greater regularity in shape (subspherical) as well. This smaller and more uniform size of sclerotia in root-infecting fungi can probably be related to the fact that they function individually as *single infective propagules*, producing on germination either hyphae or hyphal strands ; sclerotia of air-borne pathogens, on the other hand, germinate to form either fruit bodies or stromata, each of which produces a large number of infective propagules (spores). All organisms, including fungi and higher plants, can be seen to exercise economy in the deployment of their reproductive resources ; the most economical size for a reproductive propagule is the *minimum effective size*, in which definition some allowance must be made for a 'safety factor' to provide for at least a certain range of variation in the habitat, individual host plant or substrate of higher plant, parasite or saprophyte, respectively. The application of this generalization to the sclerotia of root-intecting fungi is quite clear ; the most economical size is the minimum size that will generate an inoculum potential adequate for infection of the roots of the average or typical host plant. All the root-infecting fungi mentioned above have as host plants either herbaceous annuals or the smaller woody perennials ; we can therefore conclude that the limited range in root-resistance to infection imposed by size-limitation in the host plants has imposed a similar size-limitation and uniformity on the sclerotia of the fungal pathogens. No such limitation is thus imposed on the size of sclerotia produced by airborne pathogens, because they do not act as infective propagules directly ; instead they produce a widely variable number of infective propagules, according to their size and reproductive capacity.

Finally, it is worth noting that fungi infecting tree roots do not usually produce sclerotia of the type defined above ; it is the infected root that serves as a food base for a new infection. For several such fungi, Campbell (1934) has proposed the term 'pseudosclerotium' to designate a volume of infected wood enclosed by a membrane of sclerotized fungal cells.

Phymatotrichum omnivorum

The morphogenesis of these sclerotia was first described by King and Loomis (1929), who showed that they arose at intervals along the mycelial strands by a process of renewed cell growth and division, which occurred in the large cells of the wide central hypha as well as in the smaller cells of the narrower investing branch hyphae. This resulted in a central mass of closely packed, colourless, thin-walled cells of large and small size intermixed. The central mass, or medulla, is bounded by a rind composed of brown, thicker-walled cells of very irregular outline in surface view. From some of these surface cells, acicular hyphae (setae), similar to those surrounding newly formed mycelial strands, are produced (Fig. 6.15).

During germination of the sclerotia, as described by Presley (1939), the new hyphae do not arise as extensions of the cell

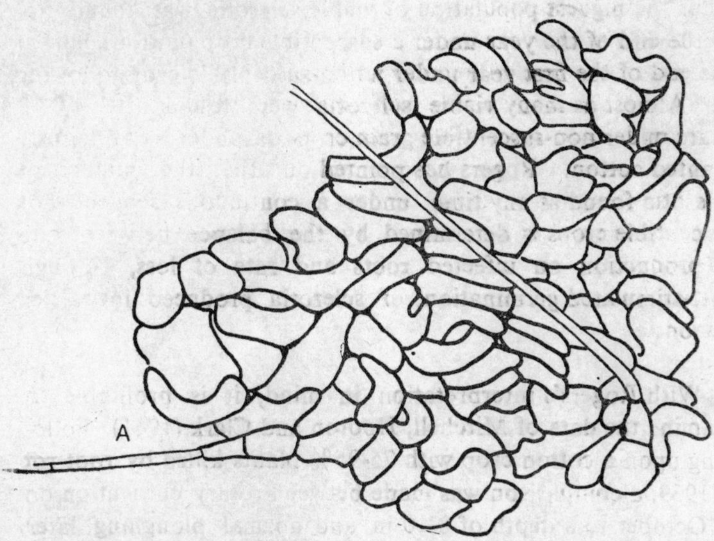

Fig. 6.15. Surface cells bounding sclerotium of *Phymatotrichum omnivorum*. Acicular hypha at *A*.

walls of the medullary cells, but are organized *inside* these cells, from which they later emerge. Ergle (1948) found that sclerotia taken from 30-day cultures of *P. omnivorum* would germinate on a sterile medium containing only inorganic salts, so that endogenous dormancy could have lasted only for a brief period ; carbon reserves were in the form of glycogen.

In their study of sclerotium longevity, King and Eaton (1934) found that spontaneous germination in a well aerated soil contributed substantially to decline in the population of viable sclerotia. Evidence from field trials suggests that disturbance of the soil by cultivation implements also encourages spontaneous germination ; cultivation temporarily improves soil aeration, though it may have another and more important effect on sclerotium populations, as we shall see shortly. But it is first necessary to appreciate the general sequence of events that determines the rise and fall of sclerotium populations in the field, as elucidated by Rogers (1942). Rogers's data show that formation of sclerotia occurs mainly after the conclusion of the parasitic phase of *P. omnivorum* on living host roots.

Thus the highest population of viable sclerotia was found not at the end of the year under a susceptible crop (cotton) but at the end of the first year under a non-susceptible crop following it. Almost as many viable sclerotia were found after 10-12 years under non-susceptible grain crops as under continuously planted cotton. Rogers has pointed out that the number of sclerotia found at any time under a continuous sequence of susceptible crops is determined by the balance between rate of production on infected roots and rate of loss, through host-stimulated germination, of sclerotia produced in earlier seasons.

With Rogers's interpretation in mind, it is profitable to examine the data of Mitchell, Hooton and Clark (1941). Following upon a cotton crop with 75-95% plants killed by root-rot in 1939, a comparison was made between rotary cultivation on 30 October to a depth of 8-10 in, and normal ploughing later in the autumn as a preparation for the spring-sown crop of cotton in 1940. In August 1940, 36% plants were dead from root-rot on the rotavated plots, as compared with 90% on the ploughed plots. In the soil of the rotavated plots, no sclerotia were found down to 12 in, though they were present below this level. On the ploughed plots, at Mitchell *et al.* state, 'sclerotia were observed in abundance in the 6- to 12-in level'. Numbers of sclerotia thus present in August 1940 would have represented the number produced on the roots of the 1939 crop *minus* the number stimulated to germinate by the roots of the 1940 crop. Since the latter figure is unlikely to have been affected by the cultivation treatments in autumn 1939, it is evident that number of sclerotia *produced* after the two cultivation treatments must have determined the difference in numbers surviving at the sampling in August 1940. As Mitchell *et al.* have pointed out, early rotavation would have chopped up the cotton root systems and interfered with further extension of infection by *P. omnivorum* along the roots and with its production of sclerotia on the mycelial strands ; ploughing later in the autumn would have interfered much less with the production of sclerotia. It is further possible that the soil-aerating effect of rotavation would have encouraged spontaneous germination of sclerotia already formed, though it seems likely that this would have been subsidiary in importance to the mechanical dismemberment of root systems by the rotavator blades. Rogers (1942) found a similar beneficial effect in reduction of root-rot to follow subsoil cultivation.

In the same field trial, Mitchell *et al.* also included as treatments the rotavation into soil of farmyard manure and sorghum fodder, respectively ; these had the same effect on sclerotium populations in August 1940 as had rotavation alone, but reduced percentage of deaths from root-rot to a yet lower level. In accompanying laboratory experiments, Mitchell *et al.* showed that amendment of soil either with farmyard manure or with fresh plant material strongly stimulated germination of sclerotia. After germination, much the same sequencc of events occurs as with some kinds of spore in the same situation ; the sclerotia exhaust their nutrient reserves by outward growth of mycelial strands, but the strands are unable to secure a sufficient share of the germination-stimulating substrate to form new sclerotia and thus replace the population wastage so induced. Death of the young fungal colonies thus initiated can be ascribed to two causes, the relative importance of which is difficult to assess : (1) microbial competition for substrate nutrients, and especially for carbon nutrients, resulting in starvation and autolysis ; (2) microbial lysis (heterolysis). As a result of other experiments similar to those in collaboration with Mitchell and Hooton, Clark (1942) stressed the heterolysis component of this effect, which he considered to be one of 'biological control'. Whether he was right in so doing is still debatable, but there is no doubt that the initiation of this sequence of events is due to a *direct* effect of substrate nutrients on the sclerotia, stimulating them to germinate ; in current practice, biological control is defined as being mediated by one or more organisms (excepting man himself) outside the host-parasite relationship. This fungus causes the violet root-rot disease of sugar beet, carrot, legumes and various other field crops, and the sclerotia are more variable in size and shape than are those produced by other root-infecting fungi. They may develop on the mycelial strands, as described by Duggar

(1915) and Faris (1921), but on crops with storage tap-roots, such as sugar beet, mangold and carrot, they are usually formed around small lateral roots ; when sectioned, the vascular elements of the rootlet can be distinguished as the longitudinal axis of the sclerotium. Sclerotia formed thus around small lateral roots are elongated and often irregular and contorted in shape.

In the course of examining sugar beet and carrots with violet root-rot over more than 20 years, Pathologists have never been able to find sclerotia produced directly on the fleshy tap-root, though it could be always found out on lateral rootlets and sometimes on the unswollen tail end of the tap root. This recalls an observation recorded by McNamara and Hooton (1933), who regularly found sclerotia of *Phymatotrichum omnivorum* on small lateral rootlets but not on the tap-root or larger lateral roots of infected cotton plants. It is most unlikely that the nutrients required for production of sclerotia are wholly supplied by the infected tissues of these small lateral roots, because the total nutrient content of the mature sclerotium must greatly exceed that of the original root before infection. It is probable that the rootlets provide a 'perch' on which sclerotia are produced by translocation through the mycelium or mycelial strands from the main body of infected tissue in the fleshy tap root. It is possible that the nutrient balance in the mycelium on the surface of the tap root is adverse to the initiation and maturation of sclerotia, but that this nutrient balance becomes altered to a favourable one as the mycelium travels away from the tap-root surface along the lateral roots.

This explanation if unsupported by other evidence could be dismissed as too speculative to commit to print, but it is possibly compatible with the results of a study scientists made on carbon and nitrogen requirements for production of sclerotia by *H. purpureum* on nutrient agar. Sclerotia are produced by *H. purpureum* on suitable nutrient agars, but they vary widely in size and also in position with respect to the colony centre ; moreover, individual sclerotia often fuse during development to give compound sclerotial masses. To overcome this difficulty, they employed whole small colonies of *H. purpureum* as sclerotium-survival units ; ten colonies were grown around the periphery of a 9 cm agar plate for an incubation period of 2 months at 25° C. After burial in soil for various periods, these small colonies were exhumed and tested for presence of viable sclerotia by inoculating small groups of young carrot plants. The basal nutrient medium contained 1.5% malt extract, providing vitamins and approx. 1% of sugars, and this was supplemented in different treatment-series with 1, 2 and 4% glucose, respectively ; the different sugar levels were combined in a factorial design with six nitrogen levels over the range 0-0.16%. Colonies were buried in soil for a sufficient period (2 months) for elimination of unorganized mycelium, so that survival could be equated with production of mature, viable sclerotia ; this was confirmed by direct observation when colonies were exhumed. In this way, scientists were able to show that production of sclerotia increased with rising sugar content of the agar medium, up to the maximum concentration tested (5%). At the lower sugar concentrations, nitrogen above a certain critical level depressed production of viable sclerotia ; for 2% sugar, this critical level was 0·01% (C/N 80) ; for 3% sugar, it was 0·04% (C/N 30) ; for 5% sugar, it was not reached at 0·16% (C/N 12·5), which was the highest nitrogen level tested. It thus appears that the C/N ratio of the medium docs not itself determine the initiation and maturation of viable sclerotia, though one of its consequences may well do so. Thus at any given sugar level, the nitrogen level will determine the amount of sugar remaining in the medium after mycelial growth has used up all the available nitrogen and has therefore come to a halt The amount of residual sugar will then determine whether sclerotia are produced and if so, then the volume of sclerotial tissue formed. It is now permissible, though admittedly speculative, to extrapolate these results to interpret production of sclerotia on an infected root system. Thus we can suppose that when *H. purpureum* is growing on a living root, vegetative growth of mycelium will continue until the available nitrogen, located chiefly in the cytoplasm and vacuoles of the host cells, is exhausted. After that, the mycelium will continue to hydrolyse both reserve starch and the carbon constituents of the plant cell walls until it has accumulated sugar to a level at which sclerotia are initiated and carried through to final maturation.

Spontaneous germination of these sclerotia was studied by Valder (1958), who employed sclerotial colonies on individual grains of rye as his survival units, and slices of living carrot-root tissue for his sclerotium-viability tests. He found that survival of the fungus on rye grains was reduced to a level of 10-40% in different experiments after some 4 months in a somewhat acid soil (pH approx. 5·3), and direct observation showed this to be due to spontaneous germination of the sclerotia. In a slightly alkaline soil (pH approx. 7·3), no spontaneous germinating was observed, and 70-80% of the rye grains carried still viable sclerotia after 4 months in the soil. When the acid soil was brought to pH 7·3 by incubation with powdered calcium carbonate, then the sclerotia survived as well as in the other, originally alkaline soil, suggesting that soil pH itself directly controlled the occurrence of spontaneous germination. In neither soil, however, did addition of 1% of ground, dried plant material (peameal and grassmeal) stimulate sclerotium germination, because neither of these supplements significantly affected sclerotium survival in either soil. Both of Valder's results were somewhat unexpected, though stimulation of spontaneous germination by a suitable degree of soil acidity can be compared with a similar effect on the resting spores of *Plasmodiophora brassicae*, noted above.

Verticillium dahliae

V. dahliae produces microsclerotia in culture and no its host plants and can thereby be distinguished from *V. albo-atrum*, which develops dark-coloured mycelium but no microsclerotia in culture. Most isolates of *V. dahliae* can further be distinguished from those of *V. albo-atrum* by their greater tolerance of growth temperatures above 23°C, and generally also by a higher optimum temperature for pathogenesis. European mycologists recognize these differences as worthy of specific rank, but in the U.S.A. both species are usually included within the single specific epithet of '*albo-atrum*'. The microsclerotia of *V. dahliae* are amongst the smallest known in root-infecting fungi; their range in length has been given as 50-200 μ by Isaac (1949), from whose paper (Fig. 6.16) has been reproduced.

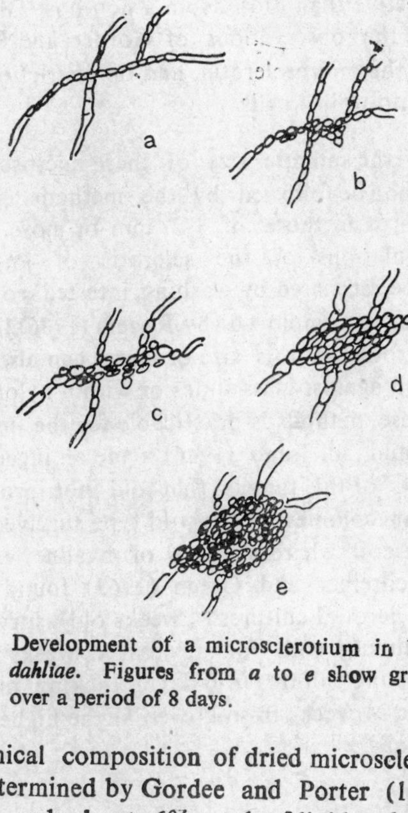

Fig. 6.16. Development of a microsclerotium in *Verticillium dahliae*. Figures from *a* to *e* show growth stages over a period of 8 days.

The chemical composition of dried microsclerotia of age 4 weeks was determined by Gordee and Porter (1961) as 80% carbohydrate and about 6% each of lipid and protein; after 40 weeks more, carbohydrate content had fallen to just below 60%, whilst lipid had increased to 22% and protein to 8%. Microscopical examination showed the microsclerotia to be made up of melanized, thickwalled cells and hyaline, thin-walled cells in an apparently random mixture. When micro-sclerotia were sonically disintegrated and the tissue debris suspended in buffer solutions, germ tubes developed only from the thin-walled cells, more than 90% of which germinated over a wide pH range (4·2—9·2). Isaac and MacGarvie (1962) found that microsclerotia taken from 3-4 week cultures on Dox's agar would germinate on water agar, provided that they were first soaked in distilled water. Schreiber and Green (1963) observed that germination of the microsclerotia in soil was stimulated by root exudates, those from a host (tomato) being more effective than those from a non-host (wheat. They also confirmed the observations of Gordee and Porter on the morphology of the microsclerotia, and on restriction of germination to the thin-walled cells.

Because of the minute size of these microsclerotia, their fate in soil cannot be followed by the methods employed for larger sclerotia, *i.e.* those of 1-2 mm or more in diameter; thus soil populations of the sclerotia of *Phymatotrichum omivorum* can be estimated by washing intested soil through a bank of sieves, as employed by Rogers (1936). The fate of individual sclerotia of this size or larger can also be followed by burying them against glass slides or within nylon envelopes. But none of these methods is practicable for the microsclerotia of *V. dahliae*, and so Isaac (1946) made an ingenious use of 'hyaline variants' of this fungus that did not produce microsclerotia. Agar colonies of the wild type survived for at least 5 months in the soil, whereas those of hyaline variants were short-lived. Schreiber and Green (1962) found that homogenized microsclerotial cultures (4 weeks old) survived in soil (0.5 mg inoculum/g soil) for at least 82 weeks; in contrast, homogenized young cultures (4-6 days) lacking microsclerotia failed to survive 21 weeks in soil, even at the highest inoculum level (7 mg/g soil).

Sclerotium rolfsii

Under some conditions, the sclerotia of *S. rolfsii* will germinate spontaneously, in the absence of any nutrient stimulus from fresh plant tissue or from root exudates. This was incidentally demonstrated by Leach and Davey (1938) by their viability test for sclerotia separated from samples of field soil by washing over a bank of sieves. The washed sclerotia were plated out in Petri dishes on the surface of moist, unsterilized, black peat soil and their germination was so regular that this technique provided a much more reliable viability test than did plating out surface-sterilized sclerotia on potato-dextrose agar.

In the behaviour of sclerotia, it was noted an interesting but quite isolated observation by Dunlap (1943) on production of sclerotia by *Phymatotrichum omnivorum*. He found that this could be completely inhibited in autoclaved Houston black clay soil, which normally supported a good production of sclerotia, by autoclaving 325-mesh dusting sulphur with the soil, at the rate of 1 part sulphur to 1000 parts air-dry soil. If the mixture was sterilized dry, no restraint on sclerotium production eventuated. But mycelial growth through the autoclaved soil-sulphur mixture was as good as that through soil alone. Recent work on production of sclerotia by *Sclerotium rolfsii* has now suggested an explanation for Dunlap's observation ; Chet, Henis and Mitchell (1966) have reported that production of sclerotia by this fungus on nutrient agar was inhibited by five sulphur-containing amino acids and also by glutathione over the range 10^{-5}—10^{-4} M, without affecting the mycelial growth. Although not restricting mycelial growth, all these substances induced production of basidia of the perfect state of *S. rolfsii, i.e. Corticium rolfsii,* which had never been seen in culture before, and only rarely on its host plants. This morphogenetic effect of L-cysteine was found to be competitively antagonized by iodoacetic acid ; at a molar ratio of 30 I cysteine : iodoacetic acid, the converse effects of these two morphogens cancelled one another out, and growth of *S. rolfsii* was normal. The effect of iodoacetic acid by itself, over the range 3×10^{-5}—10^{-4} M, was to induce precocious production of sclerotia, which grew to larger than normal size. The morphogenetic action of the inducer was found to be limited to hyphal tips at the colony margin.

Following up this discovery, Chet and Henis (1968) found that radioautograms of *S. rolfsii* grown on ^{14}C iodoacetic acid showed a selective accumulation of radioactivity first on future sites of sclerotium production and subsequently in the developing sclerotia. In contrast, radioactivity of ^{14}C L-cysteine was quite evenly distributed through the mycelium and in the few sclerotia that were formed. So the high molar ratio (30 : 1) of L-cysteine required to neutralize the morphogenetic activity of iodoacetic acid could be explained by its even distribution throughout the mycelium, whereas that of iodoacetic acid was more economical, being restricted to sites of sclerotium development. Chet and Henis further reported that chelating agents and potassium iodate, in addition to indoacetic acid, induced production of sclerotia ; they have suggested that all these inducers operate by modifying a repressor, which they think to be a sulphydryl-containing, copper-linked protein entity. They regard the action of L-cysteine as one of increasing either rate of production or of activity of this repressor, or of slowing down the rate of its turnover.

It seems likely that similar morphogenetic effects will be demonstrated with *Phymatotrichum omnivorum*, and perhaps with some other sclerotium-producing fungi as well. Any method for repressing production of sclerotia on infected crop plants that can be developed as an economic proposition for agricultural use certainly deserves further exploration.

Sclerotium cepivorum

In its host range, *S. cepivorum* is restricted to species of *Allium*. The behaviour of its sclerotia also conforms with the view that it is a highly specialized parasite, because they germinate freely in natural soil only when stimulated by root exudates of *Allium* species, or by volatile effluents from these plants. Coley-Smith (1959) conducted a survival-test with the sclerotia (0.2–0.5 mm diam.) buried in soil in nylon-mesh bags outside in the field ; the percentage of viable sclerotia, determined by ability to germinate on malt-extract agar, after 4 years burial in soil was 70, 75 and 80 at depths in soil of 3, 6 and 9 in, respectively. Percentage infectivity, estimated by inoculation of onion seedling roots, was 100 after 3 years and 75 after 4 years. In some experiments, addition of fresh organic material to the soil reduced the population of sclerotia, but in others it failed to do so. In his next paper, Coley-Smith (1960) reported that sclerotia taken directly from 6-week-old cultures were endogenously dormant, inasmuch as they failed to infect onion seedlings. This endogenous dormancy began to be broken after I month's burial in soil, had nearly disappeared after 2 months and had completely vanished after 3 months ; it could be broken by abrading the sclerotial rinds but, not surprisingly, this treatment impaired infectivity. Germination was stimulated both by root exudates and by root extracts of *Allium* species, but not by those of non-host plants. Percentage germination of sclerotia adjacent to onion roots was highest amongst those in contact with the apical region, which is known to be most active in exudation. The mode of germination by these sclerotia has been figured by Coley-Smith (Fig. 6.17).

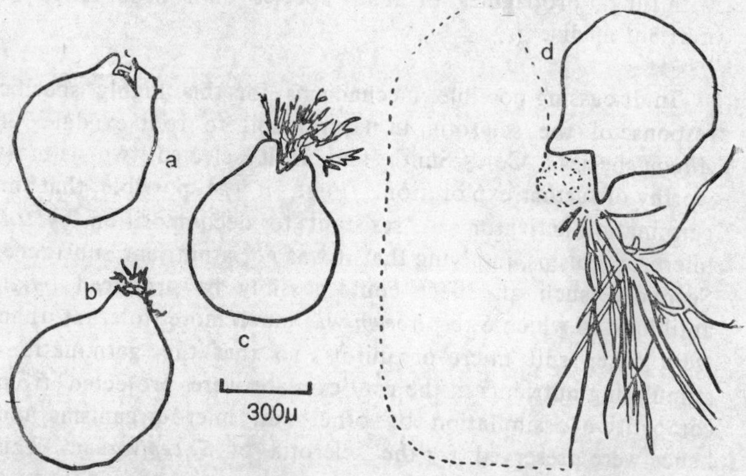

Fig. 6.17. Stages in germination of the sclerotium of *Sclerotium cepivorum*, showing rupture of the rind (*a*), followed by extrusion of a plug of mycelium (*b, c, d*).

This conclusion concerning the specificity of germination-response by the sclerotia to root exudates and extracts of *Allium* species was confirmed in a more extensive investigation by Coley-Smith and Holt (1966), who tested sixty-three plant species from a wide range of families. Germination-response to root exudates of non-hosts did not exceed the maximum percentage of spontaneous germination (14%) shown by the particular isolate of *S. cepivorum* employed in these tests. In a comparative test with eleven fungal isolates, percentage spontaneous germination showed a general correlation with percentage germination-response, indicating that isolates varied in the degree to which exogenous dormancy was enforced by soil fungistasis. Response to 1% sugar$\pm$nitrogen as casein hydrolysate in natural soil did not significantly exceed percentage spontaneous germination. Comparable tests with a number of nutrient solutions and plant extracts, but under sterile conditions on agar, silica gel, filter paper and sterilized soil, were made by Coley-Smith, King, Dickinson and Holt (1967) ; under these conditions, germination response was both high and quite unspecific, as has been the general experience with fungal propagules of many species on a wide range of nutrient media.

In discussing possible mechanisms for this highly specific response of the sclerotia in natural soil to root exudates of *Allium* species. Coley-Smith and Holt selected two as most worthy of further exploration. First, it was possible that the germination-activator was resistant to decomposition by soil micro-organisms, implying that it was not a nutrient substance. Secondly, such an effect could possibly be produced by an antibiotic to which *S. cepivorum* was much more tolerant than were other soil micro-organisms, so that the germination-stimulating nutrients in the root exudate were protected from competitive assimilation by other soil micro-organisms and hence were preserved for the sclerotia of *S. cepivorum*. This second hypothesis was thus much the same as that advanced by Cook and Schroth (1965) to explain the germination-response of chlamydospores of *Fusarium solani* f. *phaseoli* in soil to antibacterial antibiotics. Evidence for or against this second hypothesis was sought by Coley-Smith, Dickinson, King and Holt (1968), who appear to have eliminated its application to the sclerotia of *S. cepivorum*. They found that germination of the sclerotia in natural soil was induced by *Allium* extracts at a concentration too dilute to exercise any antibiotic effect on a number of common soil bacteria that they tested ; they further failed to find any evidence for the view that *undamaged* roots of *Allium* species exert any significant antibiotic effect on the rhizosphere microflora. Previous claims for such an effect were ascribed to insufficient precautions against damage to roots under investigation ; injured roots do excrete antibiotics at a significant concentration. Coley-Smith *et al.* therefore concluded that they were left with the first of the two hypotheses proposed by Coley-Smith and Holt, *viz.* that the germination-activator is a substance resistant to microbial decomposition.

The most recent advance in this investigation has been reported by King and Coley-Smith (1968), who found that sclerotia were induced to germinate in natural soil by volatile effluents from growing onion and leek seedlings, and also by air bubbled through diluted onion and garlic juices and by distillates and distillands of garlic juice. They also noted that diallyl sulphide, which possesses little antibiotic activity, was stimulatory, and suggest that this and other germination-activators produce a germination-stimulus sufficiently strong to override the effect of soil fungistasis. This explanation is compatible with all the evidence on behaviour of these sclerotia, which appear to be the most highly specialized and efficient of any yet investigated.